Food Proteins

Improvement through Chemical and Enzymatic Modification

Robert E. Feeney and
John R. Whitaker, EDITORS
University of California, Davis

A symposium sponsored by
the Division of Agricultural
and Food Chemistry at
the First Chemical Congress of
the North American Continent,
Mexico City, Mexico,
Dec. 3–4, 1975.

ADVANCES IN CHEMISTRY SERIES **160**

AMERICAN CHEMICAL SOCIETY
WASHINGTON, D. C. 1977

Library of Congress CIP Data

Food Proteins.

(Advances in chemistry series; 160 ISSN 0065-2393)

Includes bibliographical references and index.

1. Proteins—Congresses.
I. Feeney, Robert Earl, 1913- . II. Whitaker, John R., 1929- . III. American Chemical Society. Division of Agricultural and Food Chemistry. IV. Series: Advances in chemistry series; 160.

QD1.A355 no. 160 (TP453.P7) 540'.8s
(664) 77-7550
ISBN 0-8412-0339-3 ADCSAJ 160 1-312 (1977)

PRINTED IN THE UNITED STATES OF AMERICA

Advances in Chemistry Series

FOREWORD

ADVANCES IN CHEMISTRY SERIES was founded in 1949 by the American Chemical Society as an outlet for symposia and collections of data in special areas of topical interest that could not be accommodated in the Society's journals. It provides a medium for symposia that would otherwise be fragmented, their papers distributed among several journals or not published at all. Papers are refereed critically according to ACS editorial standards and receive the careful attention and processing characteristic of ACS publications. Papers published in ADVANCES IN CHEMISTRY SERIES are original contributions not published elsewhere in whole or major part and include reports of research as well as reviews since symposia may embrace both types of presentation.

CONTENTS

PREFACE

As requirements for high quality protein increase, it will become more and more necessary to improve the properties of seed proteins, fish protein concentrate, single cell protein, leaf protein, etc. to meet the world's protein needs. Cereals and legumes genetically selected for high yields of protein generally have lower nutritional value. To satisfy the nutritional needs we can (a) return to lower yielding, higher nutritive value varieties, (b) continue to seek genetically selected varieties that are both high yielding and high in nutritional values, or (c) chemically or enzymatically modify the proteins from the higher yielding varieties to give them the desired nutritional and functional properties. The last two methods will probably be needed since the exploding world population will not permit us to depend on the lower yielding varieties of plants. The day may come when we can no longer afford to convert low quality protein to higher quality protein via the cow, pig, or chick with the relatively low efficiency yield.

The intentional modification of food constituents to improve quality has long been practiced. In food manufacture lipids and carbohydrates are often intentionally modified to impart certain desired characteristics to foods. Important examples are the hydrogenation of polyunsaturated lipids to increase the melting point, and the acetylation and carboxymethylation of carbohydrates to change their textural characteristics. Similar examples are not available for food proteins.

Modifications of food proteins to produce different functional properties, to prevent processing damages, or to improve their nutritional values are not used commercially to a large extent at this time. We hope that this symposium will provide information which will attract interest to this problem.

This symposium brings together experts in the field of chemical and enzyamtic modification of proteins. The authors discuss what is presently being done in this area; however, each of them has been challenged to set forth further possibilities in this area and to bring forth new ideas which may at this time appear to be "blue sky" thinking but which will challenge the food scientist and the technologist to strive to implement the ideas. The history of mankind indicates over and over again that it is for lack of ideas that progress is stalled; given the ideas, man can usually rise to the challenge of implementation. We hope this will be true for modification of proteins for food use.

Davis, Calif.
January 1976

ROBERT E. FEENEY
JOHN R. WHITAKER

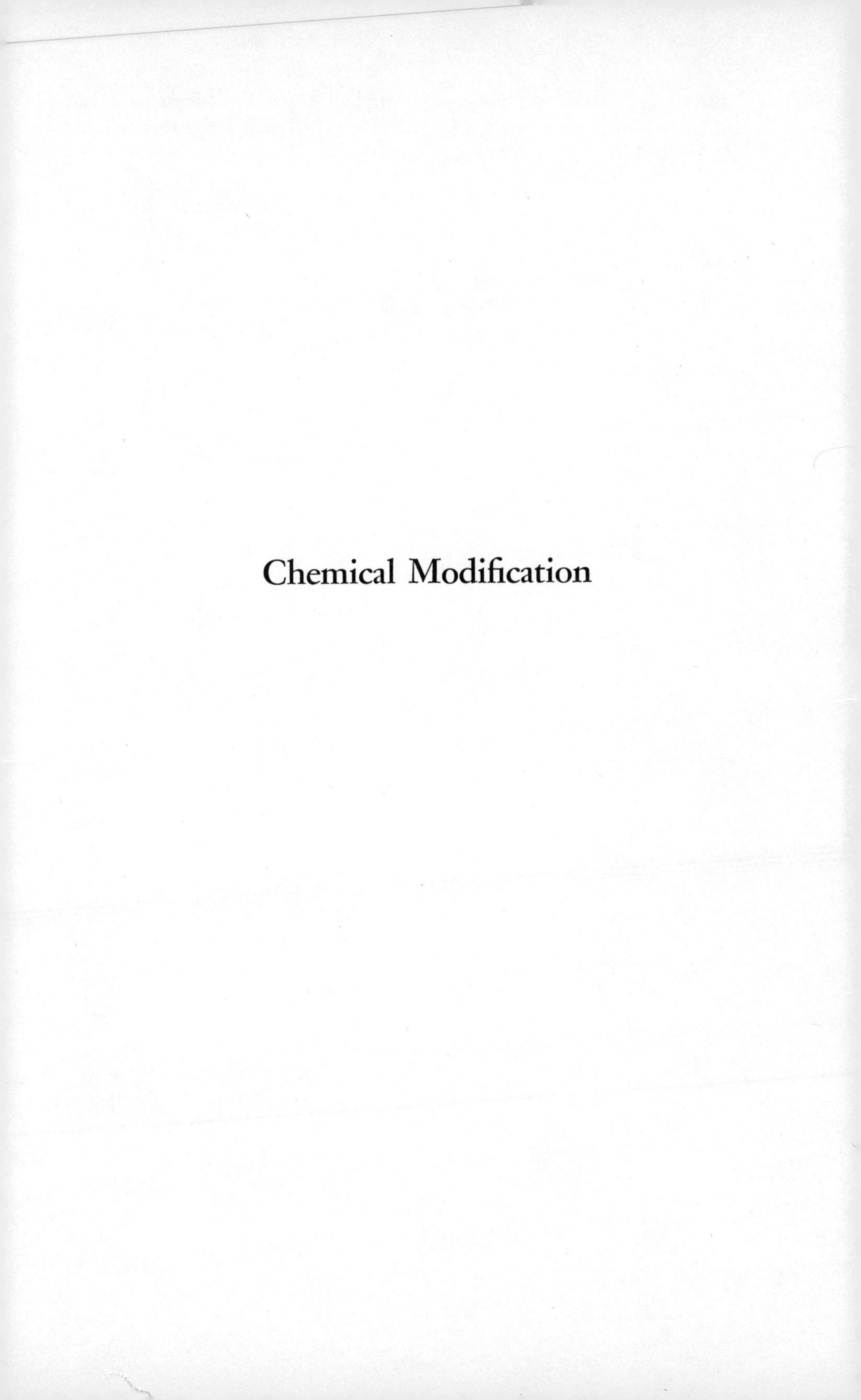

Chemical Modification

1

Chemical Modification of Food Proteins

ROBERT E. FEENEY

Department of Food Science and Technology, University of California, Davis, Calif. 95616

Alteration of the chemical structure of food proteins can be caused by deteriorative reactions arising during processing and storage, and by modifications with chemical reagents intentionally done to alter the properties of the proteins. The deteriorative type includes the Maillard reaction and the alkaline degradations leading to compounds like lysinoalanine. Although the intentional chemical modification of food proteins is currently applied to only a limited extent, it offers opportunities for improving food proteins and extending their availability from nonconventional sources. Chemical modifications of protein side chains can: (a) improve the nutritional quality, (b) block deteriorations, (c) improve physical states (e.g., texturization), and (d) improve functional properties (e.g., whipping capacity). Whereas many such modifications are theoretically possible, careful considerations of the safety and acceptability are required.

The modification of the properties of proteins by treatment with chemicals is widely used in fundamental studies (*1*). Since small changes in the chemical structures of proteins can result in large changes in their physical and biological properties, the chemical modification approach has been successfully applied to many biochemical problems. Most of these studies have been designed to obtain information concerning the mechanisms of action of biologically active proteins like enzymes.

Although the addition of chemicals to foods has long been practiced, the intentional chemical modifications of food proteins are still largely found only in the patent literature and are practiced to only a very limited extent. Most of the exceptions are those instances where food proteins have been chemically modified for the purpose of producing an industrial, rather than a food, product. Obvious barriers to the chemical modifications of food proteins for human usage have been such factors as the esthetic, cultural, legal, and medical aspects. Although the literature contains many references to the chemical modification of food proteins, nearly all of these are similar to those of many of the authors' researches (*1*), which are fundamental studies on the proteins from foods rather than studies directed at solving immediate practical needs.

Some chemical modifications of food proteins will be discussed in succeeding articles in this volume as well as in this chapter. The primary focus of this article, however, will be on the potentialities of chemical modification for improving the functionality, nutritional quality, and acceptability of food proteins.

General Perspective of the Chemistry and Chemical Reactions of Proteins

Historical Aspects. Alterations of the chemical, physical, and biological properties of proteins by chemically changing their structure have been an objective of protein chemists for many years. One of the first things discovered about proteins was how easy it was to change their properties by treatment with chemical reagents. In fact it was usually too easy, and most of the early changes were caused by labilities to the chemical reagents and the reaction conditions employed, rather than by the actual modifications. Lability of proteins is still, of course, one of the primary difficulties in handling proteins. Chemical modification of proteins, however, is now one of the most useful tools in biochemistry—a result of the application of modern knowledge of proteins, new chemical reagents, and much more sophisticated analytical techniques. Much of the history of protein chemistry is intimately related to the methods developed for determining structure and for analytical uses.

The very early study of protein chemistry was concerned mainly with the chemistry of the individual amino acids. A major turning point was the work of many investigators, in particular Emil Fischer and his colleagues, on the synthesis and properties of small peptides. Overlapping with this chemistry and synthesis of peptides, and continuing into the present time, are the analytical methods for the amino acid side chains in proteins. In fact, the primary motivating factor in earlier studies of the chemical modification of proteins was the objective of determining quantitatively the amounts of amino acids that are components of proteins. Many of these methods are very harsh because the intent was to obtain an analytical value, and not to conserve the integrity of the protein. Some of the earlier methods are still very much in use and can also be used for modification of proteins. Such an analytical method, still used to some extent, is the modification of amino groups with nitrous acid.

About the start of World War II there was increased emphasis on the chemical modification of proteins. Part of this undoubtedly stemmed from support of basic research aimed at learning more about the fractionation and use of blood plasma proteins and the properties of stabilization of proteins in foods, but some of the impetus came from the direct use of chemical modification for changing the properties of proteins. Certain studies of toxins and toxoids were also conducted, and the effects of various war gases and related substances on proteins and enzymes were studied.

Pharmaceutical and Industrial Uses of Chemical Modification. One of the main uses of chemical modification by the pharmaceutical industry has been in the modification of bacterial toxins and viruses. Formaldehyde has been used to kill, inactivate, or so change the virus or toxin as to make it incapable of inducing its toxic or pathological response while still retaining its ability to produce an immunogenic response when injected into an animal or man. Toxoids are bacterial toxins which have been so treated. The pharmaceutical industry is also currently using chemical modification to make insolubilized enzymes, or to attach enzymes to insoluble supports for enzymatic conversions of drug intermediates.

Many chemicals have also been used in leather and fiber technology, and almost all of these directly derivatize the constituent proteins. Some of the techniques used are like those with glutaraldehyde for forming cross-linkages and thereby stabilizing the substances. Others are used for the purpose of temporarily or permanently softening or solubilizing protein, as by reduction of the disulfide bonds.

Literature Related to Chemical Modification. There is extensive literature on chemical changes caused by purposeful chemical modification. This literature has been directed primarily at fulfilling the rapidly developing needs in fundamental areas of protein research. Consequently nearly all the publications are directed at fundamental biochemistry and chemistry. A general textbook and research monograph on the subject was written by Means and Feeney (*1*). A recent monograph has also appeared (*2*). The most extensive and detailed general coverage is in two different editions of "Methods in Enzymology" (*3, 4*). Recent review articles include those by Heinrikson and Kramer (*5*) and by Thomas (*6*). Articles on specialized subjects by Knowles (*7, 8*), Feeney et al. (*9*), and Feeney and Osuga (*10*) have also appeared recently. In contrast, reviews related to foods or food proteins are largely restricted to deteriorations or chemical modifications occurring inadvertently (*9, 11, 12, 13*).

There is also extensive literature on specialized techniques for synthesizing or elongating polypeptide chains (*14, 15*), as well as for degrading or shortening these chains (*16, 17*). At present, such techniques are often used for fundamental studies on protein structure–function relationships, or for analytical purposes, e.g., the Edman degradation for the determination of the sequences of amino acids in proteins (*16*).

General Properties of Proteins

Proteins are usually composed of almost all of the 20 common amino acids, or derivatives of them, plus varying amounts of attached substances, such as carbohydrates. The sequences of the amino acids in the protein structure govern the conformation (three-dimensional structure) of the protein in any particular environment. Changing either the sequence or types of amino acids, or the environment in which the protein is placed, can have extensive effects on the structure and properties of the protein.

Types of Bonds in Proteins. Two kinds of bonds are usually present. The main covalent bonds are the peptide bonds between the amino acid residues and the disulfide bonds which are the cross-links. Both of these are subject to chemical modification. With very few exceptions, methods for chemical modifications must not affect the peptide bonds. In addition, there are certain other types of cross-linkages found in specialized proteins such as in the structural proteins collagen and elastin.

Table I. Forces in Proteins

Force	*Distance Acting (Å)*	*Strength (Kcal/mol)*
Covalent	< 3	100
Electrostatic	< 20	1–10
H bond	< 3	3–5
van der Waals and hydrophobic	< 5	1–3

Noncovalent interactions exist among various parts of a peptide chain and between the peptide chain and the solvent (water or solution). These have a pronounced effect on the protein's conformation. The noncovalent forces usually cited are (a) electrostatic interactions, (b) hydrophobic interactions, (c) van der Waals interactions, and (d) hydrogen bonding (Table I). Chemical changes of the amino acid side chains of a protein usually affect one or more of these structure-determining forces. The object of many modifications is to change these forces in a purposeful way, and to avoid changes which might interfere with the purpose of the modification.

In the conformation of most proteins, the more hydrophilic amino acids such as lysine, glutamic acid, and serine are more often on the surface of the molecule, whereas the more hydrophobic amino acids such as phenylalanine and valine are in the interior part of the protein. The hydrophobic amino acids literally "exclude the water" in their mutual compatability with one another. Adding hydrophobic groups to the surface of the molecule by chemical modification could possibly disrupt the protein's entire organization by causing this new hydrophobic area to be folded into an internal section.

Electrostatic interactions are extremely important, particularly at longer ranges, and these are easily changed by modification. Individual hydrogen bonds are relatively weak, but there may be many of these bonds, and they can contribute extensively to the total energy determining a protein's conformation. van der Waals interactions also are weak contributors individually, but collectively may be very significant.

Effects of Environment. Proteins vary in stability all the way from those which are so unstable that solutions of them cannot be poured from one container to another without extensive denaturation, to proteins which are so stable that the solution can be boiled for ten min at neutrality without affecting the structure. Smaller proteins are generally much more resistant to denaturation than the large complex proteins, or at least they are able to resume their original structure after harsh treatments. It is essential that proteins have such large differences in properties, because some must function as relatively inelastic, hard coverings —such as skin or hair—whereas others must be extremely pliable, like some of the biologically active proteins which undergo reversible structural changes as part of their function.

The external environment has a great effect on the structure of most proteins. Chemical modification can change the response of a protein to its environment, and indeed this may be a purpose of many modifications that would be useful for food proteins. Denaturation is any change from the natural or so-called native state (*18*). It is also frequently defined as a change from an ordered to a disordered state. Operationally, many denaturations are irreversible, although the reversibility of the individual processes leading to the extremes of denaturation is a fact. In other words, denaturations are reversible, and it is only the interplay of many factors and subsequent reactions occurring after denaturation that make it practically impossible to return to the native state.

All of these interplaying factors are critical in the handling and the use of food proteins. A very important part of processing, storage, and use of food proteins is the control of denaturation, in some cases preventing it at certain stages, and in other cases accelerating it. Food processing techniques are usually designed to avoid denaturation in products that must remain soluble (e.g., most dried milk products), but either the processing techniques or methods of food preparations must be directed toward increasing the denaturation of some resistant proteins in order to increase the digestibility (e.g., legumes). It is obvious, therefore, that chemical modifications that alter the way a protein is denatured will have a great effect on the value of the protein as a food product.

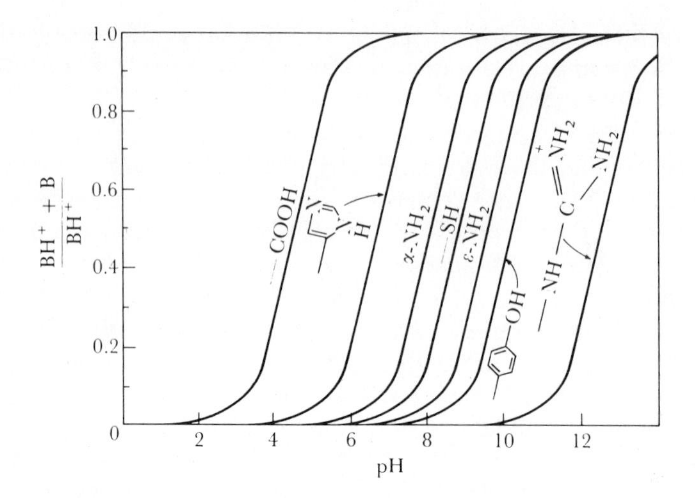

Chemical Modification of Proteins

Figure 1. Theoretical ionic states of amino acid side chains as a function of pH (1)

Relative Chemical Reactivities of Amino Acids. The relative reactivity of a protein's side chain to a chemical reagent depends upon the properties of the side chain and the reagent, and upon the environment at the time during which they are together. In the case of a protein, the ability of the reagent to approach physically the side chain is also dependent upon the structure of the protein. In some instances the chemical reagent will be completely incapable of approaching a side chain which is either tucked into a crevice or pocket, or located in the interior of the protein. In other instances, the nature of the charges of the other amino acid residues which are the neighbors of this group will have a profound effect on whether a particular chemical reagent approaches this group for the modification. Changing the reaction conditions can influence the effects of the neighboring groups.

The reactivity of an individual amino acid side chain along the protein structure will therefore depend not only upon the intrinsic reactivity of the side chain, but also upon the conditions of the reaction. For example, the majority of the side chains are chemically modified when they are good nucleophiles, and they are only nucleophiles in the nonprotonated condition. This is because the p*K* (Figure 1) of the side chain is a strong factor in determining whether the side chain is modified by a particular chemical reagent. Therefore, an amino acid side chain, such as the ε-amino groups of lysine, is readily modified only at a pH greater than eight, where the amino group is at least partially unprotonated.

The chemical reagents used in most protein modifications are either electrophiles, oxidizing chemicals, or reducing chemicals. It is obvious that many of the common reagents in organic chemistry cannot be used to modify proteins because of the lability of the peptide bond and the possible denaturing effects. As a consequence, with the exceptions of certain reagents which generate free radicals, such as photogenerated reagents (7), amino acid residues with hydrocarbon side chains (such as alanine, valine, leucine, or phenylalanine) are not usually able to be modified chemically.

Many of the side chain groups of amino acids in proteins are much more reactive in the protein than in an isolated amino acid state, although the reverse is usually the case. In these particular proteins there is probably a composite effect of many factors in the protein structure; this greatly increases the reactivity of such a particular amino acid side chain to a chemical reagent. In many cases this hyperreactivity can be utilized in chemical modifications. Many such highly reactive groups are found to be related to the active sites of enzymes. Modifications of active sites will be discussed in more detail below.

Amino Acid Side Chains That Can Be Chemically Modified. Table II lists both the amino acid side chains which can be chemically modified by commonly used reagents, and the types of modifications which can be effected easily with such amino acid residues. This is a very abbreviated listing, and more extensive texts (*1, 2*) and reviews (*5, 6, 10, 19*) should be consulted for other existing and workable chemical reactions and reagents for proteins.

Probably the amino acid side chains of proteins most often modified are the ε-amino group of lysine and the sulfhydryl group of cysteine, or its oxidized product, the disulfide group of cystine.

Table II. Amino Acid Side Chains Commonly Chemically Modified

Side chain	*Commonly used modifications*
Amino	Alkylation, acylation
Carboxyl	Esterification, amide formation
Disulfide	Reduction, oxidation
Guanidino	[a]
Imidazole	Oxidation, alkylation
Indole	Oxidation, alkylation
Phenolic	Acylation, electrophilic substitution
Sulfhydryl	Alkylation, oxidation
Thioether	Alkylation, oxidation

[a] Guanidino groups are usually not chemically modified because they are protonated below pH 10. They condense, however, with dicarbonyls, the main type of reagent used.

Objectives of Chemical Modification

Chemical modifications practiced in laboratories doing fundamental research are done for many different purposes. Many of these purposes are so interrelated that it would be difficult to classify them under separate headings, but certain general divisions can be made.

Direct Determination of the Contents of Amino Acids in Proteins. Amino acid contents of proteins are presently determined, with few exceptions, on an acid hydrolysate of the protein. In some instances, however, a particular amino acid may be at least partially destroyed by the conditions for acid hydrolyses. In other instances, a particular derivative of an amino acid of the protein may be hydrolyzed by the acid condition to regenerate the original amino acid. An analysis of such a chemically modified protein would thus need to be done on the protein before acid hydrolysis. In still other instances, there may be a need to perform a large number of analyses of different proteins for a single amino acid. The availability of a simple, direct chemical method for determination of such a single amino acid would be desirable.

Because an analysis for amino acids in an acid hydrolysate of a protein gives no information about the state of that particular amino acid in the original protein, chemical methods that do not distort the conformation of the protein can thus provide information on the state of the side chain of that particular amino acid in the protein in solution or in a crystal state. Thus, poorly reactive (sometimes called "masked") sulfhydryl groups of cysteine show different degrees of reactivity, depending upon the chemical agents used (Table III).

Methods used for the specific determination of the number of amino acids of a certain type in a protein are beyond the scope of this article and not pertinent to the subject. Other references should be consulted

Table III. Sulfhydryl-Group Reactivities (*1*)[a]

	Ovalbumin		*Serum albumin*		*β-Lactoglobulin*	
Reagent	*Native*	*Denatured*	*Native*	*Denatured*	*Native*	*Denatured*
p-Mercuribenzoate[b]	2.8	4.0	0.45	—	1.9	—
Iodine[c]	2.9	4.0	0.45	0.6	2.1	—
N-Ethylmaleimide[d]	0.4–0.6	3.8	0.45	0.45	0	1.9
5,5′-Dithiobis(2-nitrobenzoic acid)[e]	0	3.8	0.32	0.29	1.0	1.8

[a] Sulfhydryls per mol protein.
[b] At pH 4.6.
[c] At pH 6.5.
[d] At pH 7.0.
[e] At pH 8.0.

for these methods (*20, 21*). Exceptions are those supplied in the previous paragraphs and those particular methods which are directly useful in the chemical modification of proteins (*10*).

Group-Specific Modifications. Very few chemical modification procedures are "entirely specific" for any one particular type of side chain, e.g., indole, imidazole, amino, etc. Table II illustrates the principal specificity of certain reagents for some amino acid side chains as well as the other principal groups that may also be modified to at least a small degree. Two of the more specific procedures appear to be the reductive alkylation of amino groups with formaldehyde and a reducing agent, and the formation of disulfides from sulfhydryl groups with Ellman's dithionitrobenzoic acid. With those reagents which have less specificity, however, a high degree of specificity can be obtained if the treatment conditions for modification are mild. Such conditions are possible when only a few groups of a total number of one type of group (such as three amino groups out of a total of ten) need to be modified by acetylation. Then few or none of the tyrosyl groups might be modified in a particular protein.

There are various means to increase the specificity of a given chemical modification. The most obvious is the careful control of pH and other conditions. Another frequently used procedure involves reversible blocking reagents, which will be discussed in later sections. The objective in using a blocking reagent is to block another amino acid side chain which would be modified by the reagent in question, and then to remove the block after the modification of the first group has been achieved.

Modification of Groups Essential for Activity. An essential group in a protein is any group that is required for the protein's function. According to this definition, an essential group could be one that not only does not participate directly in the protein's function but could even be remote from those other essential groups participating in the protein's function (active-site groups). The remote group, however, could be essential to maintaining the protein's correct conformation.

There is no unanimously accepted definition of an active-center group in a biologically active protein. One more or less generally accepted definition is that active-center groups are all those groups which are in the region where the activity proceeds, such as in or around the active-site where amino acids are located in an enzyme. Thus some of these active-center groups would not participate in the catalytic action directly, although they might be important for binding or for merely providing a pocket or hole in which the substrate can fit. Modification of any of these groups could abolish the activity of the enzyme. In contrast, an active-site group can be considered to participate directly in the function.

One method of locating an active-center is the substrate protection of the active-site or -center of an enzyme. The object is to perform the chemical modification in the presence of a substrate, or a substrate analog, so that the active-site or -center is occupied by the substrate or substrate analog; therefore modification of groups in this area cannot occur. Consequently, when the substrate or analog has been removed, the active-site or -center is exposed, and an enzyme is now obtained which can be further modified in the active-site or -center by the same type of reagent as previously used, but one which can be quantitated, such as one containing a radioactive label. It might then be possible to identify which residue(s) was in the catalytic pocket.

The technique can be modified by using either a true substrate which may occupy only part of the active-center, or a substrate analog which could have either a larger group or groups with different charges. These procedures also have many other uses in fundamental studies.

Active-site-selective reagents are, as the name indicates, reagents which are specific, or relatively specific, for the active-site of the biologically active protein. Some of the reagents that have been used for many years are the substrates themselves, or pseudosubstrates. A substrate can be used when there is a procedure available for covalently attaching the substrate, or part of the substrate, while it is in the active-center. Covalent attachment has been done with a number of enzymes which form Schiff bases by adding a chemical reducing agent to the solution of the enzyme and its substrate (Figure 2). A pseudosubstrate can be a reagent which is similar to the substrate but which is poorly attacked by an enzyme, although the initial part of the catalytic process occurs. Such a reagent is the well-known diisopropylfluorophosphate, DFP.

Affinity reagents resemble substrates but have an additional group. They are "double-headed," one "head" of the molecule closely resembling a normal substrate, and the other "head" containing a group capable of making a covalent linkage with an amino acid side chain. The latter group may be chemically reactive or potentially so after an activation process (Figure 3).

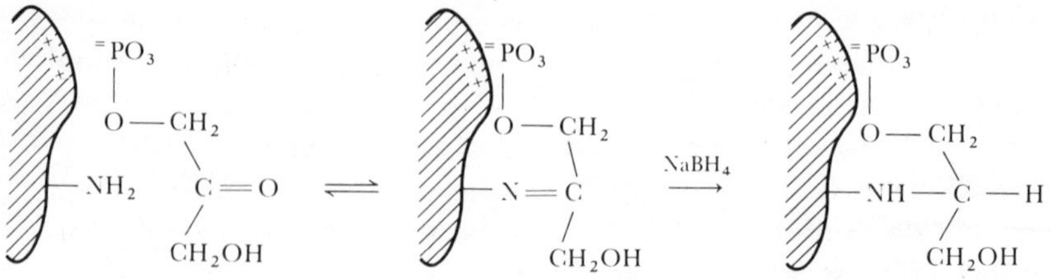

Chemical Modification of Proteins

Figure 2. Active-site labelling of rabbit muscle aldolase (1)

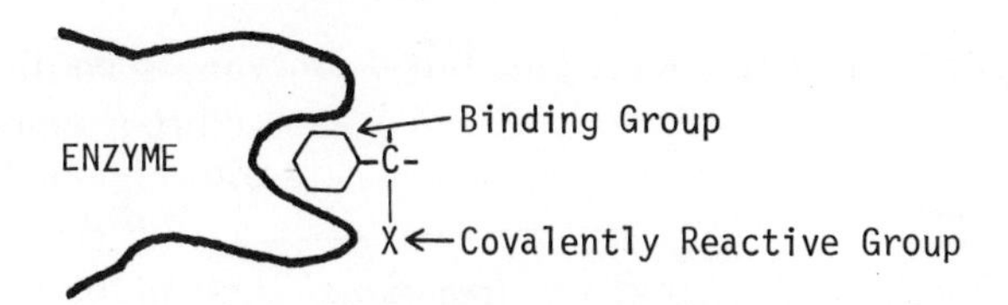

Figure 3. Diagram of principle of affinity labelling of a reactive site. In affinity labelling (1) there is (a) a binding group that resembles the type of substance (substrate, antigen, etc.) with which the protein normally interacts specifically; and (b) an additional group, a covalently reactive group, capable of forming a covalent bond in the reactive site. Affinity reagents are usually classified into three different types: general affinity, photoaffinity, and "suicide" affinity (see *Figure 5*).

The general or substrate-affinity reagent has a high rate of reaction of the X group merely because of the high relative binding of the reagent when it is used at limited concentrations. The high association at the binding site on the enzyme preferentially brings the active group of the affinity reagent into the active-center. Such a reagent will frequently inactivate an enzyme, or, when a reagent serves as a hapten with a covalently reactive group attached, it can inactivate the appropriate antibody molecule. When the proteins are examined under conditions in which the substrate or hapten would normally dissociate from the protein, the covalently linked part of the reagent is still attached to the protein. The use of an affinity reagent is thus effective in mapping the active-centers.

General affinity reagents can be relatively simple in structure. For example, a series of alkyl cyanates of differing chain lengths was used successfully to differentiate the sizes and types of binding pockets of a series of proteolytic enzymes (Table IV, Figure 4) (*22*). Elastase, which has specificity for bonds of small hydrocarbon side chains, was preferentially modified by the smaller hydrocarbon-containing reagent, while α-chymotrypsin was modified by the longer reagent.

A photoaffinity reagent is similar to the general or substrate affinity reagent with the important exception that the X group is unreactive until it is photogenerated (*7*). Photoaffinity reagents are useful not only because they are generated when in the active-center of the protein, but also bcause the photogenerated products may contain such highly reactive groups as carbenes and nitrenes which can react with hydrocarbon side chains of amino acids that are not modified by the reagents commonly used for chemically modified proteins.

A "suicide" reagent is a substrate, or a substance similar to a substrate, which is acted upon by the enzyme. Catalytic conversion pro-

Table IV. Effect of Octyl and Butyl Isocyanate on Proteases[a]

			% Activity remaining at reagent–enzyme molar ratio of 50:1	
Enzyme	*Enzyme conc* (M × 10^6)	*Reaction pH*	*Octyl isocyanate*	*Butyl isocyanate*
Chymotrypsin	1.9	7.6	0	19
Trypsin	2.0	7.6	85	85
Elastase	1.9	7.6	94	5
Carboxypeptidase A	6.4	7.6	90	
Carboxypeptidase B	3.0	7.6	100	
Pepsin	2.3	2.0	100	
Papain	4.8	3.0	0	

[a] Adapted from Brown and Wold (*22*).

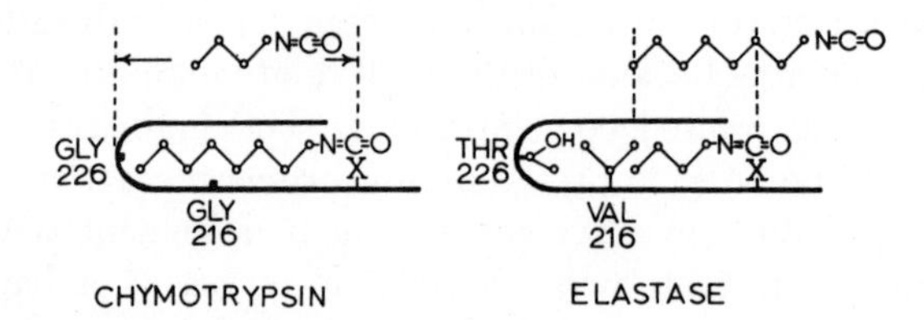

Figure 4. The effect of length of hydrocarbon side chains of alkyl isocyanates on their interactions with three related proteolytic enzymes. Schematics of proposed interactions of alkyl isocyanates with the binding pockets of trypsin, chymotrypsin, and elastase (22).

duces a chemically reactive group which was absent in the original substance (*23, 24*). The enzyme thus generates in its own vitals a substance which is capable of inactivating it at that point. Hence, the colorfully descriptive name "suicide" (*25*) has been used. One such example of modification by a suicide reagent is the irreversible inactivation of an essential histidine residue of the enzyme β-hydroxydecanoyl thioester dehydrase by $\Delta^{(3,4)}$ decynoyl *N*-acetyl cysteamine (NAC) (*23, 26*) (Figure 5). This suicide inhibitor of the enzyme, unlike the normal β,γ-olefinic substrates which undergo cis-trans conversions, is converted into the highly reactive conjugated allene.

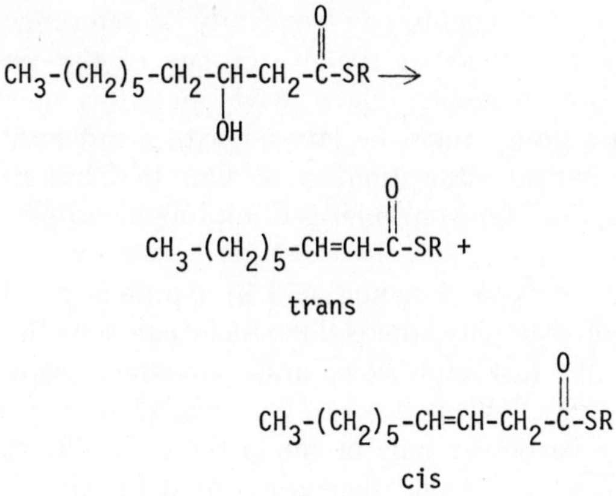

$$CH_3\text{-}(CH_2)_5\text{-}CH_2\text{-}\underset{\displaystyle OH}{CH}\text{-}CH_2\text{-}\overset{\displaystyle O}{\overset{\|}{C}}\text{-}SR \longrightarrow$$

$$CH_3\text{-}(CH_2)_5\text{-}CH{=}CH\text{-}\overset{\displaystyle O}{\overset{\|}{C}}\text{-}SR \; +$$

trans

$$CH_3\text{-}(CH_2)_5\text{-}CH{=}CH\text{-}CH_2\text{-}\overset{\displaystyle O}{\overset{\|}{C}}\text{-}SR$$

cis

Scheme 1

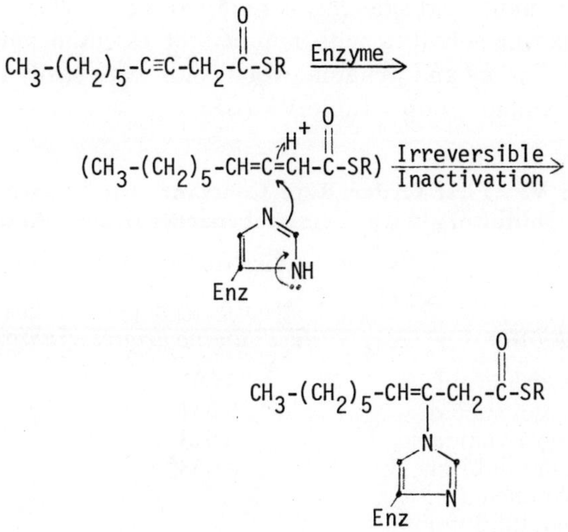

$$CH_3\text{-}(CH_2)_5\text{-}C{\equiv}C\text{-}CH_2\text{-}\overset{\displaystyle O}{\overset{\|}{C}}\text{-}SR \xrightarrow{\text{Enzyme}}$$

$$(CH_3\text{-}(CH_2)_5\text{-}CH{=}C{=}CH\text{-}\overset{\displaystyle O}{\overset{\|}{C}}\text{-}SR) \xrightarrow[\text{Inactivation}]{\text{Irreversible}}$$

H⁺, N, NH, Enz

$$CH_3\text{-}(CH_2)_5\text{-}CH{=}\underset{\displaystyle N}{C}\text{-}CH_2\text{-}\overset{\displaystyle O}{\overset{\|}{C}}\text{-}SR$$

N, Enz

Scheme 2

Science

Figure 5. The irreversible inactivation of β-hydroxydecanoyl thioester dehydrase by $\Delta^{(3,4)}$ *decynoyl* N*-acetyl cysteamine. The enzyme catalyzes the reversible interconversion of hydroxydecanoyl thioesters with their α,β-trans and β,γ-cis counterparts (Scheme 1). The acetylenic analog is converted by the enzyme into the highly reactive conjugated allene, which alkylates a histidine residue in the active center (Scheme 2)* (23).

The absolute number of essential residues, either of the same or different types of amino acids, can frequently be determined. One of the simplest and best methods is the use of one of the several types of active-site labeling reagents which have an easily monitored group attached. Such a group could be labeled with a radioactive element or with a distinct optical absorption band. One that has frequently been used with the "serine"-type enzymes is diisopropylfluorophosphate labeled with ^{32}P.

Since the same type of amino acid in a protein can have very different degrees of reactivity, these differences can sometimes be used to classify them into fast and slow, and sometimes even intermediate, degrees of reactivity. When this can be done, there may be only a few amino acids of a particular type in one of these classifications. If one of the amino acids in a classification is essential for the function of the protein, then it may be possible to do a kinetic analysis relating the loss of activity to the modification of residues in that particular class. It is then frequently a simple procedure to determine whether there are one, two, or more amino acid side chains required for activity. This was used to relate losses of activities with numbers of essential amino groups in ovomucoids. Turkey and penguin ovomucoids were easily shown to have one essential amino group (Table V) (*27*).

Table V. First-Order Rate Constants for Treatment of Inhibitors with Trinitrobenzenesulfonic Acid[a]

	*First-order rate constant (*k *min^{-1})*	
Inhibitor	*Modification of "fast" amino groups*	*Loss of trypsin-inhibitory activity*
Turkey ovomucoid	0.087	0.081
Penguin ovomucoid	0.094	0.092
Cassowary ovomucoid	0.043	0.059
Colostrum inhibitor	0.058[b]	0.122
Duck ovomucoid	[c]	0.051
Lima bean inhibitor	[d]	0.092
Chicken ovomucoid	0.041	0.006[e]
Tinamou ovomucoid	0.042	

[a] Adapted from Haynes et al. (*27*).
[b] Two amino groups are implicated. One could be part of the reactive site; the other could be in or near the active center. The modification of either could cause inactivation.
[c] Duck ovomucoid inhibits two trypsins simultaneously. It was not possible to separate the fast and intermediate amino groups.
[d] Difficulties encountered in distinguishing among rates of three differently reacting amino groups.
[e] Chicken ovomucoid has an arginine at the reactive site and therefore no essential amino group.
[f] Tinamou ovomucoid inhibits chymotrypsin, not trypsin. No loss in activity against chymotrypsin was observed.

Modifications to Introduce Special-Purpose Groups. An investigator sometimes wishes to introduce a group into a protein structure for the sake of having the group present in the protein, and not for any modification of the properties of the protein. These groups may serve as labels for tracing the protein through some physical or biological series of events or as labels to monitor changes in the conformation of the protein. These reagents have various names, such as environmentally sensitive labels (for detecting changes in conformation when a protein interacts with its environment or with some other substance or substrate), radioisotopic label, colored or fluorescent label (for following a protein through some process), or isomorphic replacement label (as a target for x-ray analysis).

A small molecule attached to a protein for the purpose of eliciting antibodies to the small molecule when the molecule-protein conjugate is injected into an animal is called a hapten. When antibodies are made for the hapten, many different types of experiments can be done, either with the free hapten and the antibody or with the hapten attached to the protein and the antibody.

Modifications to Change Physical Properties. Modifications have been used extensively to change physical properties of proteins, although they are used more frequently to modify specific active-center residues for changing biochemical or chemical functions. Changing physical properties can, of course, result in changes in the biochemical functions also. Indeed, the purpose of changing the physical properties is frequently to study the specific effects of the physical change on the biochemical functions.

Modifications which alter the charges on amino acid side chains usually effect profound changes in the properties of the protein. The most obvious change is one in the isoelectric point of a protein, but the changes in charge can also affect the conformation of the protein and thus its overall functional activity. There have been many applications of changing charge in proteins in order to study both fundamental and practical phenomena. One of the common procedures for separating subunits of a protein is using a reversible reagent which will change the charge and cause the monomeric constituents to repel one another, and thus dissociate the polymers. Still another application of effectively changing charge is in the study of interactions of two proteins with each other. When ovomucoids were modified by procedures that made the proteins more acidic, they reacted much more rapidly with the proteolytic enzyme they inhibited (Figure 6) (*28*).

The addition of hydrophobic groups to proteins should have many different effects on protein properties, depending upon the location of the additions and the type of hydrophobic groups. The most profound effects would be on the general conformation of the protein, the solubility, and the interfacial characteristics. Many studies have been made on the properties of polyhydrophobic amino acid peptides such as polyalanines (*29*). Hydrophobic residues such as cyclopentyl groups have also been attached to amino groups of lysine by reductive alkylation (*30*).

The attachment of hydrophilic groups to protein could, in most instances, increase the solubility of the protein by increasing its affinity for water, and by changing its interaction with other substances. However, it might not have such extensive effects as the attachment of hydrophobic groups. The less probable effects from hydrophilic groups as compared with hydrophobic groups are easily understood when it is

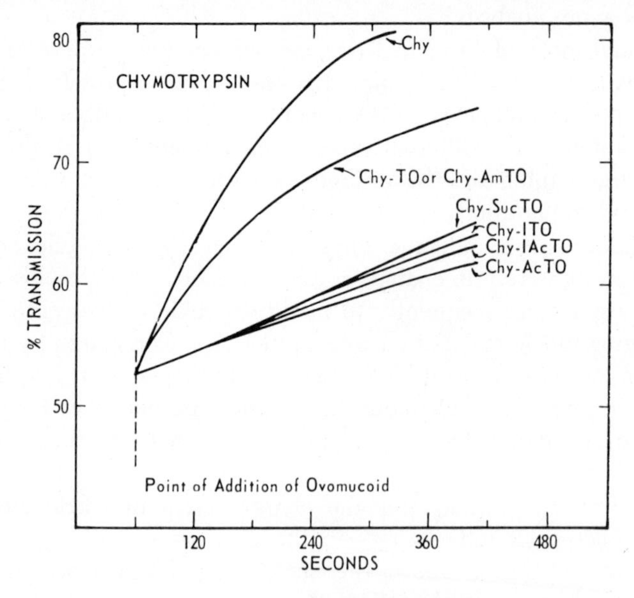

Archives of Biochemistry and Biophysics

Figure 6. Effect of chemical modification of turkey ovomucoid on its inhibition of bovine α-chymotrypsin (by delay time assays). Abbreviations used: Chy, chymotrypsin; TO, turkey ovomucoid; AcTO, acetylated turkey ovomucoid; AmTO, amidinated turkey ovomucoid; SucTO, succinylated turkey ovomucoid; ITO, iodinated turkey ovomucoid; IAcTO, iodinated and acetylated turkey ovomucoid. To a mixture of enzyme-buffer and substrate (benzoyl tyrosine ethyl ester, plus m*-nitrophenol as indicator) was added the inhibitor solution within 18–25 sec and the enzyme activities recorded on a chart at 395 mμ. The weight ratio for chymotrypsin and the turkey ovomucoids was 22:15. The percent change in transmission is proportional to the amount of enzyme activity. At the time of the additions of the inhibitor, the enzyme was hydrolyzing the substrate* (28).

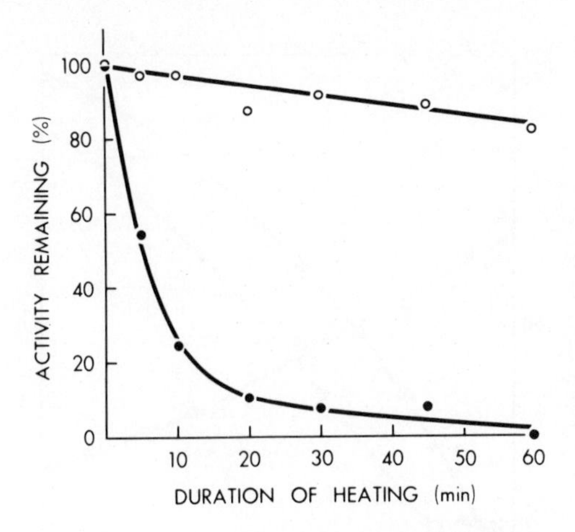

Archives of Biochemistry and Biophysics

Figure 7. Loss of activity on heat treatment (60°C) of sweet potato β-amylase (●) and its conjugate with dextran (○) (31)

remembered that the outer surface of the protein is already relatively hydrophilic. Nevertheless, when polysaccharide dextran was conjugated with the enzyme sweet potato β-amylase, the product was much more stable in the presence of heat than was the original protein (Figure 7) (*31*).

When a protein is reduced so as to modify the disulfides and form sulfhydryls, this change, with very few exceptions, causes such large and drastic effects on the protein's conformation that biological activities are absent in the product. With many proteins, the disulfide bonds can be reformed by reoxidation with atmospheric oxygen. Correct pairings of sulfhydryls can be accelerated by the presence of mixtures of low molecular weight organic mercaptans and disulfides (*32*). Many different ways of reducing and reoxidizing, and many different types of reactions, can be included in this sequence of events. For example, when turkey ovomucoid is reduced and then reoxidized, its "double-headed" inhibitory capacities against trypsin and chymotrypsin are not regained to the same degree of reoxidation (Figure 8) (*33*). In addition, all the activity is regained before all the sulfhydryls are reoxidized to disulfides. When one of four disulfides of the pancreatic trypsin inhibitor is reduced by sodium borohydride, all the activity is still retained (*34*).

Cross-linkages can be formed intramolecularly or intermolecularly by means of bifunctional reagents (*1, 35*). This is currently a rapidly expanding field for many different purposes. When the cross-linkages are intramolecular, the protein molecule is frequently more resistant to deformation when exposed to different environmental conditions. An

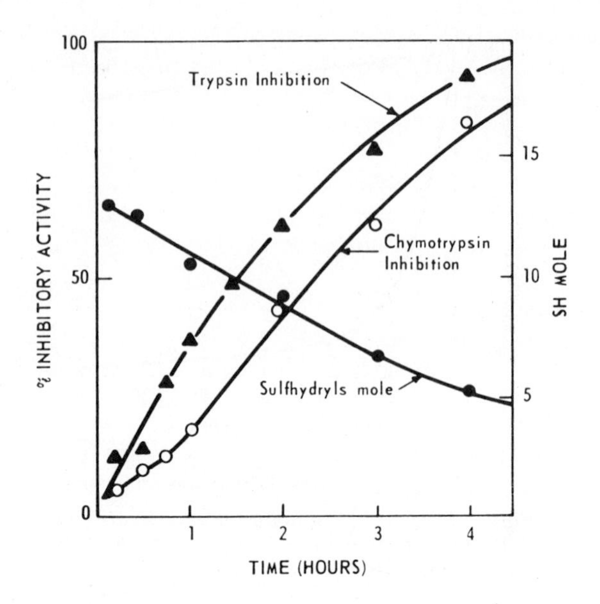

Biochimica et Biophysica Acta

*Figure 8. Reformation of disulfides and inhibitory activities on reoxidation of reduced turkey ovomucoid, a dual inhibitor of bovine trypsin, and α-chymotrypsin with nonoverlapping sites. The disulfides of turkey ovomucoid were reduced with mercaptoethanol and the product purified. Reoxidation was made with protein concentrations of 60 μg/ml in 0.006*M *Tris buffer at pH 8.2 at room temperature. (Turkey ovomucoid contains eight disulfides, all of which were reduced to 16 sulfhydryls in this experiment. Of these 14 were found at the first sampling and five at the time the experiment was discontinued.)* ▲, *Trypsin inhibition;* ○, *α-chymotrypsin inhibition;* ●, *sulfhydryls per mole* (33).

intramolecularly cross-linked protein could thus be more stable to denaturation. Sophisticated uses of cross-linkages for intramolecular modification have included some which also incorporate affinity labeling. In these instances it might be possible to show areas of the protein that are near one another topologically in the native active protein molecule.

There are many different applications of intermolecular cross-linkages of proteins. A currently active area is the mapping of the positions of proteins on ribosomes (*36*).

Attachment of proteins to solid supports has been a development primarily of the attachment of enzymes to solid supports to make immobilized enzymes (*37*). The attachments can be made by many different biochemical reactions. The selection of the appropriate reaction depends on the relative importance of modifying different types of side chains on the protein. Immobilized enzymes are discussed in more detail in other chapters of this volume.

Typical Reactions for the Chemical Modification of Proteins

Typical reactions have been conveniently classified on the basis of the type of reagent used (*1*). The following is an abbreviated list of these modifications.

Acylation Reagents. Acetylation with acetic anhydride:

$$\text{(P)}\text{—}NH_2 + O(\text{—}C(=O)\text{—}CH_3)_2 \xrightarrow{pH > 7} \text{(P)}\text{—}NH\text{—}C(=O)\text{—}CH_3 + CH_3COO^- + H^+$$

Carbamylation with cyanate:

$$\text{(P)}\text{—}NH_2 + HNCO \xrightarrow{pH \gtrsim 7} \text{(P)}\text{—}NH\text{—}C(=O)\text{—}NH_2$$

$$\text{(P)}\text{—}S^- + HNCO + H_2O \underset{}{\overset{pH\ 6\ to\ 8}{\rightleftharpoons}} \text{(P)}\text{—}S\text{—}C(=O)\text{—}NH_2 + OH^-$$

Guanidination with alkyl acetimidate:

$$\text{(P)}\text{—}NH_2 + H_2N^+{=}C(\text{—}NH_2)\text{—}OR \xrightarrow{pH > 9.5} \text{(P)}\text{—}NH\text{—}C(={}^+NH_2)\text{—}NH_2 + ROH$$

Alkylating Agents. Alkylation of several groups with iodoacetic acid:

$$\text{(P)}\text{—}S^- + ICH_2COO^- \xrightarrow{pH \geq 7} \text{(P)}\text{—}S\text{—}CH_2COO^- + I^-$$

$$\text{(P)}-\text{Im(NH)} + ICH_2COO^- \xrightarrow{pH > 5.5} \text{(P)}-\text{Im}-N-CH_2COO^- + I^- + H^+$$

$$\text{(P)}-S-CH_3 + ICH_2COO^- \xrightarrow{pH\ 2\text{–}8.5} \text{(P)}-S^+(CH_2COO^-)(CH_3) + I^-$$

$$\text{(P)}-NH_2 + ICH_2COO^- \xrightarrow{pH > 8.5} \text{(P)}-NH-CH_2COO^- + I^- + H^+$$

Reductive alkylation with formaldehyde:

$$\text{(P)}-NH_2 + 2H_2CO + \tfrac{1}{2}\ NaBH_4 \xrightarrow[pH\ 9]{0°C,} \text{(P)}-N(CH_3)_2 + \tfrac{1}{2}\ NaH_2BO_3 + \tfrac{1}{2}\ H_2O$$

Esterification. Esterification with methanol:

$$\text{(P)}-C(=O)OH + CH_3OH \xrightarrow{0.02 \text{ to } 0.1\ M\ HCl} \text{(P)}-C(=O)OCH_3 + H_2O$$

Amidation. Amide formation with amine and condensing agent:

$$\text{(P)}-C(=O)O^- + NH_2-R \xrightarrow[R'-N=C=N-R'']{pH \sim 5} \text{(P)}-C(=O)NHR$$

Oxidation. Strong oxidation of sulfhydryls and disulfide with performic acid:

$$\text{(P)}-SH + 3H-C(=O)O-OH \longrightarrow \text{(P)}-SO_3H + 3H-C(=O)OH$$

$$\text{(P)}-S-S-\text{(P)} + 5H-C(=O)O-OH + H_2O \longrightarrow 2\ \text{(P)}-SO_3H + 5H-C(=O)OH$$

Purification and Analysis of Chemically Modified Proteins

Selection of Methods. The purification and analysis of a chemically modified protein can be as formidable a task as the chemical modification itself (*10*). An important general rule to follow in the selection of the proper procedures for the purification and fractionation of modified proteins is that the selections must be on an operational basis. Any method of purification should be selected with regard to: the type of chemical modification, the manner in which modification is done, and the purpose of the modification. It is frequently necessary to change the original chemical modification to give a product which can be purified more easily. Similar reasonings may also be applied to the selection of techniques for analyses of the products.

Fortunately, sometimes purifications can be accomplished by very simple methods such as precipitation by salt or changing pH. Sometimes the extents of modification can also be determined by simple methods such as gel electrophoresis or determination of amino groups. More often, however, the methods required are much more difficult.

Problems Encountered in the Purification and Analysis of Chemically Modified Proteins. One of the main problems plaguing the chemical modification expert is the heterogeneities of the products. With chemically modified proteins heterogeneity may make purification nearly impossible and analysis merely a reflection of an average value for the heterogeneous population of molecules. Heterogeneity can be caused by incomplete modifications as well as by side reactions of either a physical or chemical nature. Careful considerations of these as well as other possible problems is mandatory in achieving satisfactory purifications and analyses. A recent review (*10*) should be consulted for a more comprehensive discussion of the purification and analysis of chemically modified proteins.

Chemical Changes in Food Proteins Caused by Deteriorative Reactions

Food proteins may undergo many different types of chemical changes during processing and storage. These changes, while usually deleterious, are sometimes beneficial. Although the deteriorative chemical changes are not the chemical modifications which are the main subject of this article, chemical treatments or additions are employed to control such deteriorative reactions. Two of the deteriorations will be discussed briefly (for a more complete discussion see Ref. *13*).

The Maillard Reaction. The Maillard reaction is probably one of the best known deteriorative reactions in the drying and storage of foods containing carbohydrates. The initial reaction is a carbonyl-amine reaction between the carbonyl group of the carbonyl compound (usually glucose, fructose, or pentose) and the amino groups of protein (*9*). Proteins may become insolubilized, giving off colored products and off-flavors. A loss of nutritive value occurs, but there is usually no gross toxicity in products still acceptable for consumption. The loss of nutritive value has been the subject of numerous investigations (*38*).

Effects of Alkali. Although alkali had been used to treat certain foods for many years, only recently has it been used widely by the texturized protein industry. Alkali-mediated degradation of proteins has long been known (*13, 39–44*). Some of the main initial reactions are apparently β-eliminations of cystines and substituted serines and threonines. The products (or their intermediates) then alkylate various other amino acid side chains to form substances like lanthionine and lysinoalanine [N^{ϵ}-(DL-2-amino-2-carboxyethyl)-L-lysine]. Possible toxicities are currently under investigation (*45, 46*), but nutritional losses could also be important.

In addition to the loss of several amino acids by alkaline treatment, and the possibility of toxic compounds being produced, alkali may also cause racemization of amino acids (*47*), which can also occur with roasting (*48*). There is need for further nutritive investigations of such racemized products, as well as for fundamental studies on the racemizations of amino acids in different proteins under various conditions (*48*).

Application of Chemical Modification to Food Proteins

Some General Objectives of Chemical Modifications of Food Proteins. The application of chemical modification to foods is a very young technology (*13*). The author prefers to consider it a technology for the future. Some of the more obvious possibilities will be considered in this section. Three general objectives are: the blocking of deteriorative reactions, improvement of physical properties, and improvement of nutritional properties.

Deteriorative reactions affecting proteins have been discussed in the previous section, and are also covered elsewhere (*13*). The general purpose here is to modify the proteins to either prevent a chemical reaction or greatly retard its rate. These objectives can be accomplished by blocking a protein group which undergoes a reaction, or by changing the conditions, thus greatly retarding the reaction.

Improvements of the physical properties include modifications to change the gross physical characteristics, such as texture, and to change the physical chemical interactions important to functional properties in cookery, such as foaming and whipping capabilities. Physical chemical laboratory methods and standards for assessing physical characteristics, unfortunately, are not well developed at this time (*49*).

Texturized protein foods have become very important in commercial food products (*50*). Some of the methods used involve chemical changes perhaps only indirectly (such as alkali-produced cross-linkages). Here is a probable place for advances in food technology in the application of chemical cross-linking agents under controlled conditions. This would include the bifunctional reagents.

Chemical modifications affecting solubility would also affect many of the protein properties in foods. These could be achieved through such techniques as changing the isoelectric point or changing the interaction with water or other substances in the food products.

Products with superior properties for use because of their functional characteristics in cooking may prove to be those which receive the most attention insofar as the application of chemical modification to food proteins goes. When low-cost proteins and proteins from different sources are used for these purposes, it is obvious that chemical procedures that can change physical properties and give them the characteristics required for these physical processes will be valuable.

Although chemical modifications may receive extensive attention for improving the physical properties of proteins because these give economically desirable properties, improvement of the nutritional quality of proteins by chemical modifications may prove to be the most important use from the standpoint of society's fundamental needs. Improvement of the nutritional quality might be brought about by increasing the digestibility of the protein, inactivating toxic or inhibitory substances, or attaching essential nutrients to the proteins. Nutrients that might be attached could include essential amino acids. The attachment of coloring or flavoring agents to proteins might also improve their acceptability.

Proteins have characteristics suitable for encapsulating other substances, particularly lipids. The purpose in such a procedure would be to protect the lipid from interaction with other food materials until the lipid would reach the digestive tract where protein could be removed.

Biological Considerations with Chemically Modified Food Proteins. Perhaps the greatest barrier to the use of chemical modification of food proteins is the biological one. The following are some of the factors to be considered.

Any chemically modified product must be acceptable organoleptically in order to be of any value. Ethnic and cultural habits might also dictate the chemical modifications used for different populations.

Losses in nutritional value would be an undesirable effect of chemical modification and would usually be avoided. However, for products incorporated into mixtures of different proteins, reductions in nutritional value of certain amino acids might be inconsequential because sufficient amounts of the destroyed amino acids could be supplied by the other proteins. This type of supplementation is common practice with other products in the food industry.

In some products, however, even large losses in nutritional value would be inconsequential. This would be the case when the chemically modified product would be used to supply certain physical properties, and the amounts of proteins supplied by the product would be relatively small compared with the total protein composition. In other instances, only a very small amount of the total protein content might be modified, although that small amount would be modified extensively. Such an example will be discussed below in the chemical treatment of certain grains or seeds, wherein apparently only a small percentage of the total proteins (the surface proteins) are modified.

Formation of toxic products by chemical modification is perhaps the area which will receive the most attention in future investigations. The possibility of forming toxic products frightens commercial research directors and higher administrators in industry. Toxicity in a protein is not a simple matter because of the complexities of protein structure and protein digestion in the digestive tract. A chemically modified protein might be toxic as:

(a) the intact protein molecule,

(b) certain peptide products produced from proteolytic enzymes in the digestive tract, or

(c) the free, chemically modified amino acids.

Toxicity of the intact protein when fed would easily rule out its use, but the latter two might be encountered only infrequently.

Chemical modifications of chemical groups other than the side chains of amino acids in the protein might also form toxic or harmful substances. Nucleic acids, carbohydrates, and lipids all contain functional groups which could be derivatized by many of the reagents used for protein modification.

In contrast to the other classifications just discussed, reversibility of modification would be a desired objective. A hypothetical example would be the use of a reagent which would block the Maillard reaction on amino groups, but which would be completely removed by the acidity in the stomach. This would be important if the blocked group were nontoxic but unavailable nutritionally. The protein would regain its complete nutritional value when the block was removed. Of course, an extrapolation of this concept is related to the removal of the blocking groups of modified amino acids after intestinal absorption, e.g., the removal of acetyl groups from ϵ-amino groups of lysine by a kidney deacylase (*51*).

Examples of Chemically Modified Food Proteins. Only a few examples can be quoted at this time, mainly because information from

Table VI. Examples of Chemically Modified Food Proteins

	Reaction	*Purpose*	*Ref.*
1.	Acylation of $—NH_2$	Blocking Maillard	*53*
2.	Dimethylation of $—NH_2$	Blocking Maillard	*56*
3.	3,3-Dimethylglutaric anhydride treatment	Egg white "stabilization"	*54*
4.	Formaldehyde treatment	Functional property of flours	*55*
5.	Formaldehyde treatment	Grain stabilization	*57*
6.	Succinylation	Solubilization, increase in emulsifying capacity	*58–61*

commercial research is not available. Some of the applications, both in the experimental stage and on a practical basis, are described in the chapter by Meyer and Williams (*42*) in this volume. A few examples will be discussed in this section, and some of the programs in the author's laboratory will be outlined.

Modifications performed for six different purposes are listed in Table VI. In each instance, the direction of the work has proceeded along different routes.

Bjarnason and Carpenter (*38, 52, 53*) studied the use of formylation, acetylation, and propionylation for blocking amino groups in food proteins. Many of these chemical modifications could be easily applied on a commercial scale. The formyl and acetyl derivatives were nutritionally utilized at least partially. The propionylated lysine was not utilized; however the propionylated lactalbumin was partially utilized (Table VII). The acylation procedure lowered considerably the extent of the Maillard reaction. Previous investigations, in support of the observations of Bjarnason and Carpenter, have shown deacylase in the kidney (*51*).

Gandhi et al. (*54*) made a general study of the effects of treating chicken egg white with 3,3-dimethylglutaric anhydride. Their work included examinations of electrophoretic diagrams as well as studies of the functional properties of the egg white for foaming, whipping, and cake baking (Table VIII). Extensive changes in properties were observed, but no attempts were made to vary conditions or reagent to achieve selectivity in the changes in properties.

Primo et al. (*55*) studied the effects of formaldehyde and iodate on the properties of flours related to breadmaking (Table IX). Their data indicated that chemical treatments might be used in lieu of the curing techniques. Since formaldehyde is known to cause cross-links (*9, 56*), the results might indicate some effects caused by cross-linking. No attempts to characterize cross-links in the flour were reported.

Primo et al. (*57*) studied the effects of formaldehyde on the stabilization of rice and grain. The results indicated that the formaldehyde treat-

Table VII. Comparison of Results from Rat Growth

Material	*% of lysine ε-NH_2 groups in sample*[a] *unreacted (a)*	*Relative activity of substrates for rat kidney ε-lysine acylase (51) (b)*
Formyl		
Lysine	—	(94)
Protein (lactalbumin)	28	—
Acetyl		
Lysine	0	(100)
Protein (BSA)	15	—
Propionyl		
Lysine	0	(0)
Protein (lactalbumin II)	2	—
Control protein (BSA)	91	—
Heat-damaged protein (BSA)	24	—

[a] For the amino acids, the zero values were based on the failure to find free lysine by thin-layer chromatography; for the proteins, the direct FDNB-available lysine values were taken as the measure of unreacted lysine.

ment stabilized the grain to physical deterioration occurring during transportation, storage, and handling. Apparently the modification is restricted to the periphery of the grain and, therefore, to a very small amount of protein. Thus, any biologically adverse reactions would be relatively minor unless highly toxic compounds were produced. The latter was not reported.

Succinic anhydride, a reagent frequently used to solubilize proteins (*1*), has been studied for its effect on the properties of: wheat flour by

Table VIII. Performance of Glutarinated Egg White in Angel Food Cake (54)

Mol DMGA/ mol EWP[b]	*Meringue specific gravity*			*Cake volume (ml)*		
	Control	*Treatment*	*Ratio T/C*	*Control*	*Treatment*	*Ratio T/C*
3	0.160	0.138	0.86	613	639	1.04
6	0.153	0.133	0.87	596	602	1.01
15	0.148	0.127	0.86	505	285	0.56
30	0.154	0.125	0.81	345[c]	230	—
60	0.154	0.119	0.77	345[c]	265[c]	—
150	0.172	0.118	0.69	240[c]	230[c]	—

[a] DMGA, 3,3-dimethylglutaric anhydride; EWP, egg white protein.
[b] Unsatisfactory cake with both control and treated egg white.

and Metabolism Experiments Using Acylated Materials (52)

Estimated partition of lysine from test materials in rat experiments

% Available for growth (c)	% Excreted Fecal (d)	Urinary (e)	Sum of c + d + e (f)
—	—	—	—
77	—	—	—
50	18	18	86
67	—	—	—
0	19	41	60
43	(−16)	41	84[b]
85	14	1	100
13	66	4	83

[b] This value is the sum of *c* and *e* alone, without the negative value for *d* subtracted.

Grant (*58*), single-cell protein concentrates by McElwain et al. (*59*), and fish protein by Groninger and Miller (*60*) and Chen et al. (*61*). Succinylated fish myofibrillar protein had rapid rehydration and good dispersion characteristics at neutral pH (*60*). Succinylation of fish protein concentrate improved its emulsifying capacity and emulsion stability (*61*).

Our laboratory has long been studying the fundamental aspects of the chemical modification of proteins. Results of certain studies are now being applied to the chemical modification of proteins to obtain information that might be useful for food processing. The dimethylation of proteins by the reductive alkylation of formaldehyde and sodium borohydride-treated proteins was initially investigated by our laboratory (*30, 62, 63*). More recently, the reductive methylation of proteins has been examined from the standpoint of possible use in foods (*56*). This program is probably a good example of changed emphasis in a research area, resulting in a change of objectives. Whereas our previous studies indicated no detectable side reactions with the usual techniques employed (*30*), it was now necessary to search more diligently for possible small amounts of nonspecific side reactions that might appear under certain conditions of application of the method for industrial use. None were found by the methods available when the procedure was done correctly, but it was observed that there was an extensive decrease in the rate of hydrolysis of the dimethylated proteins by α-chymotrypsin. When casein

Table IX. Effects of Curing, Iodate, and Formaldehyde on Rheological Properties of Dough: *Brabender farinogram* (55)

Characteristics	*Un-treated sample*	*Cured sample*	*Iodate-treated sample*	*Formal-dehyde-treated sample*
Water absorption (%)	55.9	55.2	55.2	55.2
Dough development time (sec)	60	75	75	105
Stability (min)	8.5	12	9	12.5
Mixing tolerance index—5 min after peak (B.U.) [a]	295	280	260	270
Degree of softening—12 min after peak (B.U.)	90	75	90	75

[a] B.U., Brabender units.

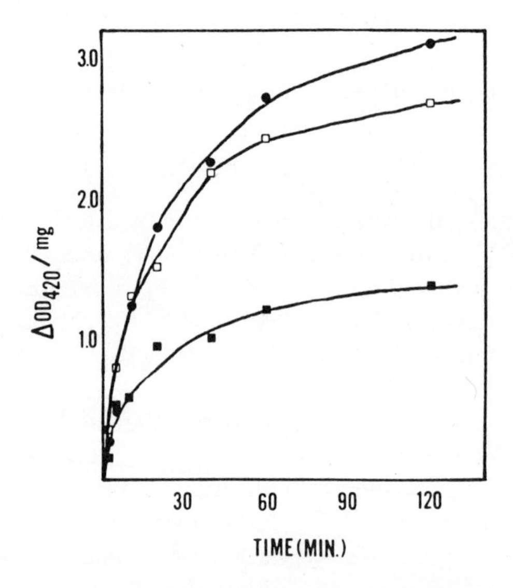

Figure 9. Digestion of casein and chemically modified caseins by bovine α-chymotrypsin. Amino groups of casein were reductively methylated and the hydrolysis by α-chymotrypsin followed with time. ●, *unmodified casein;* □, *reductively methylated casein—33% modified;* ■, *reductively methylated casein—52% modified. The casein concentrations were 0.2 mg/ml in 0.02*M *borate buffer at pH 8.2 and the α-chymotrypsin was 3.2 μg/ml. The hydrolysis was followed by the procedure of Lin et al.* (62) [*Figure from Ref.* 56].

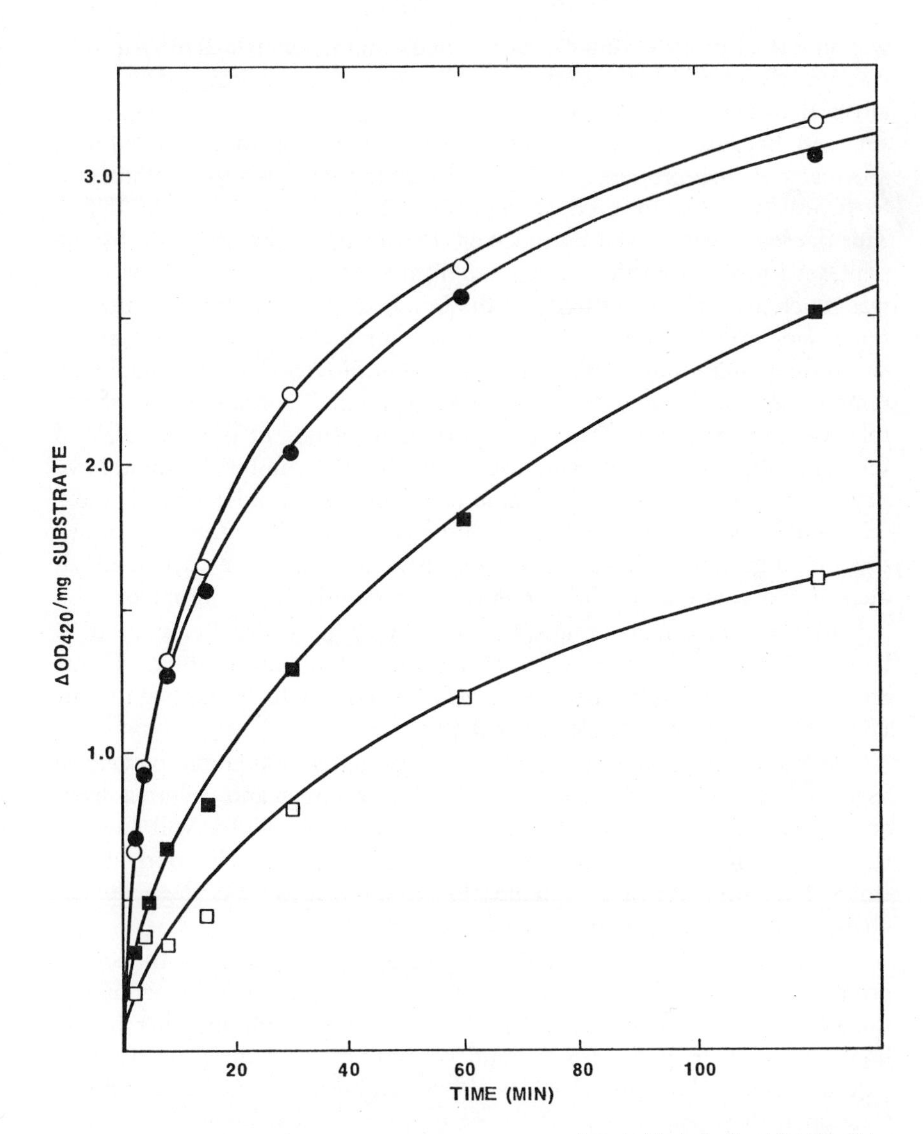

Journal of Agricultural and Food Chemistry

*Figure 10. Inhibitory effects of peptides from chymotryptic hydrolysates of modified bovine serum albumin (BSA) on hydrolysis of BSA by bovine α-chymotrypsin. BSA was acetylated and then reductively methylated. Eight ml (0.87 mg/ml) of this product in 0.012*M *Tris buffer and 0.012*M *$CaCl_2$, pH 7.7, was incubated at 37°C for 24 hr with 200 μg of α-chymotrypsin. The solution was then boiled for 10 min. Aliquots of this mixture of peptides were then added in the indicated proportions of BSA to the peptides. The various ratios of BSA to peptides in assay were: 17:1 (●); 3.5:1 (■); 1.7:1 (□). BSA alone (○). The BSA concentration was 0.2 mg/ml in 0.02*M *borate buffer at pH 8.2 and the α-chymotrypsin was 3.2 μg/ml. The hydrolysis was followed according to a modification of the procedure by Lin et al.* (62) [*Figure from Ref.* 56].

was modified to 50% dimethylation of its amino groups, the hydrolysis was greatly diminished (Figure 9). Because dimethylation should not affect the amino acid side chains susceptible to attack by α-chymotrypsin, and because of the obvious importance of this to the nutritional adequacy of such products, it was necessary to reexamine the method and to investigate this lowered susceptibility to α-chymotrypsin in detail. This decrease was caused by the inhibition of the α-chymotrypsin by the peptides formed from the dimethylated proteins (Figure 10). Thus there was no change in susceptibility of the bonds in the dimethylated protein, but rather a decrease in activity of the hydrolyzing enzyme as a result of product inhibition. Although it was not proven that this would be unimportant in the digestive system of an animal, considerations of the relative rates of removal and transport of materials in the animal, and the concentrations present, would indicate that this inhibition would have little, if any, importance. This same program is still under investigation, with emphasis on the use of the different types of alkyl groups to block the amino groups. Animal feeding experiments with these various alkylated amino compounds and proteins are presently being initiated.

Blocking of amino groups by dimethylation is also being used to help understand changes in proteins subjected to dry heating. Transamidation of carboxyls with lysines, or of amides with lysines, can occur (*38*). Dimethylation should prevent this.

Work in progress is employing the reductive alkylation procedure used for methylation to the covalent linkage of larger and different sized and shaped hydrophobic compounds to proteins. These will include the longer chain alkyl derivatives as well as some cyclic derivatives. Imidoesters also attach compounds to amino groups (*1*); these do not change the charge appreciably and are being tested.

A third related program is the attachment of fatty acids to the amino groups of lysine by formation of an amide linkage. Little is known about the presence of amidases that might remove these groups either in the digestive tract or in the tissues, and consideration of these possibilities will necessarily be important. If these products are easily digestible, they open the door to many different types of groups being attached to proteins to change the hydrophobicity.

A fourth way of attaching hydrophobic groups is the attachment of polyhydrophobic amino acids to proteins. There is much literature on this subject, and it should be possible to exploit the fundamental information available.

The attachment of hydrophilic groups to proteins can be done by methods similar to those used for the attachment of hydrophobic groups, and similar problems exist. Those under consideration in our laboratory include the reductive alkylation of sugar aldehyde groups to the amino groups of proteins (*64*). This reaction, of course, involves the first step of the Maillard reaction, namely, formation of the Schiff base, but with the simultaneous addition of the reducing agent, the intermediate is locked in as a reduced alkyl-amine, so there should be no possibility of the numerous breakdown products formed in the Maillard reaction.

A second procedure involves the coupling of the amino groups of proteins to cyanogen bromide-treated carbohydrates (*31*).

Various other methods can be used for coupling sugar to proteins, but much further work is necessary before these can be considered as possible methods for food uses.

Conclusions

A special committee (*65*) has recommended process research related to methodologies for restructuring protein materials or combining proteins with other materials. The chemical modification of food proteins offers a means for economically producing high quality foods from both unconventional and conventional sources. Plants not now cultivated for man's food might be used to produce valuable foods. Almost everyone can accept the concept of the improvement of "natural foods" by pretreating them for easier digestion, for more nutrition, and for greater palatability. The fundamental scientific information is available to serve as models for the development of chemical methods for food uses, but this is as far as the field has progressed at this time. There are strong indications that certain simpler methods might be suitable. These include acetylation or methylation of amino groups and esterification of carboxy groups by ethylation. Extensive studies, however, still must be done on the applications to food proteins before a satisfactory method will be generally available.

A major obstacle in the application of chemical modification to food proteins is the potential health hazard. The author believes, however, that this is an artificial barrier which exists only because inadequate financial support has been available for the necessary research. This research obviously must include extensive testing for the health hazards. Enzymatic procedures (*66*) offer advantages in regard to potential health hazards (perhaps because of the fear the public has for the word "chemical") and can sometimes fill the need, but chemical procedures can frequently offer more possibilities for different products at low cost.

If the major obstacle is not the return for a financial investment, but rather the size of the investment itself, the problem then seems to be one of finding the investment. Perhaps the only groups that can support the needed long-term research are national governments, international bodies, or multinational corporations in cooperation with these bodies (*13, 67, 68*).

Acknowledgment

The author gratefully acknowledges advice and criticisms from many sources. Particular appreciation goes to Chris Howland for editorial assistance and to Clara Robison and Gail Nilson for stenographic assistance. Background researches for this article were supported by FDA Grant FD 00568-02.

Literature Cited

1. Means, G. E., Feeney, R. E., "Chemical Modification of Proteins," Holden-Day, San Francisco, 1971, p 254.
2. Glazer, A. N., Delange, R. J., Sigman, D. S., "Chemical Modification of Proteins. Selected Methods and Analytical Procedures," North-Holland, Amsterdam, 1975, p 205.
3. "Methods in Enzymology," C. H. W. Hirs, Ed., Vol. 6, Academic, New York, 1967.
4. "Methods in Enzymology," C. H. W. Hirs, S. N. Timasheff, Eds., Vol. 25, Academic, New York, 1972.
5. Heinrikson, R. L., Kramer, K. J. in "Progress in Bioorganic Chemistry," E. T. Kaiser, F. J. Kézdy, Eds., Vol. 3, Wiley and Sons, New York, 1974, p 141.
6. Thomas, J. O. in "Companion to Biochemistry. Selected Topics for Further Study," A. T. Bull, J. R. Lagnado, J. O. Thomas, K. F. Tipton, Eds., Longman, London, 1974, p 87.
7. Knowles, J. R., *Acc. Chem. Res.* (1972) **5**, 155.
8. Knowles, J. R. in "MTP International Review of Science, Biochemistry Series One, Chemistry of Macromolecules," H. Gutfreund, Ed., Vol. 1, University Park, Baltimore, 1974, p 149.
9. Feeney, R. E., Blankenhorn, G., Dixon, H. B. F., *Adv. Protein Chem.* (1975) **29**, 135.
10. Feeney, R. E., Osuga, D. T. in "Methods of Protein Separation," N. Catsimpoolas, Ed., Vol. 1, Plenum, New York, 1975, p 127.
11. Mauron, J. in "Protein and Amino Acid Functions. International Encyclopaedia of Food and Nutrition," E. J. Bigwood, Ed., Vol. 11, Pergamon, New York, 1972, p 417.
12. Ford, J. E. in "Proteins in Human Nutrition," J. W. G. Porter, B. A. Rolls, Eds., Academic, New York, 1973, p 515.
13. Feeney, R. E. in "Proteins for Humans: Evaluation and Factors Affecting Nutritional Value," C. E. Bodwell, Ed., Avi, Westport, Conn., 1976, p 233.
14. Gish, D. in "Protein Sequence Determination. A Sourcebook of Methods and Techniques," S. B. Needleman, Ed., Springer-Verlag, New York, 1970, p 276.
15. Stewart, J. M., Young, J. D., "Solid Phase Peptide Synthesis," Freeman, San Francisco, 1969, p 103.
16. Edman, P. in "Protein Sequence Determination. A Sourcebook of Methods and Techniques," S. B. Needleman, Ed., Springer-Verlag, New York, 1970, p 211.
17. Stark, G. R., *Adv. Protein Chem.* (1970) **24**, 261.
18. Tanford, C., *Adv. Protein Chem.* (1968) **23**, 121.
19. Sigman, D. S., Mooser, G., *Annu. Rev. Biochem.* (1975) **44**, 889.
20. Hvidt, A., Johansen, G., Linderstrøm-Lang, K. in "A Laboratory Manual of Analytical Methods of Protein Chemistry (Including Polypeptides). The Composition, Structure, and Reactivity of Proteins," P. Alexander, R. J. Block, Eds., Vol. 2, Pergamon, New York, 1960, p 101.

21. Scoffone, E., Fontana, A. in "Protein Sequence Determination. A Sourcebook of Methods and Techniques," S. B. Needleman, Ed., Springer-Verlag, New York, 1970, p 185.
22. Brown, W. E., Wold, F., *Biochemistry* (1973) **12**, 828.
23. Rando, R. R., *Science* (1974) **185**, 320.
24. Rando, R. R., *Acc. Chem. Res.* (1975) **8**, 281.
25. Walsh, C. T., Schonbrunn, A., Abeles, R. H., *J. Biol. Chem.* (1971) **246**, 6855.
26. Bloch, K., *Acc. Chem. Res.* (1969) **2**, 193.
27. Haynes, R., Osuga, D. T., Feeney, R. E., *Biochemistry* (1967) **6**, 541.
28. Simlot, M. M., Feeney, R. E., *Arch. Biochem. Biophys.* (1966) **113**, 64.
29. Haschemeyer, R. H., Haschemeyer, A. E. V., "Proteins: A Guide to Study by Physical and Chemical Methods," Wiley, New York, 1973, p 445.
30. Means, G. E., Feeney, R. E., *Biochemistry* (1968) **7**, 2192.
31. Marshall, J. J., Rabinowitz, M. L., *Arch. Biochem. Biophys.* (1975) **167**, 777.
32. Saxena, V. P., Wetlaufer, D. B., *Biochemistry* (1970) **9**, 5015.
33. Sjöberg, L. B., Feeney, R. E., *Biochim. Biophys. Acta* (1968) **168**, 79.
34. Kress, L. F., Laskowski, M., *J. Biol. Chem.* (1967) **242**, 4925.
35. Wold, F. in "Methods in Enzymology," C. H. W. Hirs, S. N. Timasheff, Eds., Vol. 25, Academic, New York, 1972, p 623.
36. Sommer, A., Traut, R. R., *J. Mol. Biol.* (1975) **97**, 471.
37. Zaborsky, O. R., "Immobilized Enzymes," CRC, Cleveland, 1973, p 175.
38. Carpenter, J. K., *Nutr. Abstr. Rev.* (1973) **43**, 423.
39. Bohak, Z., *J. Biol. Chem.* (1964) **239**, 2878.
40. Ziegler, K., Melchert, I., Lürken, C., *Nature* (1967) **214**, 404.
41. Gross, E., ADV. CHEM. SER. (1977) **160**, 37–51.
42. Meyer, E. W., Williams, L. D., ADV. CHEM. SER. (1977) **160**, 52–66.
43. Sternberg, M., Kim, C. Y., Schwende, F. J., *Science* (1975) **190**, 992.
44. Provansal, M. M. P., Cuq, J. A., Cheftel, J., *J. Agric. Food Chem.* (1975) **23**, 938.
45. DeGroot, A. P., Slump, P., van Beek, L., Feron, V. J. in "Proteins for Humans: Evaluation and Factors Affecting Nutritional Value," C. E. Bodwell, Ed., Avi, Westport, Conn., 1976, in press.
46. Woodard, J. C., Short, D. D., Alvarez, M. R., Reyniers, J. in "Protein Nutritional Quality of Foods and Feeds," M. Friedman, Ed., Vol. 2, Marcel Dekker, New York, 1975, p 595.
47. Neuberger, A., *Adv. Protein Chem.* (1948) **4**, 297.
48. Hayase, F., Kato, H., Fujimaki, M., *J. Agric. Food Chem.* (1975) **23**, 491.
49. Ryan, D. S., ADV. CHEM. SER. (1977) **160**, 67–91.
50. "Fabricated Foods," G. Inglett, Ed., Avi, Westport, Conn., 1975.
51. Leclerc, J., Benoiton, L., *Can. J. Biochem.* (1968) **46**, 1047.
52. Bjarnason, J., Carpenter, K. J., *Br. J. Nutr.* (1969) **23**, 859.
53. Bjarnason, J., Carpenter, K. J., *Br. J. Nutr.* (1970) **24**, 313.
54. Gandhi, S. K., Schultz, J. R., Boughey, F. W., Forsythe, R. H., *J. Food Sci.* (1968) **33**, 163.
55. Primo, E., Barber, S., Benedito de Barber, C., Ribo, J. in "The Contribution of Chemistry to Food Supplies," I. Morton, D. N. Rhodes, Eds., Butterworths, London, 1974, p 357.
56. Galembeck, F., Ryan, D. S., Whitaker, J. R., Feeney, R. E., *J. Ag. Food Chem.* (1977), in press.
57. Primo, E., Barber, S., Benedito de Barber, C., personal communication, 1975.
58. Grant, D., *Cereal Chem.* (1973) **50**, 417.
59. McElwain, M., Richardson, T., Amundson, C., *J. Milk Food Technol.* (1975) **38**, 521.
60. Groninger, H., Jr., Miller, R., *J. Food Sci.* (1975) **40**, 327.

61. Chen, L., Richardson, T., Amundson, C., *J. Milk Food Technol.* (1975) **38**, 89.
62. Lin, Y., Means, G. E., Feeney, R. E., *J. Biol. Chem.* (1969) **244,** 789.
63. Lin, Y., Means, G. E., Feeney, R. E., *Anal. Biochem.* (1969) **32,** 436.
64. Gray, G. R., *Arch. Biochem. Biophys.* (1974) **163,** 426.
65. Scrimshaw, N. S., Wang, D. I. C., Milner, M., "Protein Resources and Technology: Status and Research Needs. Research Recommendations and Summary," **NSF-RA-T-75-037,** Cambridge, Mass., 1975.
66. Whitaker, J. R., ADV. CHEM. SER. (1977) Am. Chem. Soc., Wash., D. C., **160**, 95–155.
67. Revelle, R. in "War on Hunger," Report of Agency of International Development, Vol. 2, No. 6, U. S. Government Publication, 1968, p 1.
68. Feeney, R. E. in "The Social Responsibility of the Scientist," M. Brown, Ed., Free Press, New York, 1971, p 228.

RECEIVED January 26, 1976.

2

Chemistry and Biology of Amino Acids in Food Proteins: Lysinoalanine

ERHARD GROSS

Reproduction Research Branch, National Institute of Child Health and Human Development, National Institutes of Health, Bethesda, Md. 20014

Lysinoalanine, the constituent amino acid of cinnamycin (from Streptomyces cinnamoneus*) and duramycin (from* Streptomyces cinnamoneus forma azacoluta*) is found in proteins with a history of alkaline treatment, and used as a supplement in food products for human consumption. It is formed under basic conditions in peptides such as nisin, nisin fragments, and subtilin, all known to contain dehydroalanine and lysine. The chemistry of lysinoalanine is closely linked to that of the α,β-unsaturated amino acid dehydroalanine. A primarily pH-dependent equilibrium exists between dehydroalanine, lysine, and lysinoalanine, with lower values of the alkaline pH range favoring the forward reaction of the addition of the ε-amino group of lysine to the unsaturation, higher ones reversing the reaction via β-elimination. The physiology of lysinoalanine—an uncommon amino acid—deserves much additional exploration.*

The addition of the ε-amino group of lysine across the double bond of dehydroalanine leads to a product known as lysinoalanine (Figure 1). The formation of this amino acid was initially observed in bovine pancreatic ribonuclease when the enzyme was subjected to alkaline conditions. More recently, lysinoalanine has been detected in food products

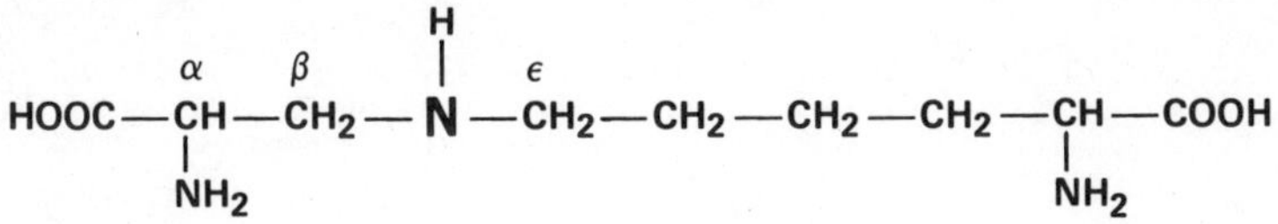

$$\mathrm{HOOC{-}\overset{\alpha}{C}H(NH_2){-}\overset{\beta}{C}H_2{-}N(H){-}\overset{\epsilon}{C}H_2{-}CH_2{-}CH_2{-}CH_2{-}CH(NH_2){-}COOH}$$

Figure 1. Structure of lysinoalanine

intended for consumption by man. The contributing sources for lysinoalanine are protein supplements secured by processing of, for instance, soya bean meal under, again, alkaline conditions. To this day, lysinoalanine has not been seen as constituent amino acid of peptides or proteins from plant or mammalian sources. However, the status of merely being an artifact changed recently when lysinoalanine was found for the first time to occur naturally in a peptide of microbial origin, namely, in cinnamycin isolated from *Streptomyces cinnamoneus.* Soon thereafter lysinoalanine was also detected in duramycin, a peptide from *Streptomyces cinnamoneus forma azacoluta.*

Nisin and Subtilin

The studies that led to the discovery of lysinoalanine in the naturally occurring peptides cinnamycin and duramycin began with the structural elucidation (*1*) of the heterodetic pentacyclic peptide nisin (Figure 2) from *Streptococcus lactis* (*2*). Unique to nisin is the presence of no fewer than three residues of α,β-unsaturated amino acids—two residues of dehydroalanine (Figure 3) and one residue of dehydrobutyrine (Figure 3)—as well as that of one residue of lanthionine (Figure 4) and four residues of β-methyllanthionine (Figure 4). The inspection of subtilin (Figure 2) from *Bacillus subtilis* (*3*) revealed the presence also

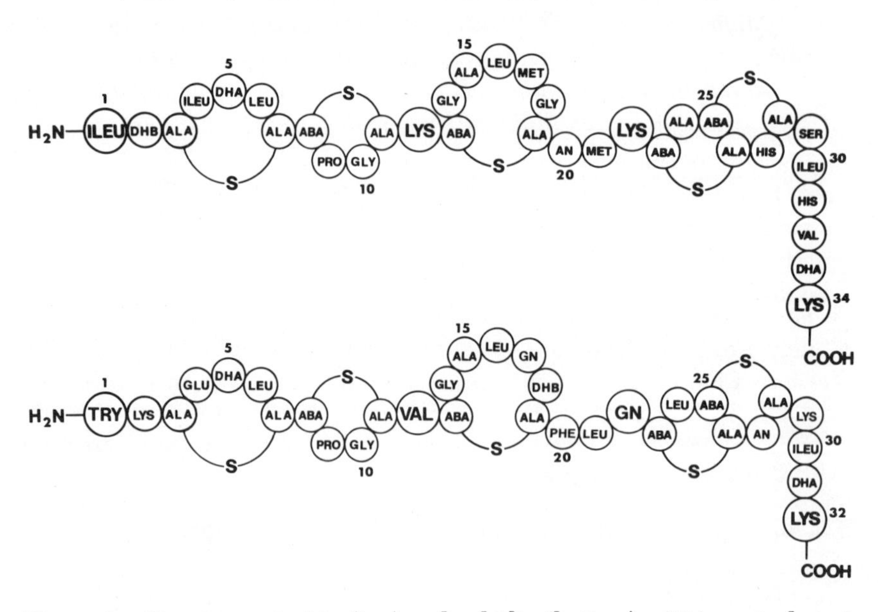

Figure 2. Structures of nisin (top) and subtilin (bottom). ABA, aminobutyric acid; DHA, dehydroalanine; DHB, dehydrobutyrine (β-methyldehydroalanine); ALA–S–ALA, lanthionine; ABA–S–ALA, β-methyllanthionine.

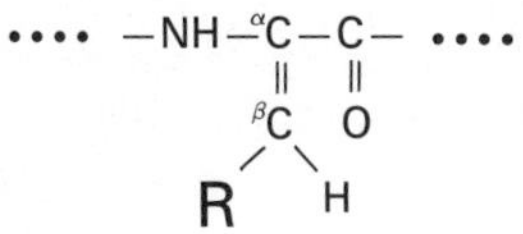

···· $-NH-{}^{\alpha}C(=^{\beta}CRH)-C(=O)-$ ····

Figure 3. α,β-unsaturated amino acids. R = H: dehydroalanine; R = CH_3: dehydrobutyrine (β-methyldehydroalanine).

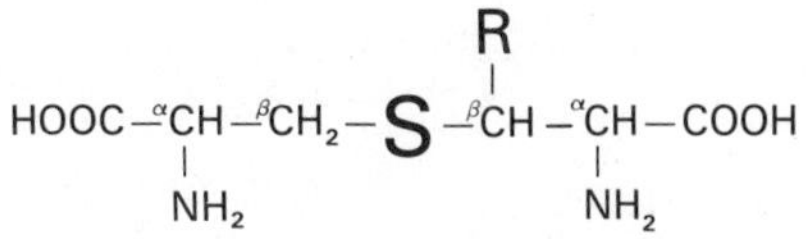

$HOOC-{}^{\alpha}CH(NH_2)-{}^{\beta}CH_2-S-{}^{\beta}CH(R)-{}^{\alpha}CH(NH_2)-COOH$

Figure 4. Structures of lanthionines. R = H: lanthionine; R = CH_3: β-methyllanthionine.

of three residues of α,β-unsaturated amino acids. As in nisin, there are two residues of dehydroalanine occupying positions identical to their counterparts in nisin (*see* Figure 2) and one residue of dehydrobutyrine, which in subtilin, however, is found in an endocyclic position, namely in ring C (*see* Figure 2).

At this point it must be admitted that little has been said about food proteins and/or amino acids found in them. The microbial sources of the peptides mentioned hardly qualify for placement on the list of food products. It must be recognized, however, that the producers of nisin and, for that matter, nisin itself are items of man's daily dietary intake as long as he eats dairy products.

Cinnamycin and Duramycin

The inspection of cinnamycin (*4*) and duramycin (*5*)—prompted by their content in lanthionine and β-methyllanthionine—for the presence of α,β-unsaturated amino acids was negative. However, cinnamycin as well as duramycin contain lysinoalanine, among other amino acids rarely seen in nature. Did the peptides at one time contain dehydroalanine and did it serve as a precursor for lysinoalanine? If the answer to this question were yes, which amino acids, in turn, are potential precursors of dehydroalanine?

Precursors of Dehydroalanine

Let us turn to answering the second question first. Potential precursors of dehydroalanine, by way of β-elimination, are the amino acids with functional groups at the β-carbon atom. Such β-elimination reactions are likely to be enzyme-catalyzed in nature. Additionally, the substrates may be suitably substituted at functions, such as the sulfhydryl group (thioether and sulfonium salt formation) or the hydroxyl group (carbohydrate attachment—glycopeptides and proteins; phosphorylation).

One β-substituted amino acid to be considered is serine (Figure 5), the proper substitution of which at the hydroxyl group will set the stage

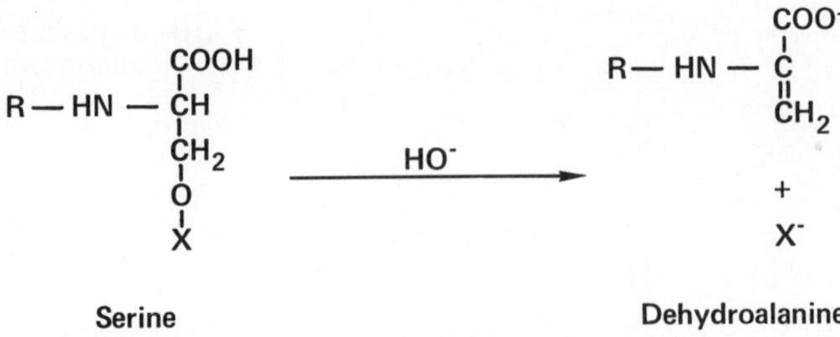

Serine → Dehydroalanine

Figure 5. Conversion (β-elimination) of serine to dehydroalanine. R, acyl or aminoacyl group in peptides and proteins; X, e.g. carbohydrate (glycoprotein).

for β-elimination and the formation of dehydroalanine. Chemically, the substitution may be provided by *O*-tosylation, and elimination may be brought about by base catalysis (*6*). Also to be considered are cysteine or cystine (Figure 6), two amino acids in which the sulfur functions in the form of the β-substituted sulfhydryl group or the disulfide bridge are the prerequisites for β-elimination under basic reaction conditions. In case of bovine pancreatic ribonuclease alkaline treatment caused β-elimination in cystine residues and the formation of dehydroalanine. To such dehydroalanine residues was then added the ε-amino group of lysine residues with the formation of lysinoalanine (*7, 8*).

Other amino acids suitably substituted at the β-carboxyl group are, in principle, also capable of undergoing β-elimination reactions. Threonine, a common constituent amino acid of proteins, must be mentioned, although there is no indication as yet of the formation of the product that would result from the elimination reaction and the addition of the

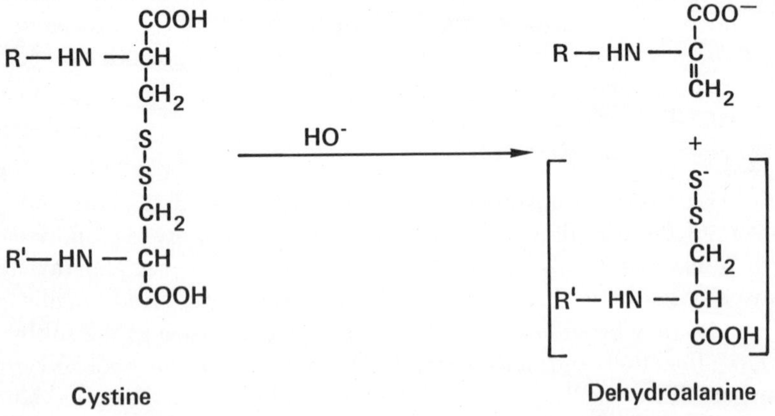

Cystine → Dehydroalanine

Figure 6. Conversion (β-elimination) of cystine to dehydroalanine. R and R′, acyl or aminoacyl group in peptides and proteins.

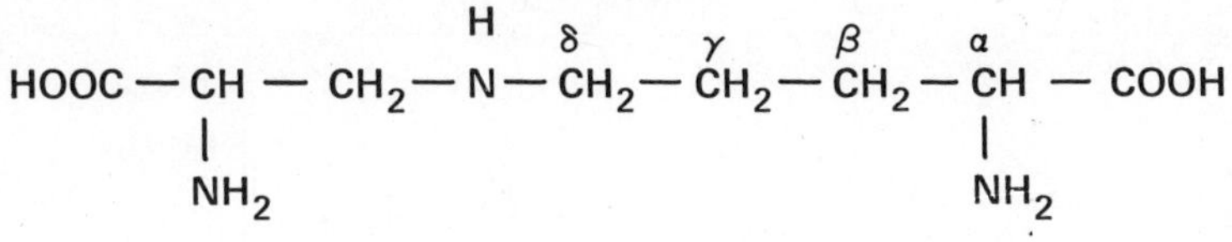

$$HOOC-\underset{NH_2}{\underset{|}{CH}}-CH_2-\overset{H}{N}-\overset{\delta}{CH_2}-\overset{\gamma}{CH_2}-\overset{\beta}{CH_2}-\overset{\alpha}{\underset{NH_2}{\underset{|}{CH}}}-COOH$$

Figure 7. Structure of ornithioalanine

ε-amino group to the resultant dehydrobutyrine (*see* Figure 3). The product expected from this sequence of reactions would be β-methyl-lysinoalanine, the methyl group being carried by the alanine moiety of the amino acid. The analogous substitution exists in the case of the lanthionines where β-methyllanthionine (*see* Figure 4) is believed to be the result of adding the cysteine sulfhydryl group across the double bond of dehydrobutyrine (*see* Figure 3).

Lanthionines are also the obvious target of β-elimination reactions. However, in view of their restricted occurrence in peptides of microbial origin, these amino acids are not of further concern in the context of this communication.

Origin and Formation of Ornithinoalanine

Since lysinoalanine is so readily formed by the treatment of protein with base, ornithinoalanine (Figure 7) also should be found in the hydrolysates of alkali-treated proteins, if only ornithine were a constituent amino acid of proteins. Ornithinoalanine may indeed become such a constituent amino acid of proteins. The alkaline conditions to which proteins are exposed provide the base catalysis required for the hydrolysis of the guanido group of arginine (Figure 8) and the formation of ornithine. The presence of ornithine in alkali-treated proteins was shown

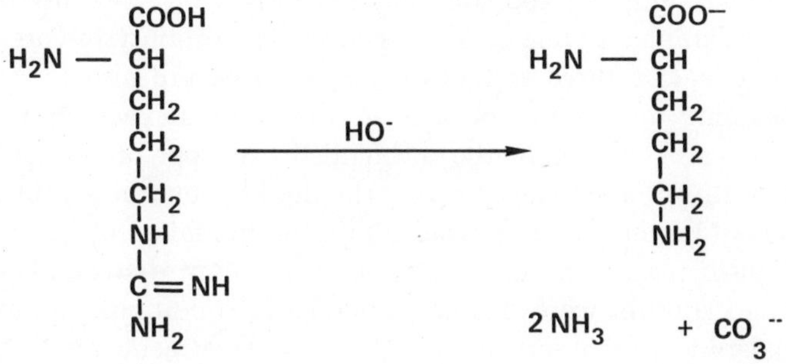

$$H_2N-CH(COOH)-CH_2-CH_2-CH_2-NH-C(=NH)-NH_2 \xrightarrow{HO^-} H_2N-CH(COO^-)-CH_2-CH_2-CH_2-NH_2 \quad 2\,NH_3 + CO_3^{--}$$

Arginine → Ornithine

Figure 8. Conversion of arginine to ornithine

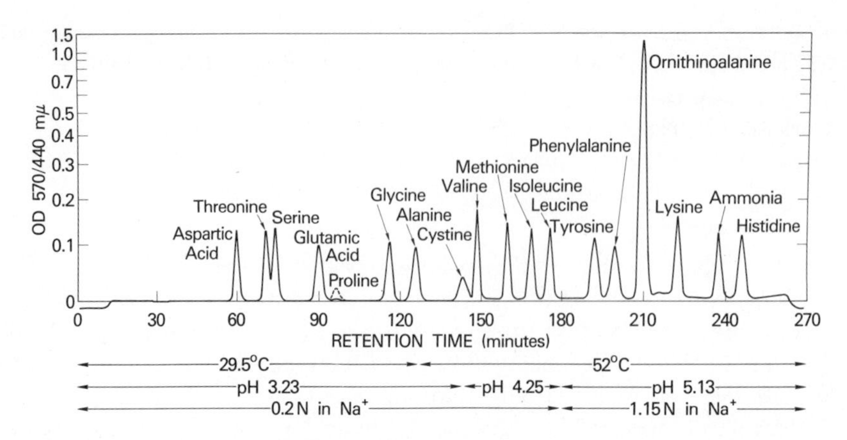

Figure 9. Elution position of ornithinoalanine in amino acid analysis. Standard amino acid mixture with an excess of ornithinoalanine; column dimension: 0.9 × 52 cm; column packing: spherical sulfonated polystyrene resin.

for sericin (*9*) and merino wool (*9*) by the detection of a new amino acid, namely, the addition product of the δ-amino group of ornithine to dehydroalanine (i.e., ornithinoalanine), in the corresponding total hydrolysates. In other cases of alkali treatment of proteins, detection of ornithinoalanine may have escaped the investigators. The elution pattern (Figure 9) reveals that the positions of lysinoalanine and ornithinoalanine are close together. Amino acid analytical systems of potent resolving power are required to distinguish between lysinoalanine and ornithinoalanine.

Formation of Lysinoalanine

Reference has been made repeatedly to α,β-unsaturated amino acids. In nisin (*1*) (Figure 2) and subtilin (*10*) (Figure 2), two residues each of dehydroalanine and one residue each of dehydrobutyrine are present. In these molecules three residues each of lysine are also found. With dehydroalanine and lysine present in the same molecules, will it be possible to verify also here the mechanistic concept of the addition of ω-amino groups of amino acids across the double bond of α,β-unsaturated amino acids (Figure 10) and demonstrate the formation of lysinoalanine?

When nisin, fragments of nisin (still containing dehydroalanine and lysine), and subtilin were treated under basic conditions, the formation of lysinoalanine was observed in each case. Treatment, for instance, of the carboxyl-terminal fragment of nisin (Figure 11) under these conditions (*11*) (1*N* *N*-ethylmorpholine, pH 10.65, 7 days, room temperature) followed by total hydrolysis and amino acid analysis showed the presence of lysinoalanine in the hydrolysate (Figure 12). The carboxyl-terminal fragment of nisin contains two residues of lysine and one residue of dehydroalanine in the penultimate position. At the time of this writing, work is still in progress to establish the extent of participation of either lysine residue in the formation of lysinoalanine. The ratio of intermolecular reaction to the two possible intramolecular reactions is as yet not known.

Considering the carboxyl-terminal–dehydroalanyllysine sequence of nisin (Figure 2) and subtilin (Figure 2), and knowing that the lysine moiety of lysinoalanine in cinnamycin and duramycin occupies the carboxyl-terminal position, it is tempting to ask the question whether the lysinoalanine residues in the latter two peptides are formed by the addition of the ε-amino group of the terminal lysine residue across a dehydroalanine residue that occupied the penultimate position. Structural studies are still in progress to settle this question unequivocally (*see* Figure 13).

The amino acid compositions of cinnamycin and duramycin differ by the exchange of one residue of arginine in the former and by a residue

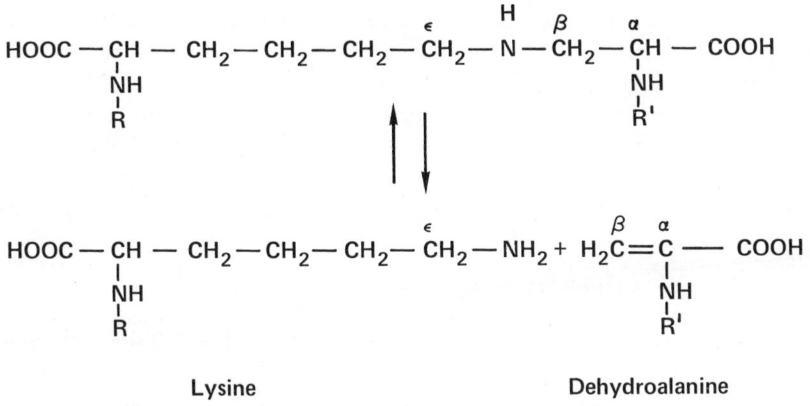

$$\mathrm{HOOC{-}CH(NHR){-}CH_2{-}CH_2{-}CH_2{-}\overset{\epsilon}{C}H_2{-}NH{-}\overset{\beta}{C}H_2{-}\overset{\alpha}{C}H(NHR'){-}COOH}$$

$$\uparrow\downarrow$$

$$\mathrm{HOOC{-}CH(NHR){-}CH_2{-}CH_2{-}CH_2{-}\overset{\epsilon}{C}H_2{-}NH_2 + H_2\overset{\beta}{C}{=}\overset{\alpha}{C}(NHR'){-}COOH}$$

Lysine Dehydroalanine

Figure 10. Formation of lysinoalanine from lysine and dehydroalanine

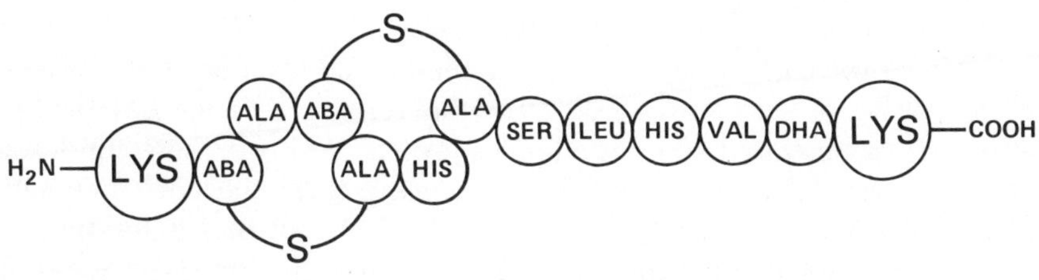

H_2N—LYS—ABA—ALA—ABA—ALA—HIS—ALA—SER—ILEU—HIS—VAL—DHA—LYS—COOH (S bridges: ABA–S–ALA)

Figure 11. COOH-terminal fragment of nisin. ABA, aminobutyric acid; DHA, dehydroalanine; ABA–S–ALA, β-methyllanthionine.

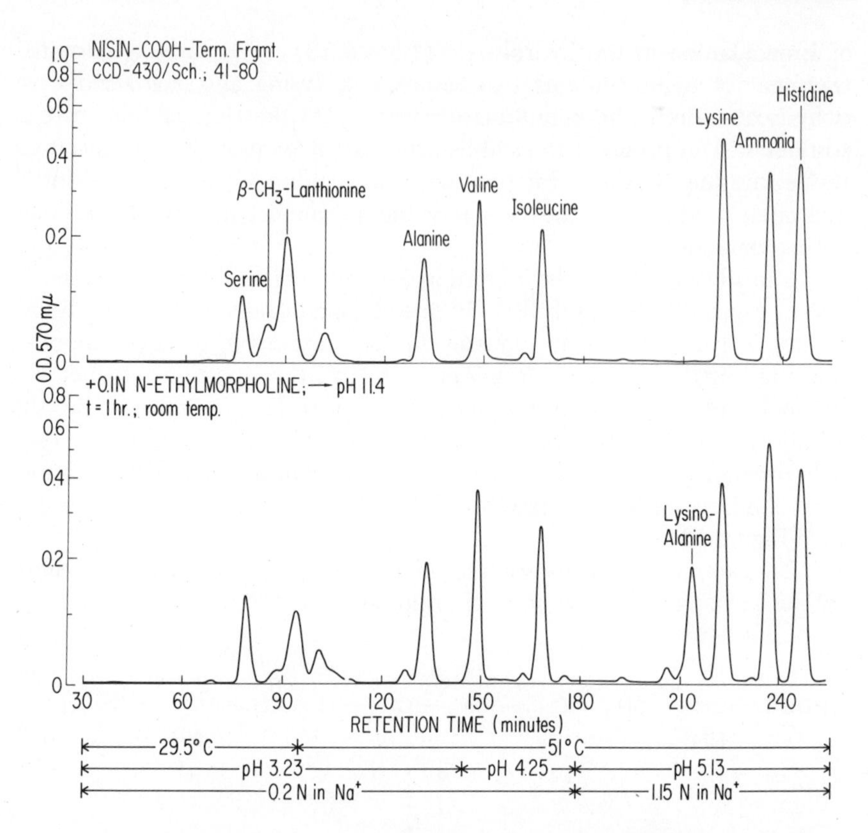

Figure 12. Formation of lysinoalanine in the carboxyl-terminal fragment of nisin. At the pH conditions chosen, β-methyllanthionine is already subject to partial β-elimination—note decrease in peak area for β-methyllanthionine comparing the amino acid analysis of treated (lower chromatogram) vs. untreated (upper chromatogram) material.

of lysine in the latter. Assuming that this exchange affects the same position in both peptides, the imino bridge of lysinoalanine is likely to link residues in identical positions in cinnamycin and duramycin.

The reaction postulated for the formation of ornithinoalanine from ornithine (derived from the hydrolysis of the guanido group of arginine) and dehydroalanine is shown in Figure 14.

Under the basic condition of the β-elimination reactions (vide supra) equilibria exist between reactants in the form of the α,β-unsaturated amino acids generated and nucleophiles present. Shifts in these equilibria are a function of a number of factors, among them the nucleophilicity of groups capable of adding across the double bonds of α,β-unsaturated amino acids. One of the most prevalent nucleophilic groups in proteins exposed to alkaline conditions is the ε-amino group of lysine, the addition of which takes place under circumstances where cystine, for instance, is subject to β-elimination. While the group leaving will give up sulfur with the formation of cysteine, the sulfhydryl group generated is apparently not a strong enough nucleophile to compete successfully in the addition reaction—unless the conformational situation in a given case is such that the side chain of certain residues is favored over that of others. The classic case of the alternative formation of lanthionine is that observed in wool (*12*). The effect of pH condition on the formation of

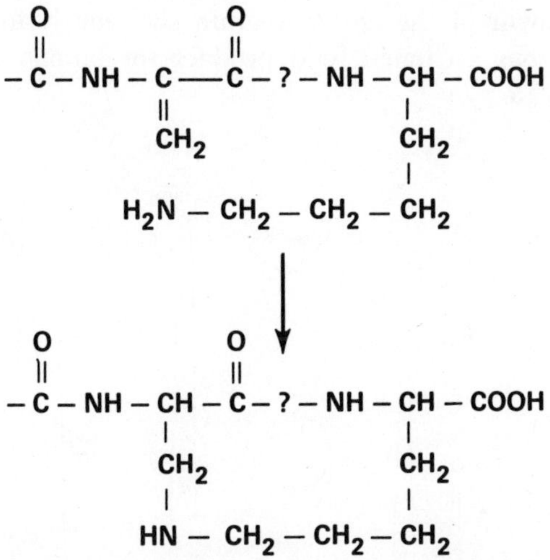

$$-C(=O)-NH-C(=CH_2)-C(=O)-?-NH-CH(COOH)-CH_2-CH_2-CH_2-CH_2-NH_2$$

$$\downarrow$$

$$-C(=O)-NH-CH(CH_2)-C(=O)-?-NH-CH(COOH)-CH_2-CH_2-CH_2-CH_2-NH-(CH_2)$$

Figure 13. Formation of lysinoalanine in cinnamycin and duramycin. Question mark indicates the uncertainty presently existing about the number of peptide bonds between the alanine and lysine moiety in lysinoalanine.

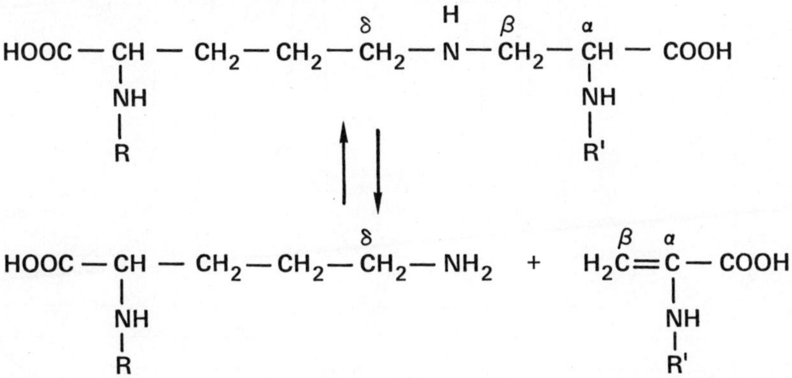

$$HOOC-CH(NH-R)-CH_2-CH_2-\overset{\delta}{CH_2}-\overset{H}{N}-\overset{\beta}{CH_2}-\overset{\alpha}{CH}(NH-R')-COOH$$

$$\uparrow\downarrow$$

$$HOOC-CH(NH-R)-CH_2-CH_2-\overset{\delta}{CH_2}-NH_2 \quad + \quad \overset{\beta}{H_2C}=\overset{\alpha}{C}(NH-R')-COOH$$

Figure 14. Formation of ornithinoalanine (see Figure 8 for the generation of ornithine from arginine).

lysinoalanine was clearly demonstrated in the case of lysine and dehydroalanine containing peptides (vide supra; nisin and subtilin) where the yield of lysinoalanine decreased with increasing concentrations in the hydroxyl ion, i.e., the reverse reaction of β-elimination occurred.

The product of the addition reaction, lysinoalanine, has been observed on numerous occasions of treatment of proteins under alkaline conditions (*7, 8, 13, 14, 15*). The amino acid is stable to the conditions of total protein hydrolysis and is easily accounted for in amino acid analysis. A segment of the chromatogram showing lysinoalanine found in the sample from a canned food product for human consumption is shown in Figure 15.

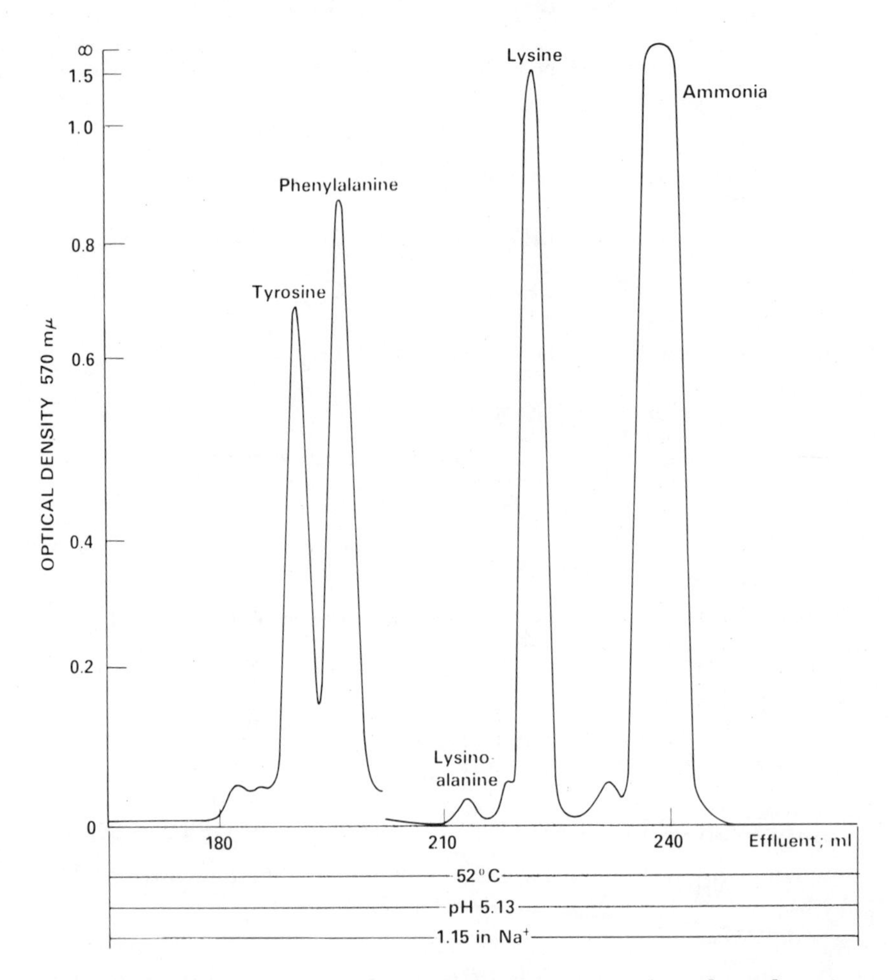

Figure 15. The presence of lysinoalanine in a canned food product for consumption by man

Physiological Aspects

The presence of lysinoalanine—an amino acid not commonly found in proteins—in food products to be eaten by man is justifiable cause for concern. Studies on the nutritive value (*15, 16*) and the toxicity (*17*) of lysinoalanine-containing proteins have already been reported. It seems highly desirable to conduct metabolic studies, initially with peptides containing appropriately radiolabeled lysinoalanine residues. Knowledge of the metabolic fate of lysinoalanine and possible breakdown products should greatly assist future investigations aimed at answering questions about the pharmacology and toxicology of lysinoalanine.

The formation of lysinoalanine in home-cooked food has been reported recently for conditions of preparation well outside the alkaline pH range (*18*). Is lysinoalanine ubiquitous in the heat- and alkali-treated preparations of man's daily dietary intake, and has it been so for ages? If the answer to this question is yes, man must have developed (induced!), since the days of having learned to tame the fire, enzymes or enzyme systems for the effective metabolic utilization of lysinoalanine. The early application of peptides with radiolabeled lysinoalanine becomes a more urgent one.

Not for one moment must we, however, overlook the prerequisite in the form of the presence of dehydroalanine for the formation of lysinoalanine. Do all residues of the α,β-unsaturated amino acid generated react by way of β-addition with ϵ-amino groups of lysine (or with other nucleophiles)? Which are the physiological effects of α,β-unsaturated amino acids in proteins of food products of man's daily diet? The answers to these questions are largely unknown.

One problem of prime importance is the reliable determination of the number of residues of α,β-unsaturated amino acids in proteins. Direct amino acid analysis subsequent to total hydrolysis of proteins is not feasible. The α,β-unsaturated amino acids are subject to degradation with the formation of amide (ammonia) and α-keto-acid. The numbers and types of α,β-unsaturated amino acids in nisin (*1*) and subtilin (*10*) and in the fragments of the two peptides were, nevertheless, determined by amino acid analysis, only, however, after the addition of mercaptan across the double bonds of dehydroalanine and dehydrobutyrine (*19*). Using benzylmercaptan, the addition products are S-benzylcysteine (from dehydroalanine) and β-methyl-S-benzylcysteine (from dehydrobutyrine). The two thioether amino acids are eluted from ion exchange columns of the amino acid analyzer free from interference by other amino acids (Figure 16).

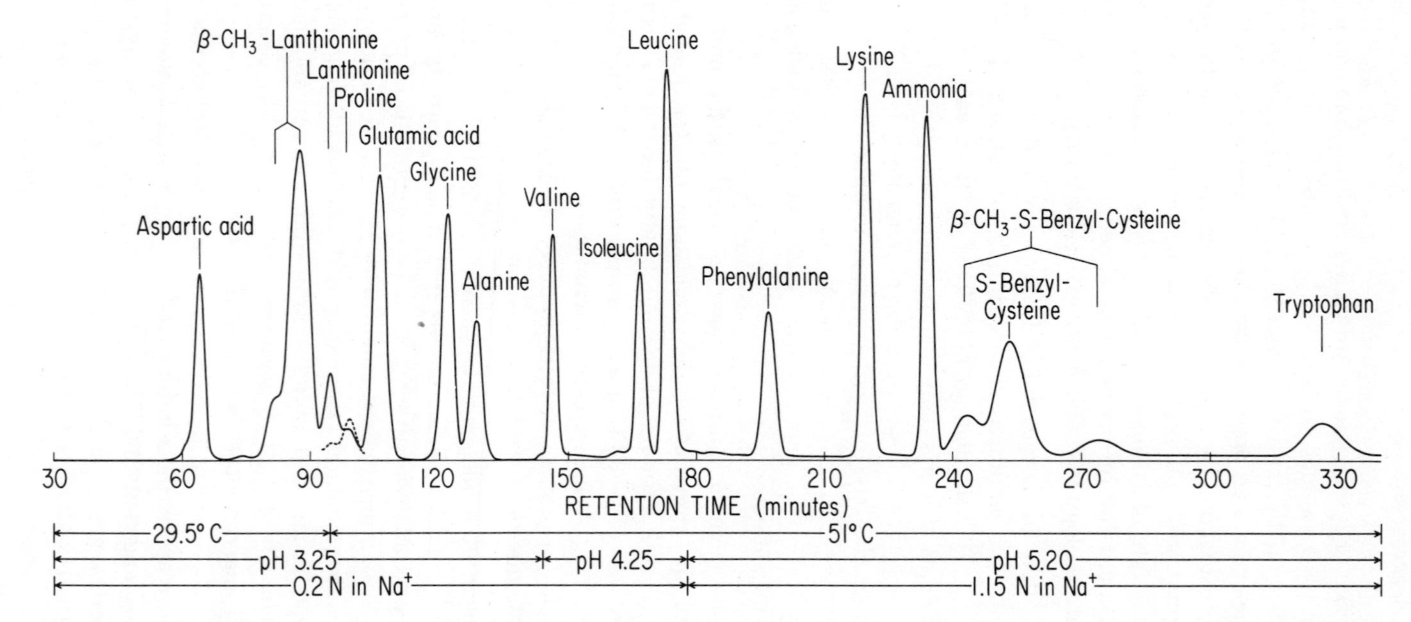

Figure 16. Determination of the type and quantity of α,β-unsaturated amino acids in subtilin—note the elution positions of the addition products of benzylmercaptan to dehydroalanine and dehydrobutyrine: S-*benzylcysteine and β-methyl-*S-*benzylcysteine (two pairs of separable diastereoisomers).*

Are α,β-unsaturated amino acids in proteins of the daily diet harmful to man? Considering nisin and its longtime intake by man in, admittedly, small quantities together with dairy products, the answer to be given will be no. The gastrointestinal system apparently provides safeguards which—presumably by proteolytic digestion—degrade nisin rapidly, thus not allowing biologically active fragments to reach the circulation. However, an important line of distinction must be drawn in this context. Parenterally administered nisin and the amino-terminal fragments of nisin (*1*) show profound physiological effects, for instance, on neonatal and neoplastic tissue. Nisin and nisin fragments induce fetal resorption in rabbits and rats, and inhibit the growth of tumor tissue. It remains to be seen whether other peptides and proteins with α,β-unsaturated amino acids are also degraded as effectively and completely by proteases of the gastrointestinal tract. It is conceivable that peptides of low molecular weight and the correct physical properties are resorbed from the gastrointestinal tract and that they are transported by the circulation to potential sites of beneficial or adverse action. A case in point is the oral administration of highly hydrophobic peptides that induce fetal resorption in rats.

The arguments offered above call for caution and the careful, however scientifically sound, evaluation of the safety of food items for human consumption. Close toxicological and pharmacological surveillance of nutritional products with additives and/or the history of chemical or physical processing is of paramount importance. The importance of this surveillance is demonstrated most convincingly by observations made recently in the field of hyperalimentation. Here the patient, deprived of the ability of oral food intake, is dependent on the intravenous infusion of solutions of essential amino acids. For solution stabilization sodium bisulfite was added. This presumed preservative, however, reacts with tryptophan (*20*) with the formation of products which affect liver tissue adversely (21).

The α,β-unsaturated amino acid dehydroalanine is a necessary precursor for the formation of lysinoalanine. The chemical events covered in the preceding discussion took place in the in vitro environment. Do α,β-unsaturated amino acids play a role in any in vivo environment other than that of microorganisms? From the latter domain have been isolated the peptides richest in α,β-unsaturated amino acids, nisin (*1*) and subtilin (*10*). No direct evidence is available at this time to document convincingly the possible physiological role of α,β-unsaturated amino acids in higher organisms. That dehydroalanine is the constituent amino acid of a plant protein has been reported for phenylalanine ammonia lyase from potato tubers (*22*). There is one case on record for the presence of dehydroalanine in a protein from a mammalian system, namely, histidine ammonia lyase from rat liver (*23*).

The study of induced fetal resorption, mentioned briefly before, was conceptually based on the potential of peptides with α,β-unsaturated amino acids to intercept essential sulfhydryl groups by way of addition across the α,β-double bond. Clearly, the unsaturation here is provided from an exogenous source. It thus remains to be established whether, and to which extent, endogenous events are linked via α,β-unsaturated amino acids. Compounds, physiologically as significant as amides and ketoacids, are conceivably to be derived from α,β-unsaturated amino acids subsequent to addition of water across the double bond, and the translation of the β-elimination reaction (vide supra) into the physiological environment would provide α,β-unsaturated amino acids from the appropriately β-substituted amino acids serine, cysteine (cystine), thereonine, the lanthionines, and perhaps lysinoalanine. The prospect of lysinoalanine formation in endogenous systems is presently the subject of an investigation relating to questions of antagonistic action towards hormones of reproductive physiology.

The chemistry of the α,β-unsaturated amino acids, inclusive of that relating to lysinoalanine, is a dynamic and fascinating one. Lysinoalanine, definitely established as the constituent of peptides of microbial origin, may well play a positive role in physiological events of higher forms of life. While the latter aspects call for continued multidisciplinary effort to provide answers to numerous questions, this immediate outgrowth of the involvement in the chemistry of α,β-unsaturated amino acids contributes to various aspects of physiology. Peptide amides, notably hormone releasing and hormone release inhibiting factors, are being synthesized on dehydroalanine resins, a variant in peptide synthesis, in which the growing peptide chain is attached to a solid support via dehydroalanine (*24*).

Acknowledgments

The enthusiasm and effort of able co-workers produced the results discussed here. John L. Morell determined the structure of nisin, H. H. Kiltz and E. Nebelin that of subtilin. H. H. Kiltz initiated the studies on cinnamycin, since then advanced by H. C. Chen and C. H. Chapin. Judith H. Brown is responsible for the information about duramycin. The studies on lysinoalanine are being continued by J. H. Brown, S. Nanno, and C. H. Chapin.

Literature Cited

1. Gross, E., Morell, J. L., *J. Am. Chem. Soc.* (1971) **93,** 4634.
2. Rogers, L. A., Whittier, E. O., *J. Bacteriol.* (1928) **16,** 211.
3. Janson, E. F., Hirschmann, D. J., *Arch. Biochem.* (1944) **4,** 297.
4. Bendict, R. G., Dvonch, W., Shotwell, O. L., Pridham, T. G., Lindenfelser, L. A., *Antibiot. Chemother.* (1952) **2,** 591.
5. Shotwell, O. L., Stodola, F. M., Michael, W. R., Lindenfelser, L. A., Dworschack, R. G., Pridham, T. G., *J. Am. Chem. Soc.* (1958) **80,** 3912.
6. Photaki, I., *J. Am. Chem. Soc.* (1963) **85,** 1123.
7. Bohak, Z., *J. Biol. Chem.* (1964) **239,** 2878.
8. Patchornik, A., Sokolovsky, M., *J. Am. Chem. Soc.* (1964) **86,** 1860.
9. Ziegler, K., Melchert, I., Lürken, C., *Nature* (1967) **214,** 404.
10. Gross, E., Kiltz, H. H., Nebelin, E., *Hoppe-Seyler's Z. Physiol. Chem.* (1973) **354,** 810.
11. Gross, E., Nanno, S., Chapin, C. H. (1976), in preparation.
12. Horn, M. J., Jones, D. B., Ringel, S. J., *J. Biol. Chem.* (1941) **138,** 141.
13. Zahn, H., Lumper, L., *Biochim. Biophys. Acta* (1966) **121,** 173.
14. Ziegler, K., *J. Biol. Chem.* (1964) **239,** PC 2713.
15. de Groot, A. P., Slump, P., *J. Nutr.* (1969) **98,** 45.
16. van Beek, L., Feron, V. J., de Groot, A. P., *J. Nutr.* (1974) **104,** 1630.
17. Woodard, J. C., Short, D. D., *J. Nutr.* (1973) **103,** 569.
18. Sternberg, M., Kim, C. Y., Schwende, F. J., *Science* (1975) **190,** 992.
19. Gross, E., Kiltz, H. H., *Biochem. Biophys. Res. Commun.* (1973) **50,** 559.
20. Kleinman, L. M., Tangrea, J. A., Gallelli, J. F., Brown, J. H., Gross, E., *Am. J. Hosp. Pharm.* (1973) **30,** 1054.
21. Grant, J. P., Cox, C. E., Kleinman, L. M., Maher, M. M., Pittman, M. A., Tangrea, J. A., Gross, E., Brown, J. H., Beazley, R. M., Jones, R. S. (1976), in preparation.
22. Hanson, K. R., Haxier, E. A., *Arch. Biochem. Biophys.* (1970) **141,** 1.
23. Givot, I. L., Abeles, R. H., *J. Biol. Chem.* (1970) **245,** 3271.
24. Gross, E., Noda, K., Nisula, B., *Angew. Chem., Int. Ed. Engl.* (1973) **12,** 664.

RECEIVED March 12, 1976.

3

Chemical Modification of Soy Proteins

EDWIN W. MEYER and L. D. WILLIAMS

Food Research, Central Soya Co., Inc., Chicago, Ill. 60639

This selective review, which deals primarily with the chemical modification of soy proteins, is further limited to nondestructive chemical reactions which alter physical and biochemical properties of importance in food systems. Soy protein products have been modified by various chemical reactions including: (a) treatment with alkalies and acids, (b) acylation, (c) alkylation and esterification, and (d) oxidation and reduction. In most instances these reactions have been applied to heterogeneous protein mixtures containing nonprotein impurities, and often to proteins of unknown prior history. Nonetheless, these reactions indicate that protein functional properties of value in food fabrication can be altered significantly through reaction with chemical reagents. It is recognized that chemically modified proteins must be critically evaluated for food safety.

During the past fifteen years a number of soy protein products have become established as useful ingredients in the manufacture of processed foods. The growing acceptance of these products for use in food manufacture has been prompted by their varied functional properties and good nutritional qualities. Examination of current utilization patterns indicates that these protein products are used in food for the extension and replacement of traditional protein ingredients and as components in newly designed foods (Table I). The use of soy protein products is expected to continue to grow because of the increasing cost and decreasing availability of traditional animal protein foods and food ingredients, and the prospect of an inadequate supply of protein-calorie-rich foods on a worldwide basis.

Much progress has been made in developing commercial soy protein products with differing functional characteristics. This has been accom-

Table I. Food Uses of Soy Protein Products

Meat foods
Meat analogs and extenders
Baked products
Dairy-type foods
Breakfast cereals and bars
Infant formulations and baby foods
Dietary and health foods
Protein hydrolyzates (HVP)
Confections

plished through protein enrichment and protein isolation, as well as through the application of mechanical, thermal, and chemical treatments, and selected combinations thereof. Nonetheless it is now recognized that limitations in the functional properties, including flavor, texture, and color, inhibit expanded utilization for the fabrication of food. Consequently, research interest in exploring various chemical and biochemical means for altering the functional properties of soy protein products is expanding.

Indeed a recent report to the National Science Foundation on "Protein Resources and Technology" recommends the study of the "modification of the functionality of oilseeds and legume proteins by physical, chemical, and enzymatic methods to facilitate development of protein materials with wide versatility and acceptability in formulated foods" (*1*).

Soybean Proteins and Protein Products

In introducing the subject of chemical modification, we will describe briefly the proteins (*2*) and protein products of the soybean (*3*).

Soybean Proteins. Protein products for food are derived from the cotyledonary tissue of the soybean seed. At the subcellular level (*4*) the proteins of the seed are distributed in cellular inclusions or protein bodies and in the surrounding cytoplasmic matrix (*2*). These proteins have often been grossly classified as storage globulins and biologically active proteins (Tables II and III). Although a number of the proteins have been separated in a reasonable state of purity (*2*), not one of the major storage globulins has been fully characterized down to the primary structure. In contrast, primary structures for two of the trypsin inhibitors, the Bowman–Birk and Kunitz inhibitors, have been elucidated through the notable efforts of Ikenaka and co-workers (*5, 6*). It now appears that at least seven of the soybean globulins have a multisubunit structure, and certain of these undergo association-dissociation reactions and sulfhydryl–disulfide interactions. While a review of the current state of knowledge about the structure and properties of the proteins of the

Table II. Major Storage Proteins of Defatted Soybean Flakes

Ultra-centrifuge Fraction	*Approx. % H_2O Sol. Protein*	*Components*	*Approx. Mol Wt*
2S	18	α-Conglycinin (?)	22,000
		Others	
7S	27	β-Conglycinin	330,000 (?)
		γ-Conglycinin	180,000
		Glycinin monomer	180,000
11S	34	Glycinin dimer	360,000
15S	6	Polymers (Glycinin tetramer, etc.)	~ 600,000

Table III. Major Biologically Active Proteins of Defatted Soybean Flakes

Ultra-centrifuge Fraction	*Approx. % H_2O Sol. Protein*	*Components*	*Approx. Mol Wt*
2S	9	Trypsin inhibitors (multiple)	8,000 22,000
6S, 7S	6	Hemagglutinins (multiple)	90,000–100,000
		Lipoxygenase (iso-enzymes)	100,000
?	?	Numerous enzyme systems: proteases, amylases, lipases, urease, etc.	—

Table IV. Potential Substrates for Chemical Modification

Soy flours and grits—full-fat and defatted
Soy protein concentrates
Soy protein isolates
Textured protein products
a. spun fibers
b. spun or extruded shreds
c. extruded products
d. compacted products
Soy protein extracts or "milks"
Soy "whey"

soybean is not within the scope of this presentation, it is important to recognize that the incomplete and fragmentary nature of such knowledge makes the interpretation of chemical reactions difficult.

Soy Protein Products. Over the years the variety of soy protein products available to the food processor has been expanding (*3*). In more recent years the introduction of such products has accelerated. The various classes of commercially available soy protein products are shown in Table IV.

The first soybean protein ingredients made commercially available for food use included full-fat and defatted soy flours and grits (*3, 7, 8*). These products contain ca. 46–59% protein ($N \times 6.25$) on a moisture-free basis and are available with various heat treatments for specific end-use. Soy protein concentrates and soy protein isolates were introduced into the market about 15 years ago (*3, 9, 10, 11*). By definition soy protein concentrates must contain no less than 70% protein ($N \times 6.25$) and isolates no less than 90% protein ($N \times 6.25$), all on a moisure-free basis. In the past several years there has been much activity in the commercialization of textured soy protein products intended for the extension and replacement of meat. These textured products may be obtained through fiber spinning, shred formation, extrusion, or compaction (*12, 13, 14, 15*). In addition, soybean "milk" solids and the heterogeneous proteins in soybean "whey" might serve as useful substrates in chemical modifications for food use. This short recitation of commercial products illustrates the type of crude protein fractions available for practical modification. Many useful functional properties have been ascribed to these new food proteins.

Chemical Modification

The chemical modification of proteins has been reviewed recently in a monograph by Means and Feeney (*16*). Other useful reviews include those of Stark (*17*), Spande and Witkop (*18*), and Cohen (*19*). Feeney (*20*) has recently reviewed chemical changes in food proteins. This particular review should be consulted for additional references of both general and specific nature.

The scope of this review is limited to those reactions with specific chemical reagents which result in the formation of new covalent bonds. It explicitly does not cover the many changes in protein structure caused by alteration of ionic, hydrogen, and hydrophobic bonding without concomitant establishment of new covalent bonds. In addition, the literature on this subject is reviewed in a very selective fashion simply to illustrate the types of chemical reactions studied and the nature and depth of such studies. Primary emphasis is placed upon investigations directed toward the modification of properties useful in food systems rather than those concerned solely with protein structure and biological properties (e.g., enzyme activity). Chemical modifications which have been suggested for altering the functional properties of soy proteins for food use are summarized in Table V. Since this information has been gleaned primarily from patent literature, it is not intended that this summary portray current commercial practice.

Table V. Altered Functional Properties through Chemical Modification

Reaction	*Change in Properties*
Alkalies (pH > 10)	
sodium hydroxide, etc.	a. Increased dispersibility and solubility
	b. Increased resistance to aggregation (heat, etc.)
	c. Increased elasticity—better fiber formation
Acylation	
acetic anhydride	a. Improved solubility in acidic foods
succinic anhydride	b. Increased solubility
	c. Lower viscosity
	d. Increased tolerance to Ca^{2+}
	e. More resistance to aggregation
Oxidation	
hydrogen peroxide (alkaline)	Reduced viscosity
chlorine	
peracid salts	
Reduction	
sulfite and related salts	a. Reduced viscosity in water dispersion
	b. Increased viscosity in salt solution
	c. Increased resistance to aggregation

Examination of the literature on the chemical modification of proteins and protein-containing products derived from soybeans reveals that the major motivation for such studies was often the development of improved products for industrial rather than food or feed utilization (*21, 22, 23, 24*). Such research activity was at the heart of the Chemurgic movement begun in the early thirties. Hence it becomes understandable that the patent literature dealing with the chemical modification of soy protein products is much more extensive than the periodical literature. This, together with our imperfect understanding of the basic structure of the soy proteins, explains why many of the reactions cited in this review cannot be chronicled in detail as to reaction course or consequence but remain highly speculative in nature. This does, however, focus attention on the many areas which need clarification.

It should be emphasized that chemically modified proteins intended for food use must be critically and extensively evaluated for their short- and long-term biological effects. Even chemical treatments not designed for modification of the protein products during processing, but applied for some processing advantage, must be studied at the molecular level to assess the biological implications of such treatment. Understandably, the design of biological experiments to evaluate these effects will be relevant to the intended food utilization.

Alkalies and Acids. The older literature dealing with the treatment of soybean proteins with alkaline substances is quite extensive since these agents have often been used for protein extraction, solubilization, and property modification, including improved solubility, increased adhesive properties, and lower viscosity in dispersion and fiber formation (*12, 21, 23, 24*). The alkali treatment of soy protein for industrial use is done under conditions which are more severe (higher temperature and higher pH, such as possibly 50°C at pH > 13) than those intended for food usage.

More recently the alkaline treatment of soy proteins became a matter of concern with the finding by Woodard and Alvarez (*25*) that feeding of severely alkali-treated industrial soy protein to rats resulted in a cellular abnormality, cytomegaly, in the kidneys. Also, de Groot and Slump (*26*) demonstrated that severe alkaline treatment of a food-grade soy protein resulted in reduced nutritional value, disappearance of lysine, cystine, and serine, and the formation of lysinoalanine, LAL (*27, 28*). Recent work reported by de Groot and co-workers (*29*) indicates that the renal abnormality in rats is caused by free or peptide-bound LAL rather than protein-bound LAL. The relationship between biological activity and peptide structure has not been established. These workers also reported that the nephrotoxicity of LAL may be species-specific since they were unable to produce the cytomegalic response in mice, hamsters, dogs, monkeys, and quail at levels up to 1000 mg/kg diet. This question is expected to be resolved in the near future. In addition, more definitive descriptions of conditions used for alkaline treatments are needed.

Lysinoalanine is formed by β-elimination reaction of cystine–cysteine and serine with the formation of dehydroalanine and the subsequent addition of the ϵ-amino group of lysine across the reactive C—C double bond (*30*). The formation of other amino acids such as ornithinoalanine (*31*), lanthionine (*32*), and β-aminoalanine (*33*) by similar mechanism has been described. Gross et al. (*34*) have pointed out that lysinoalanine occurs in the antibiotics cynnamycin and duramycin. There is evidence that LAL occurs in food proteins and products which have not been alkali-treated (*29, 35, 36*), and hence other mechanisms for the formation of LAL must be operative. Feeney (*20*) has pointed out the importance of establishing the impact of alkaline treatment with each protein in question since such proteins may undergo significant changes under quite varied conditions.

Acid hydrolyzates of soy protein products, either singly or in combination with cereal grain proteins, are used extensively as flavoring agents in food (*12, 22*). These are commonly grouped in the familiar term "hydrolyzed vegetable protein." Milder acid treatment has been described for the deamidation of proteins, but this has not been exploited in altering the functional properties of soy proteins. Finley (*37*) has described a mild acid treatment of wheat gluten to increase its protein solubility in fruit-based acidic beverages. There is need to resolve whether such improvement results from deamidation of the glutamine and asparagine residues or from the concomitant cleavage of peptide bonds.

Acylation. The acylation of soybean proteins with various acyl halides and anhydrides of low and high molecular weight was initially evaluated for its impact on properties of industrial value such as viscosity in dispersion, adhesion, foaming, and detergency (*21, 23, 24, 38*).

It is now evident that acylation with anhydrides (such as acetic and succinic anhydrides) and with lactones (such as β-propiolactone) is being proposed (*39*) for improving the solubility of soy protein isolates at acidic pH, particularly for the preparation of coffee whiteners. Moreover, this chemical modification has been evaluated for altering the food-use properties of several milk proteins (*40*), egg protein (*41*), wheat protein (*42*), fish protein (*43*), and single-cell protein (*44*).

Little attention has been given to the biological consequences of such acylation for food use. Creamer et al. (*40*) reported the result of growth and toxicity trials in rats and mice fed acetyl or succinyl casein and acetyl milk whey protein. These proteins had lower protein efficiency ratios than casein, and added L-lysine hydrochloride did not correct this deficiency. However, a trial with acetyl casein in mice showed no overt toxicity. These workers (*40*) concluded that none of the milk proteins as modified with anhydrides can be regarded as suitable for inclusion in acidic food products. It is obvious that a thorough biological evaluation is needed to assess whether acylated soy proteins are suitable for food use.

Insofar as soy protein for food is concerned, no definitive study of the course of the acylation reaction has been made. Hoagland (*45, 46*), in a study of the acylation of β-casein, found that various acyl groups altered the calcium sensitivity of the protein and its ability to associate and to form aggregates. This work indicates that the predominant reaction is the acylation of the ϵ-amino group of the lysine residues. A later study by Evans et al. (*47*) of the properties of the succinyl, maleyl, and glutaryl derivatives of β-casein revealed that carboxyacylation is important in reducing casein–casein interactions and protein aggregation. These are basic changes of importance in modifying various novel food proteins, including soy protein, to fit the need in a diversity of food systems, both fluid and solid.

Alkylation and Esterification. A number of alkylating, arylating, and related reagents have been used in studies concerned with the composition, structure, and conformation of various soybean proteins and protein fractions (*2*). Such studies are not within the scope of this review. In the modification of soy protein properties for food there is no evidence that either alkylation or esterification has been considered to any serious extent.

McKinney and Uhing (*48*) examined the carboxymethylation of a commercial soy protein with sodium chloracetate in an alkaline medium. They found that the treated protein exhibited minimum solubility at a lower pH than did the untreated protein. In addition, the treated protein had increased resistance to reaction with formaldehyde, and, when exposed in the wet state, an increased resistance to putrefaction. These workers speculate as to the course of the reaction but give little evidence of the extent of reaction and reactive sites modified.

Soy protein products have been treated with formaldehyde and a variety of other aldehydes and ketones, both simple and complex (*21, 23, 24*). These efforts were directed toward production of insolubilized protein for adhesives, films, coatings, polymers, etc.

Soy protein products undergo the typical non-enzymatic browning reaction (Maillard reaction) in the presence of reducing sugars. The course of this reaction is similar to that of other proteins which have been studied in some detail. This subject has been reviewed recently by Feeney, Blankenhorn, and Dixon (*49*) as a special aspect of the carbonyl-amine reactions of proteins.

Some years ago, Fraenkel-Conrat and Olcott reported on the esterification of proteins with low-molecular-weight alcohols (*50*). Included in this study were a number of food proteins including casein, gluten, gliadin, and egg albumin. The esterified proteins showed altered solubility behavior, as expected, from blocking of carboxyl groups. Although esterification has been applied to industrial soy proteins (*21*), it is doubtful, because of the sensitivity of the esters to hydrolysis, whether it would be of real value in altering food-use properties.

Oxidation and Reduction. The patent literature concerned with the use of soy protein products for industrial purposes describes many oxidative treatments (*21, 23, 24*). The reagents most often cited include hydrogen peroxide, sodium peroxide, barium peroxide, and various oxidizing salts. Although this work gives little insight into the nature and extent of protein modification, it does indicate the issue of developing altered properties. Such oxidative treatments of soy protein have been applied mainly for color improvement or bleaching, increased solubility, and altered viscosity in alkaline dispersion.

More recently, Johnson and Anderson (*51*) have described a process for preparing isolated oilseed proteins for food wherein the source material is extracted in an aqueous alkaline medium containing hydrogen peroxide. The process is said to be applicable to a variety of oilseed and grain products. It is claimed that the oxidative treatment results in improved protein yields. In another patent Melnychyn and Wolcott (*52*) describe a process for preparing an isolated soy protein with reduced dispersion viscosity and improved flavor. The reagents used in this process include oxidizing salts such as potassium bromate and iodate, sodium chlorate and chlorite, and ammonium persulfate, and also chlorine and bromine. In neither instance is the nature or extent of protein alteration defined. These modifications may include halogenation of aromatic residues in addition to oxidation of sensitive protein sites such as methylsulfhydryl, sulfhydryl, and disulfide groups.

One of the most important applications of hydrogen peroxide in the food industry is reported to be preservation (*53*). This simple oxidant has been recognized as an alternative antimicrobial agent in milk systems (*54*). Such recognition has prompted a number of studies of the alteration of milk proteins by hydrogen peroxide (*55*). Hydrogen peroxide has also been evaluated in the production of a peanut protein concentrate (*56*) and for the detoxification of rapeseed flour for food use (*57*).

The course of the reaction of hydrogen peroxide and several other oxidizing reagents with proteins has been reviewed by Means and Feeney (*16*). The vulnerable protein sites include those of cysteine and methionine, and in certain instances cystine, tryptophan, and tyrosine. These reactions are of interest to the protein biochemist concerned with protein structure and bioactivity, yet the fundamental concern here is whether oxidation results in beneficial alteration of properties for food and whether such treatment results in diminished nutritive value.

In recent years, several reports have dealt with the decreased nutritional value of oxidized food proteins (*53, 57, 58, 59*). Although oxidation per se of the sulfur amino acids is a factor in reduced nutritive quality, Cuq et al. (*58, 59*) have demonstrated that oxidation of the intact protein causes a sharp reduction in the release of methionine by in vitro enzymatic digestion (*60*). This may be the answer to the enigma of the biological availability of methionine sulfoxide in intact proteins. Obviously, further work is needed to resolve this matter.

It has now been recognized that protein oxidation may arise through reaction with lipid hydroperoxides (*60*). This is of practical significance with the increasing production of processed composite foods and their extended storage in reaching the ultimate consumer. The nutritional significance of methionine oxidation in casein systems by autoxidizing lipids was described by Tannenbaum et al. (*61*). Autoxidizing lipids have been implicated in the generation of undesirable flavors in soy protein products (*12, 62*). The role of oxidizing indigenous and added lipids deserves further study concerning their impact on the flavor and, more importantly, the nutritional quality of food ingredients and food products.

Sodium sulfite and related sulfiting substances have been favored reagents in the extraction and preparation of protein products from defatted soybean meal for industrial, feed, and food purposes (*21, 23, 24*). The use of sodium sulfite or sulfur dioxide in the production of soy protein isolates has been described in a number of patents (*3, 23, 63*). Although not always explicitly stated, the purpose of such use has been to improve the extractability of the protein and to inhibit the aggregation of the isolated protein. Such treatment results in proteins with improved solubility and reduced viscosity. The use of sulfite also affords the beneficial effect of retarding the growth of micro-organisms during the wet processing of the protein.

The chemistry of the reaction of sulfites with proteins has recently been reviewed by Means and Feeney (*16*). It is presumed that the described course of the reaction applies to soy proteins, at least in a general sense. Because of the long history of the use of sulfur dioxide and sulfites in food and beverage processing, and in food preservation, the literature dealing with their use and food safety is quite extensive. Much of this literature has been reviewed in a U.S. Food and Drug Administration (FDA) re-examination of the Generally Recognized As Safe (GRAS) status of sulfiting agents for food (*64, 65*). Although this re-examination has not been completed, these agents (sodium sulfite, sodium bisulfite, sodium metabisulfite, potassium bisulfite, potassium metabisulfite, and alkali sulfites) continue to be recognized as safe by the FDA. Earlier it had been suggested that bound sulfur dioxide, such as that reacted with protein, may be more toxic than the free sulfites as demonstrated by significantly decreased growth rate in rats (*66*). Gibson and Strong (*67*) repeated this work and found that sulfite-treated casein produced growth rates in rats equal to that of untreated casein. Although evidence to date indicates that proteins treated with sulfiting agents are safe for use in food, further information relating conditions from laboratory studies to conditions used in commercial practices is desirable.

The thiols, mercaptoethanol and diethiothreitol, have been used extensively for the reductive cleavage of disulfide bonds and for the maintenance of a reductive environment in the isolation and characterization of specific soybean proteins (*2*). However, these reagents have not been used for the preparation of protein products for food. The odor and flavor of the low-molecular-weight thiol compounds are not compatible with usage in a broad range of food items, yet these substances contribute to the desirable odor and flavor of selected foods.

Other Reactions. Various soy protein products have been treated with numerous organic reagents in efforts to develop properties useful in adhesives, plastics, films, coatings, textile fibers, and the like (*21, 23, 24*). These purposefully modified proteins have not been recommended for food use.

On the other hand, it has now been recognized that the treatment of soybeans or soybean proteins with chemical substances in processing for component fractionation or isolation may result in the unrecognized production of toxic reaction products. A classic and historic example is the defatting of soybeans with trichloroethylene (TCE) to produce defatted soybean meal for animal feeding. During the late 1940s and early 50s it was determined that the feeding of TCE-extracted soybean oil meal (TESOM) resulted in many cases of fatal aplastic anemia in cattle (*68*). After much investigation it was concluded that the toxicity resulted from reaction of the protein with TCE to produce protein-bound S-dichlorovinylcysteine (*69*). Further, the corresponding synthetic amino acid was shown to produce the typical syndrome of aplastic anemia when orally administered to experimental animals. This experience underscores the need to examine critically the biochemical and biological consequences of both intentional and incidental protein modification caused by exposure to chemical reagents. In addition, the TCE toxicity problem indicates that specific reagent activity with proteins must be considered with regard to environmental conditions including concentration, temperature, moisture, pH, time, etc. An enigma in the TESOM problem was that all TCE-extracted soybean meals did not cause aplastic anemia. This was found to be related to temperature during fat extraction.

Fortunately, TCE-extracted meals were prepared for animal feeds, and apparently in recent years there have not been any proven cases of toxicities to humans from the purposeful chemical modification of food proteins.

Functional Properties of Soy Proteins

Any discussion of the modification of the functional properties of soy proteins for food use is incomplete without some comment on the nature and measurement of such properties. Various practical functional properties useful in processed food production have been ascribed to soy protein products (*12, 70, 71, 72*). These properties have become evident in the evaluation of soy proteins for the production of food products having recognized acceptance characteristics. Traditional knowledge of a protein's physical and chemical properties is much less significant in innovative food items. The properties ascribed to soy protein products include: (a) emulsion formation and stabilization, (b) fat absorption, (c) water absorption and retention, (d) viscosity generation, (e) gellation, (f) chunk, shred, fiber, and lamininate formation, (g) film formation, (h) dough formation, (i) adhesion and cohesion, (j) elasticity, (k) aeration, and (l) foam formation and stabilization (*70, 71, 72*).

The increasing effort to evaluate the food potential of a diversity of novel proteins has focused attention on methods to evaluate their potential utility in food fabrication. This attention has revealed that simple bench-top methods have little relevancy to complex food systems. Pour-El, in a review of the measurement of the functional properties of soy protein products (*73*), has concluded that "what is most lacking is a set of standardized methods which have been tested with numerous products and correlated with actual food tests."

Viewing the broad problem in a more fundamental sense makes it obvious that there is a serious lack of knowledge concerning the relation of basic protein structure to functional behavior. Such understanding would provide a rational basis for developing new proteins with predetermined specific properties for food use.

Conclusion

A review of the patent and periodical literature reveals that the intentional chemical modification of soy protein products for food use has received little attention. Such work as has been described is concerned with crude protein-containing fractions and heterogeneous protein mixtures. The full nature and extent of these chemical reactions have not been defined. In spite of the lack of definitive chemical modification studies, there is evidence of the beneficial alteration of gross properties related to potential food use.

Thus there exists a challenge in modifying soy proteins and other novel proteins through chemical reactions to make them more suitable for the extension and replacement of existing food proteins in processed foods and for the fabrication of new foods. Each modified protein intended for food use must be critically evaluated for short- and long-term biological effects related to the fundamental matters of food value and food safety. This need may be considered a deterent to the exploration of the value of chemical modification of proteins, yet such modification may play a significant role in achieving an extensive food utilization of novel proteins in the long years ahead.

Literature Cited

1. Milner, M. (Coordinator), "Protein Resources and Technology, Status and Research Needs," Report to the National Science Foundation, Oct. 6, 1975.
2. Wolf, W., *in* "Soybeans: Chemistry and Technoloy," Vol. 2 of "Proteins," A. K. Smith, S. J. Circle, *Eds.*, Avi, Westport, Conn., 1972.
3. Circle, S., Smith, A., *in* "Soybeans: Chemistry and Technology," Vol. 1 of "Proteins," A. K. Smith, S. J. Circle, *Eds.*, Avi, Westport, Conn., 1972.
4. Wolf, W., Baker, F., *Cereal Sci. Today* (1972) **17,** 125.
5. Odani, S., Ikenaka, T., *J. Biochem.* (1973) **74,** 697.
6. Koide, T., Ikenaka, T., *Eur. J. Biochem.* (1973) **32,** 417.
7. Horan, F., "Defatted and Full Fat Soy Flours by Conventional Processes," *in* "Processing of International Conference on Soybean Foods," Publication **ARS-71-35,** USDA, Wash., D. C., 1967.
8. Meyer, E., "Soybean Flours and Grits," *Proc. SOS/70 Int. Congr. Food Sci. Technol.*, 3rd, 1971.
9. Meyer, E., *J. Am. Oil Chem. Soc.* (1971) **48,** 484.
10. Meyer, E., "Soya Protein Isolates for Food," *in* "Protein as Human Food," R. A. Lawrie, *Ed.*, Butterworths, London, England, 1970.
11. Meyer, E., Williams, L. D., "Soy Protein Concentrates and Isolates," *in* "Proceedings World Soybean Research Conference," Univ. of Illinois, Champaign, Ill., 1975.
12. Smith, A., Circle, S., "Protein Products as Food Ingredients," *in* "Soybeans: Chemistry and Technology," Vol. 1 of "Proteins," A. K. Smith, S. J. Circle, *Eds.*, Avi, Westport, Conn., 1972.
13. Horan, F., *in* "New Protein Foods," Vol. 1A of "Technology," A. Altschul, *Ed.*, Academic, New York, N.Y., 1974.
14. Ashton, M., Burke, C., Holmes, A., *in* "Scientific and Technical Surveys," No. **62,** British Food Mfg. Inc. Research Assoc., Letherhead, Surrey, England, 1970.
15. Burke, C., "Textured Vegetable Proteins, II," "Survey of U.K. and U.S. Patent Literature," Scientific and Technical Surveys, No. 63, British Food Mfg. Ind. Research Assoc., Leatherhead, Surrey, England, 1971.
16. Means, G., Feeney, R., "Chemical Modification of Proteins," Holden-Day, San Francisco, 1971.
17. Stark, G., *Adv. Protein Chem.* (1970) **24,** 261.
18. Spande, T., Witkop, B., Degani, Y., Patchornik, A., *Adv. Protein Chem.* (1970) **24,** 97.
19. Cohen, L., *Annu. Rev. Biochem.* (1968) **37,** 695.
20. Feeney, R., *in* "Proteins for Human Consumption," C. E. Bodwell, *Ed.*, Avi, Westport, Conn., in preparation.
21. Circle, S., *in* "Soybeans and Soybean Products," Vol. I, K. S. Markley, *Ed.*, Interscience, New York, 1950.
22. Burnett, R., *in* "Soybeans and Soybean Products," Vol. II, K. S. Markley, *Ed.*, Interscience, New York, 1951.
23. Burnett, R., *in* "Soybeans and Soybean Products," Vol. II, K. S. Markley, *Ed.*, Interscience, New York, 1951.

24. Bain, W., Circle, S., Olson, R., *in* "Synthetic and Protein Adhesives for Paper Coating," MONOGRAPH SER. NO. **22,** Technical Assoc. Pulp and Paper Inc., New York, 1961.
25. Woodard, J., Alvarez, M., *Arch. Pathol.* (1967) **84,** 153.
26. de Groot, A., Slump, P., *J. Nutr.* (1969) **98,** 45.
27. Bohak, Z., *J. Biol. Chem.* (1964) **239,** 2878.
28. Patchornik, A., Sokolovsky, M., *J. Amer. Chem. Soc.* (1964) **86,** 1860.
29. de Groot, A., Slump, P., van Beek, L., Ferson, V., *in* "Proteins for Human Consumption," C. E. Bodwell, *Ed.*, Avi, Westport, Conn., in press.
30. Asquith, R., Booth, A., Skinner, J., *Biochim. Biophys. Acta* (1969) **181,** 164.
31. Ziegler, K., Melchert, I., Lurken, C., *Nature* (1969) **214,** 404.
32. Hupf, H., Springer, R., *Z. Lebenm. Unters. Forsch.* (1971) **146,** 138.
33. Asquith, R., Carthew, P., *Biochim. Biophys. Acta* (1972) **278,** 8.
34. Gross, E., Chen, H., Brown, J., "Abstracts of Meetings," 166th National Meeting, ACS, 1973, BIOL 80.
35. Sternberg, M., Kim, C., Plunkett, R., *J. Food Sci.* (1975) **40,** 1168.
36. Sternberg, M., Kim, C., Schwende, F., *Science* (1975) **190,** 992.
37. Finley, J., *J. Food Sci.* (1975) **40,** 1283.
38. Meyer, E., Circle, S., U. S. Patent **2,932,589,** 1960.
39. Melnychyn, P., Stapley, R., U. S. Patent **3,764,711,** 1973.
40. Creamer, L., Roeper, J., Lahrey, E., *N. Z. J. Dairy Sci. Technol.* (1971) **6,** 107.
41. Gandhi, S., Schultz, J., Boughey, F., Forsythe, R., *J. Food Sci.* (1968) **33,** 163.
42. Grant, D., *Cereal Chem.* (1973) **50,** 417.
43. Groninger, H., Jr., Miller, R., *J. Food Sci.* (1975) **40,** 327.
44. McElwain, M., Richardson, T., Amundson, C., *J. Milk Food Technol.* (1975) **38,** 521.
45. Hoagland, P., *J. Dairy Sci.* (1966) **49,** 783.
46. Hogland, P., *Biochemistry* (1968) **7,** 2542.
47. Evans, M., Irons, L., Jones, M., *Biochim. Biophys. Acta* (1971) **229,** 411.
48. McKinney, L., Uhing, E., *J. Am. Oil Chem. Soc.* (1971) **36,** 49.
49. Feeney, R., Blankenhorn, G., Dixon, H., *Adv. Protein Chem.* (1975) **29,** 135.
50. Fraenkel-Conrat, H., Olcott, H., *J. Biol. Chem.* (1945) **161,** 259.
51. Johnson, R., Anderson, P., U. S. Patent **3,127,388,** 1964.
52. Melnychyn, P., Wolcott, J., U. S. Patent **3,630,753,** 1971.
53. Slump, P., Schreuder, H., *J. Sci. Food Agric.* (1975) **24,** 657.
54. Roundy, Z., *J. Dairy Sci.* (1958) **41,** 1460.
55. Cooney, C., Morr, C., *J. Dairy Sci.* (1972) **55,** 567.
56. Chandrasekhar, M., Ramanna, B., Jaganath, K., Ramanthau, S., *Food Technol.* (1971) **25,** 596.
57. Anderson, G., Li, G., Jones, J., Bender, F., *J. Nutr.* (1975) **105,** 317.
58. Cuq, J., Provansal, M., Guilleux, F., Cheftel, C., *J. Food Sci.* (1973) **38,** 11.
59. Cuq, J., Provansal, M., Besanson, P., Cherter, C., Work Documents, *Int. Congr. Food Sci. Technol.*, IV, Madrid, Topic 7a.
60. *Nutrition Rev.* (1973) **31,** 220.
61. Tannebaum, S., Barth, H., Le Roux, J., *J. Agric. Food Chem.* (1969) **17,** 1353.
62. Cowan, J., Rackis, J., Wolf, W., *J. Am. Oil Chem. Soc.* (1973) **50,** 426A.
63. Hanson, L. P., "Vegetable Protein Processing," Noyes Data Corp., Park Ridge, N.J., 1974.
64. *Food Technol.* (1975) **29,** 117.
65. "GRAS (Generally Recognized As Safe) Food Ingredients—Sulfiting Agents," NTIS Publication **PB-221 217,** U. S. Dept. Commerce, Wash., D.C., 1972.

66. Bhagat, B., Lockett, M., *Food Cosmet. Toxicol.* (1964) **2**, 1.
67. Gibson, W., Strong, F., *Food Cosmet. Toxicol.* (1974) **12**, 625.
68. Pritchard, W., Rehfeld, C., Mizuno, N., Sautter, J., Schultze, M., *Am. J. Vet. Res.* (1956) **17**, 425.
69. McKinney, L., Weakley, F., Eldridge, A., Campbell, R., Cowan, J., Picken, J., Biester, H., *J. Am. Chem. Soc.* (1957) **79**, 3932.
70. Wolf, W., *J. Agric. Food Chem.* (1970) **18**, 969.
71. Mattil, K., *J. Am. Oil Chem. Soc.* (1971) **48**, 477.
72. Mattil, K., *J. Am. Oil Chem. Soc.* (1974) **51**, 81A.
73. Pour-El, A., "Measurement of Functional Properties of Soy Products," Proceedings World Soybean Research Conference, Univ. of Illinois, Champaign, Ill., 1975.

RECEIVED January 26, 1976.

4

Determinants of the Functional Properties of Proteins and Protein Derivatives in Foods

DALE S. RYAN

Department of Food Science, University of Wisconsin–Madison, Madison, Wisc. 53706

The improvement of food proteins involves either an increase in the nutritional value of the protein or the improvement of the functional properties of the protein. The development of methodologies for altering the functional properties of proteins would increase the usefulness of many nonconventional protein resources. The functional properties of food proteins reflect both the molecular structure of the protein and the intermolecular interactions in which the protein participates. An understanding of the forces responsible for both protein structure and protein interactions should lead to an increased understanding of the molecular basis for functionality. The evidence for the roles of van der Waals forces, hydrogen bonding, ionic bonding, and hydrophobic interactions in stabilizing protein structures is examined. The influence of protein structure, protein–protein interactions, protein–lipid interactions, disulfide bonding, and chemical modification on the functional properties of food proteins is discussed.

Criteria for evaluating what is and what is not an improvement in a food protein are at present poorly defined but can be divided roughly into two areas: improvement of nutritional quality and improvement of functional properties. In this context the term functional properties refers to such properties as emulsifying capacity and stability, foam stability, water absorption, gelation, dough-forming properties, etc. The goal of this paper is to clarify the molecular basis of protein functionality. I want to emphasize, however, that the other criteria for "improvement" of food proteins (i.e., increased nutritional value) cannot be ignored and may in many cases rule out the use of modifications of different kinds even when they give improved functional properties.

The functional properties of food proteins is a subject of considerable interest and importance. For example, in the last two decades much effort has gone into the development of various nonconventional sources of protein. The vast majority of this effort, however, has been concentrated on various aspects of production of economical protein resources and not on the actual utilization of these resources. The result is that many nonconventional protein resources, while they can now be produced fairly easily, can in many cases be produced only in a form lacking desirable functional properties. The point is simply that a protein, even though it may have excellent amino acid balance and all other prerequisites for a nutritionally superior protein, will have no impact on human nutrition unless it has the functional properties necessary for its incorporation into food systems. A specific example of this problem is the case of fish protein concentrate, the utilization of which is severely restricted by its very limited solubility in water. There are, of course, many other examples, but for present purposes it is sufficient to emphasize that there is a real need for the development of chemical, physical, or enzymatic methods that increase the usefulness of proteins in food systems.

Modification of the functional properties of a food protein is illustrated most dramatically by the development of procedures for the production of extruded plant proteins (textured vegetable protein). This rather simple procedure allows the conversion of a rather insoluble powder such as a soybean extract to a textured state with vastly altered functional properties, thereby increasing enormously the potential usefulness of the protein in a variety of food systems. This paper is not intended primarily to catalog the success or failure of various approaches to the modification of functionality but rather to explore two basic questions: (a) What are the limitations in our current knowledge which prevent us from developing a variety of methodologies for alteration of specific functional properties? (b) What kind of general conceptual framework will allow the most clarity in our understanding of protein functionality and the quickest advances in our abilities to improve functionality intentionally?

The largest single difficulty in the development of methodologies for functionality alteration is our present inability to predict how a particular chemical or structural change might affect a particular functional property. It is not at all clear, for example, which structural features of a protein are in fact responsible for the ability of a protein to act as an emulsifying agent. The same is true of most other functional properties of importance in foods (water-binding capacity, foamability, gel-forming ability, and so on). Until protein structure can be meaningfully correlated with specific functional properties, the design of experiments to alter protein functionality intentionally will be largely a hit-and-miss operation.

There are, of course, many possible approaches to the study of the correlation between protein structure and functionality. An approach which could be rewarding is to see functional properties of food proteins as fundamentally related to the way in which a particular protein interacts with solvent and with other constituents of a food system. That is, basically if you can understand and predict how a protein will interact with other molecules in a complex system such as a food, then you should have the necessary information not only to make reasonable predictions about the functional properties of the protein but to make educated guesses as to how a particular alteration of the protein by chemical or other means might change the functional properties. It should also be clear that to understand how a protein interacts with its environment you must begin with an understanding of the forces responsible for these interactions. Several advantages of considering functionality in this larger context will be apparent as we continue, but one obvious advantage of beginning with molecular forces and working towards functionality is that it gives a relatively well understood (though still controversial) base from which to begin a discussion of a more poorly understood area.

It is important to emphasize that, although the following discussion will be limited to noncovalent interactions and their eventual role in protein functionality, a variety of covalent interactions can also play very important roles in protein functionality.

van der Waals Forces

The most general and nonspecific forces which can eventually be expressed as functional properties of food proteins are van der Waals interactions. A detailed description of these forces can be quite complex. These are short-range forces whose magnitude is inversely proportional to approximately the sixth power of the distance between the molecules (*1, 2, 3*). They are frequently attributed to polar interactions between induced dipoles in adjacent atoms and therefore are related to the polarizabilities of the molecules or atoms involved. This accounts for the relative nonspecificity of these interactions; that is, no particular class of chemical structures is necessary. Van der Waals forces exist among all atoms including, for example, otherwise inert atoms such as gaseous argon where the dimer has been shown to exist, or between argon and molecular hydrogen (*4*).

The contribution of van der Waals interactions to the stability of protein structures is most clearly indicated by the packing density of amino acid residues in the interior of globular proteins (*5, 6*). The calculation of individual residue volumes from crystallographic data has been described by Richards and co-workers and involves the construction of Voronoi polyhedra around each atom with dimensions determined by the distance to and the van der Waals radii of adjacent atoms (*7, 8*). The summation of individual atomic volumes gives the volume of specific amino acid side chains. Calculations of this kind on proteins of well-defined three dimensional structure have shown the protein interior to be close-packed (*5, 7*). That is, the arrangement of side chains appears to have been designed to maximize the number of possible van der Waals contacts. The average volume of amino acid side chains in nine different proteins has been determined by Clothia (*5*), and the volumes correspond very closely to the volume occupied by the side chains in crystals of amino acids (Table I). This suggests that the folding of a protein molecule may be more similar to a phase transition to a "molecular crystal" than an association to a micelle-like or oil drop structure.

Table I. Volume Occupied by Amino Acid Side Chains in the Interior of Proteins (5)

Residue	*Average Volume in Protein (Å³)*	*Side Chain Crystal Volume (Å³)*
VAL	141.7	143.4
ALA	91.5	96.6
GLY	66.4	66.5
THR	122.1	124.3
ILE	168.8	169.7

Nature

The contribution of van der Waals forces to protein–protein interactions can also be determined by calculation of individual volumes of residues which are involved in the area of contact between the two proteins (*9*). In the case of three protein–protein complexes (the insulin dimer, trypsin–pancreatic trypsin inhibitor complex, and the α-β subunit interaction of oxyhemaglobin) these interface residues have volumes very close to the volume calculated for crystalline amino acids (*9*). It seems, therefore, that both the protein interior and surfaces involved in protein–protein contact are arranged to maximize packing density and therefore maximize van der Waals interactions. Two points should be mentioned, however, about the magnitude of van der Waals forces. First, it must be remembered that van der Waals interactions are individually extremely weak. The lifetime of a molecule held together solely by van der Waals forces, such as the gas phase dimer of argon, is limited by the collision frequency of the dimer (*4*). That is, the association energy is so small that on the average every collision with another molecule of any kind will result in dissociation of the van der Waals interaction complex. Secondly, atoms which make van der Waals interactions in the interior of a folded protein would also make van der Waals interactions with water molecules in the denatured protein molecule. The net contribution to the stability of the folded structure then is the difference between the energy of these two interactions, and may for that reason be very small. In spite of these considerations, however, the apparent maximization of van der Waals contacts suggests that some constraints on protein folding and on complex formation are imposed by van der Waals interactions.

The Hydrogen Bond

A second force which must be considered as a possible determinant of protein functionality is the hydrogen bond. The role which hydrogen bonding plays in protein stability and functionality has an interesting history. That the number of possible hydrogen bonds tends to be maximized during protein folding was suggested very early and was in fact one of the main strengths of the Pauling–Corey α-helix and β-sheet structures (*10*). More recently, the net contribution of hydrogen bonding to the stability of intra- and intermolecular structures has been questioned, and it now appears that at least in aqueous environments mono-functional hydrogen bonded structures are thermodynamically unstable (*11, 12, 13*).

In general, hydrogen bonds are much stronger than van der Waals interactions (*14*) and are also much more specific. They can be thought of as an electrostatic (polar) interaction between a partial positive charge on a hydrogen atom and a partial negative charge on an electronegative acceptor atom.

$$\text{donor-H}^{\delta+} \;.\;.\;.\;.\;.\; {}^{\delta-}\text{acceptor}$$

On the basis of a very simple model of this kind, many of the properties of the hydrogen bond can be predicted, although more elaborate models must be developed to explain in detail the properties of hydrogen bonded systems (*11*).

As was true in the case of van der Waals interactions, the stability of hydrogen bonds depends on the difference in free energy between protein–protein hydrogen bonds and protein–solvent hydrogen bonds. Since the donor and acceptor atoms found in proteins (e.g., N–H and O=C) are capable of hydrogen bonding with water, the net free energy of a hydrogen bond must take this into consideration.

$$\begin{gathered}\text{protein donor} - \text{H} \;.\;.\;.\;.\; \text{OH}_2 + \text{HOH} \;.\;.\;.\; \text{protein acceptor} \\ \rightleftharpoons \\ \text{protein donor} - \text{H} \;.\;.\;.\; \text{protein acceptor} + \text{H}_2\text{O} \;.\;.\;.\; \text{HOH}\end{gathered}$$

The thermodynamic parameters governing an equilbrium of this kind have been examined in the case of the hydrogen bond stabilized dimer of *N*-methylacetamide.

$$\begin{array}{c} CH_3 \\ | \\ C{=}O \\ | \\ HN \\ | \\ CH_3 \end{array} \rightleftharpoons \begin{array}{c} CH_3 \\ | \\ C{=}O \cdots\cdots\cdots \\ | \\ HN \\ | \\ CH_3 \end{array} \begin{array}{c} CH_3 \\ | \\ C{=}O \\ | \\ -N \\ | \\ CH_3 \end{array}$$

Presumably this is a reasonable model for a peptide bond, and formation of the dimer and higher polymers can be detected easily by examination of the N–H stretching frequency by ir spectroscopy. The degree of association of *N*-methylacetamide is affected by the solvent (Table II). In aqueous solutions the association is extremely weak, with positive values for the unitary free energy change associated with hydrogen bond formation. That is, monofunctional hydrogen bonds in aqueous solution do not appear to be stable (*15*).

In addition to this kind of result, the development of convincing explanations for urea and GuHCl-induced protein denaturation on a basis other than competition with native hydrogen bonds (*16*) has left many workers suggesting that hydrogen bonds may not contribute much to the stability of protein structures. However, it is certainly possible that systems involving multiple hydrogen bond formation or systems involving hydrogen bond formation in a nonaqueous local environment may be quite stable. In addition, it is inevitable that in protein folding or in protein–protein complex formation many protein–solvent hydrogen bonds will be broken. So even though the net contribution of hydrogen bonds to the stability of a complex structure may be small, there is a clear thermodynamic advantage to forming protein–protein hydrogen bonds to compensate for the lost protein–solvent hydrogen bonds. In this way, hydrogen bonds may contribute certain constraints to the structure of a

Table II. Thermodynamics of Interamide Hydrogen Bond Formation by *N*-Methylacetamide at 25°C (*15*)

Solvent	*Association Constant for Dimerization* (k_2)	$\Delta F°$ *(kcal mol⁻¹)*	$\Delta H°$ *(kcal mol⁻¹)*	$\Delta S°$ *(gibbs mol⁻¹)*
Carbon tetrachloride	4.7 (5.8)	−0.92	−4.2	−11
Dioxane	0.52 (0.58)	0.39	−0.8	− 4
Water	0.005 (0.005)	3.1	0.0	−10

Journal of the American Chemical Society

protein or a protein–protein complex without necessarily contributing to the overall thermodynamic stability of these structures.

Ionic Interactions

Electrostatic interactions between molecules containing oppositely charged groups is another possible force that can ultimately be expressed in functional properties. There are several amino acid residues in proteins that are charged at pH values near neutrality. Aspartic acid and glutamic acid and, at higher pH values, cysteine and tyrosine bear negative charges. Arginine, lysine, and histidine usually bear positive charges. The p*K*'s of these groups can vary considerably of course depending on the local environment. One interesting example of this variability is the active site of alpha-lytic protease which has been shown to contain a histidine residue whose p*K* is lower than the p*K* of an adjacent carboxylic acid (*17*).

The extent to which charge–charge interaction forces contribute to the stability of protein structures and protein interactions is not clearly understood. It is frequently suggested that only marginal stabilization energy can be expected from these forces because even a relatively low ionic strength environment should provide a sufficient concentration of counter ions to swamp out any charge–charge force. While this may be true in general, a few specific examples of well studied charge–charge interactions should demonstrate that they can play important roles in stabilization of protein structures.

For example, a salt bridge between the α-amino group of isoleucine 16 of chymotrypsin and the carboxylate anion of aspartic acid 194 is known to stabilize a specific conformation of the protein. The two charged groups are buried in a hydrophobic environment, and the ionic bond has been shown to have a stabilization energy of 2.9 kcal/mol (*18, 19*). A second and perhaps more interesting example is the role which ionic forces play in substrate binding and catalysis of ribonuclease A (*20, 21*). The ionic contribution to the binding of substrates is highly dependent on ionic strength, but even at an ionic strength of 0.2 it can be as high as 1.8 kcal/mol. At an ionic strength of 0.02, the ionic contribution to binding can be as high as 7–9 kcal/mol (*20*). It is generally thought that a dianionic penta-coordinated phosphate occurs in the catalytic reaction pathway of ribonuclease A and that several positively charged groups of the enzyme are involved in the stabilization of this intermediate. A lower limit for the estimated electrostatic stabilization of this transition state is 4.3 kcal/mol, and it is probably closer to 8 kcal/mol (*20*). This is sufficient to account for a rate enhancement of 10^6; and since the total rate enhancement is 10^8 (*21*), electrostatic forces play a major role in the catalytic mechanism of this enzyme. Many other examples could be mentioned (e.g., substrate binding to trypsin, salt bridge between the α and β subunits of hemoglobin), but this is sufficient to demonstrate that charge–charge interactions can play a role in the stabilization of protein structure and perhaps also in the determination of protein functionality.

Hydrophobic (Apolar) Bonding

Ever since the demonstration that hydrogen bonds are not the sole or even the primary determinant of protein structure, attention has focused on the still controversial hydrophobic or apolar bond (*12, 22*). It has been suggested frequently in the last several years that hydrophobic interactions are the predominant contributors to the stability of protein structure, but there is still no general agreement on this point and in fact there is at least some experimental evidence which indicates that the contribution of hydrophobic bonds may be much less than is widely assumed (*23*). Part of the problem in evaluating the contribution of hydrophobic bonds is the lack of clarity as to just what a hydrophobic bond is. In 1969 W. P. Jencks defined hydrophobic bonding as "an interaction of molecules with each other which is stronger than the interaction of the separate molecules with water and which cannot be accounted for by covalent, electrostatic, hydrogen-bond, or charge transfer forces" (*12*). This is sort of an "all others" category, a category defined only in reference to what it is not. While a definition of this kind is a measure of our collective ignorance, it is probably the best that can be done. There simply is not sufficient hard evidence to base a definition on a more mechanistic level. A detailed description of hydrophobic bonding in terms of the entropy of displaced solvent, for example, must still leave room for so-called nonclassical hydrophobic bonds which appear to be enthalpy-driven reactions (*16*). The difficulty with attempts to explain hydrophobic bonding in detail is that experimentally it is impossible to separate hydrogen bonding and strictly entropic effects. Approaches which avoid a strict separation of enthalpy and entropy terms may in time provide a better foundation for our understanding of hydrophobic bonding.

Quantitative estimations of the stability of hydrophobic bonding in proteins is not at all easy. One approach has been to approximate the free energy gained by folding an amino acid side chain away from water as the free energy change associated with the transfer of the side chain from water to a 100% ethanol solution. Solubility data of this kind have been compiled (Table III) (*24*) and do in fact suggest that a large

Table III. Free Energy Change for Transfer from Water to 100% Ethanol at 25°C (24)

	$\Delta G_{transfer}$	
	Whole molecule (kcal/mol)	*Side chain only (kcal/mol)*
Glycine	+4.63	0
Alanine	+3.9	−0.73
Valine	+2.94	−1.69
Leucine	+2.21	−2.42
Isoleucine	+1.69	−2.97
Phenylalanine	+1.98	−2.65
Proline	+2.06	−2.60
Methionine	+3.33	−1.30
Tyrosine	+1.76	−2.87
Threonine	+4.19	−0.44
Serine	+4.59	−0.04
Asparagine	+4.64	+0.01
Glutamine	+4.73	+0.10

Journal of the American Chemical Society

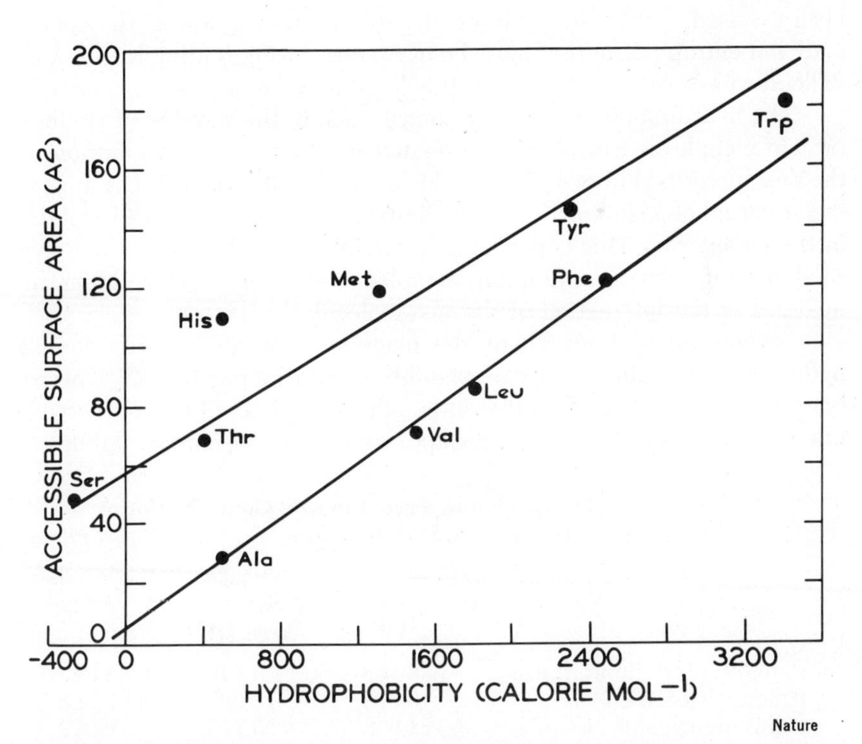

Nature

Figure 1. Accessible surface areas of residue side chains plotted against hydrophobicity—free energy change for the transfer from 100% ethanol to water (25)

negative free energy change accompanies transfer of amino acid side chains to an environment resembling a protein interior. The free energy of transfer from water to ethanol of a single isoleucine side chain, for example, is almost three kcal/mol.

A second approach to quantitative estimation of hydrophobic bonding is based on an empirical relationship between the accessible surface area of a solute and its free energy of transfer from ethanol to an aqueous solution (25) (*see* Figure 1). The larger the free energy associated with transfer from ethanol to water, the more hydrophobic is the solute and the larger the accessible surface area of the molecule. For amino acids a reduction in accessible surface area of 1 Å^2 corresponds to a gain in hydrophobic free energy of approximately 25 cal/mol. This is really an elaboration of the previous approach since it is still based on the free energy of transfer of amino acids from organic to aqueous solvents.

As shown in Table IV, considerations of this kind yield extremely large estimates of hydrophobic free energies (5). In the case of the higher molecular weight proteins, these values are significantly larger than previous estimates. The reliability of these estimates has not yet been assessed, but as we shall see shortly, the magnitude of the conformational entropy term that must be overcome during folding is extremely difficult to estimate, and energies this large may be necessary.

The hydrophobic free energy contribution to the stability of protein–protein complexes can also be estimated making similar assumptions to the ones described above (9). In this case as well (Table V) it appears that hydrophobic interactions contribute greatly to the overall stability of the complexes. This is particularly interesting in the case of the trypsin–pancreatic trypsin inhibitor complex since very few residues are involved in the interaction of the two molecules.

Several reservations about the magnitude of the contributions of hydrophobic bonding to protein stability should be mentioned. One frequently mentioned result is that although hydrophobic bonding between low molecular weight compounds appears to have maximum stability at

Table IV. Hydrophobic Free Energy Contribution to Protein Stability (5)

Protein	*MW*	*Buried Surface Area (Å^2)*	*Hydrophobic Free Energy kcal/mol*[a]
Pancretic trypsin inhibitor	6,158	5,740	143.5
Ribonuclease S	13,690	13,290	332.3
α-Chymotrypsin	24,608	25,770	644.2
Carboxypeptidase	34,409	38,610	965.3

[a] 25 cal mol^{-1} per Å of buried surface area.

Nature

Table V. Hydrophobic Free Energy of Protein–Protein Interactions (9)

Complex	K_D *(M)*	*Total Free Energy (ΔG) (kcal/mol)*	*Buried Surface Area (Å²)*	*Hydrophobic Free Energy*[a] *(kcal/mol)*
Insulin dimer	10^{-5}	7	1130	28
Trypsin–pancreatic inhibitor	10^{-13}	18	1390	35
Hemoglobin αβ-dimer	$< 10^{-8}$	> 11	1720	43

[a] 25 cal/mol/Å² of buried surface area.

Nature

about 18–20°C (*26*), the maximum stability of proteins varies over a much wider range. Another result involves the changes in accessibility of polar (electronegative) and nonpolar (relatively nonelectronegative) side chain atoms which occur during folding. The average polar atom in an amino acid side chain is three to four times more accessible to water molecules than the average nonpolar side chain atom, at least in native globular proteins, but the reduction in accessibility to water molecules which occurs during folding (and this is really the figure related to the magnitude of hydrophobic bonding) is only twice as large for nonpolar as for polar side chain atoms (*27*). It would be a mistake to overstress this kind of indication, but it certainly does raise some questions about the actual contribution of hydrophobic bonds to protein stability. Finally, although the transfer of hydrophobic solutes from organic solvents to aqueous solution causes a reduction in the volume occupied by the solute, no such effect is observed during protein denaturation (*23*). There is essentially no change in the total volume accompanying denaturation of ribonuclease A (*23*). It is difficult to know exactly what this means, but certainly one reasonable explanation is that hydrophobic bonding is not the dominant force it is often assumed to be.

These, then, are the major noncovalent forces that can be invoked to explain the stability of protein structures, protein–protein complex formation, and the functional properties of proteins which depend on interaction with other molecules. Before looking at how these forces can express themselves in functional properties, it will be advantageous to clarify how much stabilization energy is actually needed to explain the protein conformations and interactions which we can actually observe.

Protein Stability and Conformational Entropy

Estimation of the stability of various conformational states of a protein molecule is not easy. However, globular proteins can exist in a variety of energetically equivalent conformation states, the relative stability of which is a complex function closely interrelated with the kinds of forces under discussion. The net stability of the so-called native state of globular proteins with respect to the randomly coiled state has been estimated in a number of cases and is much less than expected from simple observation of the interactions possible in the folded native state which are not possible in the randomly coiled state. The net stability of lysozyme at 25°C, for example, is less than 10 kcal/mol. The reason for this relatively low value, of course, is the loss in conformational entropy which is associated with a random coil to folded transition. This energy must be compensated for before a conformation will have any net stability. Estimation of the magnitude of the conformational entropy term or protein folding is very difficult. Brant, Miller, and Flory have estimated the conformational entropy associated with a transition to a randomly coiled polypeptide backbone (no side chains) to be 10 cal/°K/mol residues (*29*). This is a very approximate value, and even more uncertain is the conformational entropy associated with amino acid side chains. An estimate for the conformational entropy of lysine side chains is 3.7 cal/deg/mol (*79*). If these are reasonable values, then the difference in conformational entropy between native and random coil states is about 400–550 kcal/mol for small globular proteins (*28*). A loss in translational and rotational entropy is also associated with the formation of protein–protein and protein–other molecule interactions. Estimation of the magnitude of this entropy term is also largely speculative, but, using the same kinds of calculations which appear to hold for small organic molecules in aqueous solution (*30*), values in the range of 20–30 kcal/mol are obtained (*9*).

With these uncertainties associated with estimation of conformational entropies, it is very difficult to draw up an energy balance sheet for a protein conformation or interaction itemizing energy required and specific sources of stabilization energy. However, in spite of these uncertainties it is a useful exercise, and, if its limitations are remembered, making up an energy balance sheet can provide some insight into the current state of our understanding of protein interactions. For the purpose of discussion, then, we will look briefly at the very well studied interaction between trypsin and bovine pancreatic trypsin inhibitor. This interaction illustrates quite well the complexity of even a relatively simple protein–protein interaction and certainly suggests some of the problems which we will encounter in attempting to study much more complex interactions on this level.

The association between trypsin and bovine pancreatic trypsin inhibitor (Table VI) has a dissociation constant of $10^{-13}M$ ($\Delta G = 18$ kcal/mol) and is one of the most stable known protein–protein associations

Table VI. Free Energy of Association of Trypsin and Bovine Pancreatic Trypsin Inhibitor

	kcal/mol
Conformational entropy	+27[a]
Formation of tetrahedral intermediate	+16[b]
Hydrophobic bonding	−35[c]
Polar interactions (hydrogen and ionic bonds)	−10 to −15
Relief of strain	−5
Bonding to tetrahedral intermediate	−5 to −10
Total free energy change (estimate)	−12 to −22 kcal/mol
Actual free energy change	−18 kcal/mol

[a] Calculated from translation/rotational partition functions (*9*).
[b] Estimated activation energy for amide substrate of chymotrypsin (*33*).
[c] 25 cal/mole/Å^2 of buried surface area.

(*31*). The conformational entropy connected with complex formation between molecules of these molecular weights has been estimated from translational and rotational partition functions to be 27 kcal/mol (*9*). So you need at least 27 kcal of stabilization energy before the complex will be stable at all. In the case of this particular trypsin–trypsin inhibitor complex there is another source of free energy which must be compensated for before the complex can be formed. X-ray crystallographic data suggest that the complex between trypsin and BPTI involves the formation of a tetrahedral structure resembling the tetrahedral intermediate thought to occur during enzymatic catalysis (*32*). The energy input necessary to form this structure is probably close to the activation energy for catalysis, which in the case of an amide substrate of chymotrypsin is about 16 kcal/mol (*33*). So a total of about 43 kcal/mol of free energy is necessary before a complex of even vanishing stability can be formed between these two proteins.

The contribution of hydrophobic bonding to the stability of the complex can be estimated as discussed above by determining the change in accessible surface area which occurs as a result of complex formation. In this case a total of 3,390 Å^2 of surface area is buried during complex formation, corresponding to a hydrophobic free energy of 35 kcal/mol (*9*). Ten hydrogen bonds or fewer can form between the two molecules when a complex is formed, and since some of these will be forming in the very nonpolar contact region, they may be quite stable (*32*). A value of 10–15 kcal/mol for the stabilization caused by both hydrogen and ionic bonds is probably not unreasonable. Another source of stabilization energy in this case which probably is not common to most other protein–

protein interactions is that formation of the tetrahedral structure allows for the relief of strain which is present in the uncomplexed inhibitor molecule (*34*). The energy associated with strained amide bonds is not large (*35*), and the contribution of this factor to complex formation is probably 5 kcal/mol or less. Finally, the tetrahedral intermediate formed during complex formation can be stabilized by hydrogen bonding to the newly formed oxyanion (*36*). This kind of bonding is thought to stabilize the tetrahedral intermediate during normal enzyme acylation (*37*), and while difficult to estimate quantitatively is probably on the order of 5–10 kcal/mol.

Approximations of this kind lead to an estimation for the free energy of complex formation of anywhere between 12 and 22 kcal/mol. Several comments should be made at this point about this balance sheet. First, the major portion of the stabilization energy comes from hydrophobic bonding, and, while this may well hold true for a large number of protein–protein and protein–other molecule interactions, no interaction can be attributed simply to one kind of force. Secondly, formation of the tetrahedral intermediate probably makes no net contribution to the stability of the complex even though this formation is compensated for to some extent by relief of strain and by better hydrogen bonding to the tetrahedral oxyanion. The noncovalent forces known to be present between the two proteins are more than enough to account for the stabilization of the complex without postulating any role for the covalent bond formed between the two proteins by the tetrahedral intermediate.

Protein Structure and Functionality

The interaction between proteases and their inhibitors is, at least by comparison with interactions which take place in food systems, a remarkably simple and straightforward association. Simple correlations of functional properties with different kinds of molecular forces cannot be made. It is possible, however, to illustrate the importance of protein structure and of protein–protein interactions as determinants of the functional properties of food proteins. I would like therefore to look at several food systems in which protein–protein or protein–other constituent interactions play a role and examine the relationship between functionality and protein structure in these systems. One of the simplest areas in which to examine this relationship is the well studied collagen–gelatin transition.

Gelatin is widely used in food products as a gelling agent, a whipping agent, an agent that increases the viscosity of various products, and a preventative against ice crystal growth in ice cream and frozen desserts. It is obtained from hot-aqueous extractions of alkali-treated collagen from bone and hide. Because collagen is the major protein component of most vertebrates, its structure and properties have been studied thoroughly (38). The fundamental subunits of collagen tissue are tropocollagen molecules. Tropocollagen is a long thin semirigid rod-shaped molecule (2800 Å × 13.6 Å; MW 300,000) composed of three super-coiled peptides (38). The thermal stability of tropocollagen is a complex function of a variety of factors including the proline and hydroxyproline content of the peptides. Raising the temperature of a tropocollagen solution past its melting temperature (T_M) results in a disruption of the forces holding the three peptides together, and the denatured peptides assume a random-coil structure (Figure 2). A subsequent reduction in the tem-

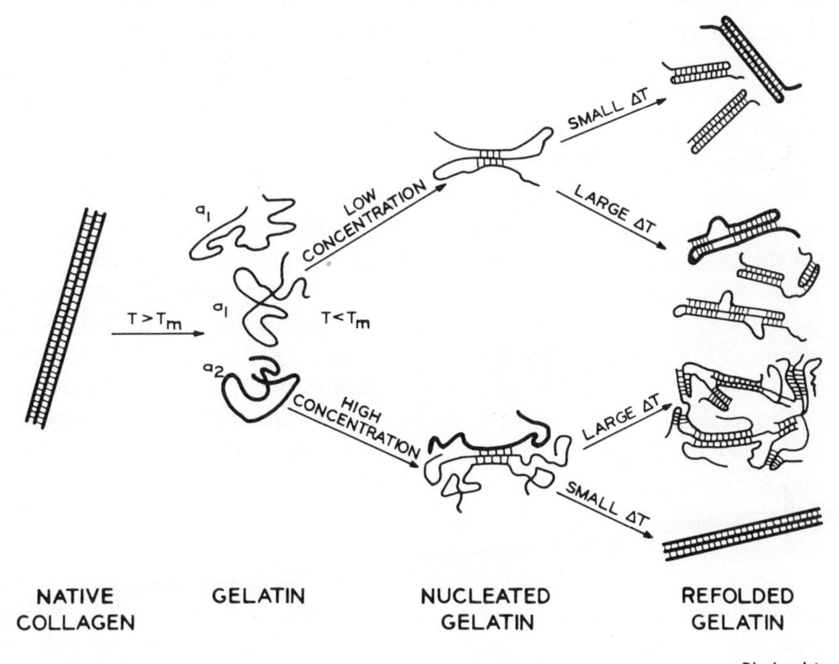

Figure 2. Suggested scheme for the concentration-dependent folding of collagen peptides below T_M (39)

perature to below the melting temperature will result in the reformation of some elements of protein structure.

At low concentrations of the denatured collagen peptides (less than 0.1 mg/ml) renaturation is a first-order process with respect to protein concentration (39). This suggests that refolding is entirely an intramolecular process resulting in formation of single-stranded structures folded back upon themselves. If the renaturation temperature is substantially below the T_M for chain unfolding, the renaturation will occur rapidly with many nucleation sites, and maximization of noncovalent interactions will not be achieved. At renaturation temperatures closer to the T_M, however, maximization of noncovalent interactions is more likely, and a more highly ordered refolded structure results. At high initial concentrations of gelatin, renaturation upon cooling is no longer a first-order process. Formation of intermolecular nucleation complexes involving two or more chains occurs, and at temperatures very near the T_M the native triple-helix structure of collagen can be reformed. While this complete reversibility of denaturation is difficult to obtain under normal conditions, chemically cross-linked collagen or collagen with a large number of naturally occurring cross-links such as *Ascaris* collagen can be much more easily refolded to give the native structure (*40*). The interchain cross-links presumably maintain the three chains in alignment even in the denatured state. Renaturation at high concentrations but well below the T_M for unfolding results in a larger number of intermolecular interactions and the formation of the familiar gelatin gel structure.

The noncovalent forces responsible for the various interactions between the peptide subunits of tropocollagen have been the subject of considerable debate. The arguments parallel in many ways the arguments we discussed in connection with the role of different noncovalent forces and protein structure in general. Hydrogen bonding has always been thought to play a major role in the stabilization of tropocollagen, although this has been questioned (*41*). Most of the detailed models for the structure of tropocollagen, however, involve the regular formation of hydrogen bonds between the peptide chains. Whether this represents a major driving force for the interaction or merely a constraint on the interaction that takes place is not clear. The ordered clustering of charged residues may also play an important role in stabilizing the structure of collagen (*42*).

Although our understanding of which particular molecular forces are responsible for these interactions is limited, it should be emphasized that the noncovalent forces responsible for the native tropocollagen structure appear to be identical to those forces responsible for the types of structures observed in refolded gelatin. Rearrangement of these forces and a redistribution between intra- and intermolecular forces, rather than any new kinds of forces, accounts for the large structural changes that can take place and that are responsible for the large changes in functional properties of the proteins which occur. The case of the collagen–gelatin transition also illustrates quite well the interplay of protein structure and protein–protein interactions in determining the functional properties of proteins. The fact that the functional properties of a protein can be altered substantially by denaturation should suggest a number of possible approaches to the intentional alteration of functionality of food proteins. This is a largely unexplored area but one which could hold much promise. It may be for example, that partially denatured proteins or mixtures of partially denatured proteins with native proteins would allow for the formation of a variety of restructured states which would be useful for the incorporation of the proteins into foods.

Protein-Lipid Interactions and Functionality

A second kind of protein interaction which can play an important role in determining the functionality of protein is the interaction of proteins with lipids. The development of our understanding of membrane structure and its importance in biological systems has led in recent years to a renewed interest in the study of protein–lipid interactions. These interactions have been recognized for a long time to play critically important roles in determining the functional properties of many food proteins (*43, 44, 45*). As with protein–protein interactions, it is not possible at present to attribute protein–lipid interactions to any single specific kind of molecular force. Electrostatic attraction can occur between the phosphate groups of phospholipids and positively charged amino acids or lipids, or negatively charged amino acid side chains if mediated by divalent calcium or other metal ions. Hydrophobic bonding is also likely to play a major role in stabilizing the interaction of both polar and nonpolar lipids with proteins. An example of the role of protein–lipid interactions in functionality which illustrates both their importance and complexity is the role of lipid–protein interactions in maintaining the structure of wheat-flour dough.

Certain polar lipids, notably stearoyl-2 lactylate, have been shown to be efficient dough improvers and even at very low concentrations result in increased mixing tolerance, increased loaf volume, finer grain, and softer crumb in breads (Table VII) (*46, 47*). They have become particularly useful in improving the functional quality of baked wheat-

Table VII. Specific Volumes of Loaves Prepared from Flour Supplemented with Indicated Protein Concentrates and Treated with Sodium Stearoyl-2 Lactylate (*48*)

		Specific Loaf-volume (cc/g)			
Agent	*Amount Added (%)*	*Nonfat Dry Milk*	*Fish Conc.*	*Cottonseed Flour*	*Chick Pea Flour*
Control	0.0	5.12	4.48	5.23	4.55
SSL	0.25	5.36	5.67	5.55	4.74
	0.50	6.32	5.99	5.70	5.01
	1.00	6.39	—	6.11	5.15

Baker's Digest

flour products that have been nutritionally improved by addition of soy flour (*48*). A water-soluble protein component of wheat flour (0.2 wt % of starting flour) has been isolated which can bind divalent calcium and then bind phospholipids or stearoyl-2 lactylate to form complexes with an ordered, crystalline structure (*49*). In this way stabilization of protein–lipid interactions should strengthen the dough; characterization of this interaction may permit the detailed study of the relationship between functionality and protein–lipid interactions in dough. In a related piece of work, Hoseney et al. suggested that free polar lipids of dough (mainly glycolipids) bind to gliadin proteins by largely hydrophilic bonds and bind simultaneously to glutenin proteins by largely hydrophobic bonds (*50*). The evidence for this lipid-mediated association between glutenin and gliadin proteins is primarily from lipid extractability studies. It is therefore extremely difficult to draw conclusions of this kind about molecular forces, but this association could provide an interesting explanation of the role of protein–lipid interactions in the formation of the gas-retaining complex of dough. The existence of hydrated protein fibrillar structures with elastic properties prior to dough development may suggest that the basic properties of dough are determined by the structure of the protein and only modified by the addition of lipids or other compounds (*51*). Many other protein–lipid interactions affect functionality—lipid–casein interactions in the formation of the casein micelle is a well studied example—but it is sufficient to indicate the role of lipids in determining protein functionality.

Disulfide Bonds and Protein Functionality

To this point the discussion has been limited to the relationship between noncovalent forces and protein functionality. Covalent forces, obviously, can also play substantial roles in determining the functional properties of a protein. Disulfide bonds are the clearest example of this fact (*52, 53*). The occurrence of disulfide bonds in proteins varies considerably, though they are most common in small, single subunit proteins which exist extracellularly. Bovine serum albumin, for example, has 17 disulfide bonds and a molecular weight of only 66,500 g/mol. Proteins of particular importance in foods very frequently contain a large number of disulfide bonds. Disulfide bonds contribute much stabilization energy to the protein conformations which allow their formation. However, the formation of disulfide bonds does not appear to result in any net stabilization of any particular protein conformation. That is, a structure with randomly formed disulfide bonds would be stabilized to the same extent as a structure with correctly paired disulfide bonds. The presence of free sulfhydryl groups will catalyze through thiol-disulfide exchange the rate

Table VIII. Effect of Added Cysteine on Loaf Volume and Mixing Data (*62*)

	Cysteine (ppm)		
	0	*80*	*120*
Series A (mixed to peak consistency)			
loaf volume (cc)	480	840	875
net energy (w-hr/lb)	3.2	1.7	1.0
mixing time (min)	51.0	11.5	7.2
Series B (mixed at energy levels half that received for peak consistency)			
loaf volume (cc)	465	885	890
net energy (w-hr/lb)	1.7	0.85	0.5
mixing time (min)	25.0	7.1	4.4

Cereal Chemistry

at which the randomly formed disulfide pairs rearrange to give the correctly matched pairs, but the noncovalent interactions appear to determine the nature of the structure finally formed (*54, 55*).

The role of intra- and intermolecular disulfide bonds in determining the functionality of wheat-flour dough is an excellent illustration of the importance of disulfide bonds. The gluten of developed dough forms a network of interconnected sheets which surround the starch granules (*56*). All of the noncovalent forces we have discussed almost certainly play a role in stabilizing this matrix, but disulfide bonds have received particular attention in this context for a variety of reasons. The functional properties of dough, for example, are particularly sensitive to reagents which attack disulfide bonds (*57, 58, 59*), and the addition of free sulfhydryl groups, oxidizing agents, or reducing agents can have very great effects on dough properties (*60, 61, 62*). The addition of cysteine, for example, results in a large reduction in dough mixing time with "short time" dough development procedures and a dramatic increase in loaf volume (Table VIII). There is no general agreement, however, as to the exact manner in which disulfide bonds affect dough structure and properties. One long-standing proposal is that the unusual viscoelastic properties of dough are caused by the ability of intermolecular disulfide bonds to be broken under conditions of physical stress and then reformed again with different partners. This disulfide stabilized matrix would permit relative movement of the matrix while maintaining a stable structure (*63, 64*). Another possibility based in part on evidence which suggests that there are very few intermolecular disulfide bonds in dough is that the rate of disulfide interchange rather than the thermodynamic stability of the disulfide bonds is responsible for dough functionality and the effects of different additives (*65*). In this case, the rate of conformational changes accompanying protein hydration determines functionality; noncovalent forces are the main determinants of the conformational changes which take place, and disulfide interchange is a necessary step in the conformational change, but dough structure does not necessarily involve polymerization of proteins by disulfide cross-linking of proteins (*65, 66*). As with noncovalent interactions, it should be possible to restructure certain proteins by appropriate rearrangement of intra- and intermolecular disulfide bonds. For example, mixing partially reduced proteins under conditions which favor intermolecular bonding and then reoxidizing will result in a restructured protein with functional properties very different from those of the native protein. Many variations of this basic idea are possible. The use of mixtures of different proteins or the incorporation of other food constituents such as carbohydrates or lipids into a disulfide stabilized matrix of this kind could result in the fabrication of a tremendous variety of products with vastly different functional properties.

Chemical Modification and Functionality Alteration

The ability to modify the functional properties of food proteins intentionally by manipulation of the forces we have discussed would clearly be advantageous in developing ways to incorporate proteins into food systems. As we have seen, both protein denaturation and the interactions of proteins with other food constituents offer potential methods for changing the functional properties of proteins. A third possible approach to alteration of the functionality of food proteins is chemical modification. Many chemical modifications of proteins have been studied and have been used in the characterization of protein structure and biological function (*67*). For a variety of reasons, however, only a small percentage of these chemical reactions would be even potentially suitable for use with food proteins. Alteration of functionality by chemical modification may in many cases be a much more serious threat to the nutritional value of a protein than would a simple structural alteration or conformational modification. Many known unintentional chemical modifications of food proteins occur during food processing, and almost without exception these unintentional chemical modifications result in a loss in nutritional value. An additional problem with the chemical modification approach to alteration of functionality is the potential toxicological problems associated either with unreacted reagent left in the food or with metabolism of chemically modified amino acid residues. In spite of these potentially serious problems several attempts have been made to use chemical modifications for alteration of food protein functionality.

The most successful attempt of this kind is probably the alteration of proteins by alkali treatments. Depending on the severity of the treatment, this can mean only a modification of the state of ionization of the protein or it can mean extensive hydrolysis of peptide bonds, formation of various kinds of cross-links, and racemization of amino acids to give the D-isomer. Since some of the problems associated with partial or extensive alkaline hydrolysis are discussed elsewhere in this symposium (*68*), they will not be discussed here. Alkaline treatments are used in the solubilization of protein-containing material in preparation for extrusion processing (*69*). In addition, hydrolysates of some proteins have improved flavor characteristics, better emulsification properties, and improved foaming ability (*70*).

The only other chemical modification approach to alteration of functionality which has received much attention is the use of various acylating

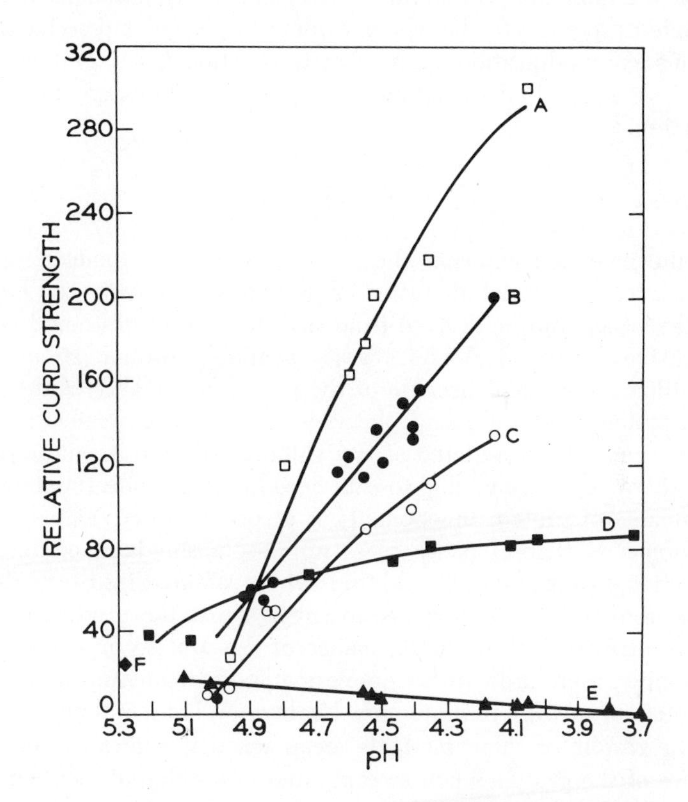

Journal of Milk and Food Technology

Figure 3. Relative curd strength (Brookfield viscometer readings) of gels prepared from 3% solutions of either fish protein concentrate extracted at high pH, succinylated FPC at various pH values, or raw skim milk. (A) Succinylated protein plus corn oil. (B) Succinylated protein. (C) Succinylated protein plus Ca^{2+}. (D) High pH FPC extract plus corn oil. (E) High pH FPC extract. (F) Raw skim milk curd prepared with rennet. (75)

agents. A variety of patents have appeared, including such things as the use of *N*-succinylated egg yolk proteins in mayonnaises (*71*), acylation of soybean protein to improve its flavor, odor, and dispersion characteristics in coffee whiteners (*72*), and acetylation of skim milk or blood serum to improve the stability of emulsions of these proteins with edible oils (*73*, *74*). The use of succinylation to increase the solubility and emulsifying characteristics of fish protein concentrate (*75*), single cell protein concentrate (*76*), wheat-flour proteins (*77*), and fish myofibrillar proteins (*78*) has also been reported. In some of these published cases improvements in functional properties were obtained (Figure 3), but in one case (*78*) some evidence was obtained which suggested that the nutritional value of the modified protein might be considerably less than that of the unmodified protein. On the whole, evidence accumulated so far suggests that chemical modification approaches to functionality alteration may be much less fruitful than either enzymatic or physical modification approaches.

Conclusion

Some general comments about the potential for developing methodologies for intentional alteration of protein functionality can be made. First, the benefits to be derived from such methodologies could be enormous. Many economical and readily available protein resources are underutilized at present because of the poor functionality of the protein in food systems. At least from the viewpoint of food science, a protein resource cannot be considered of any value at all until it can be put into a form that people are willing to eat. Secondly, our understanding of the determinants of protein functionality is at present very limited. Part of the problem is that in many cases no simple standardized tests exist for functional properties. In addition, where standardized tests do exist, there is frequently little evidence to suggest that the results of the test actually correlate with the performance of the protein in a food system. For example, there are various simple methods for measuring the ability of a protein or mixture of proteins to bind water, but it is difficult to know what the results of these methods mean when it comes to actual performance of the protein when incorporated into different meat emulsions or other food products. Therefore, a definite need exists for the development of simple standardized tests for protein functionality which relate to actual performance in food systems. This would increase immeasurably our ability to design intentional modifications of functionality.

Of the intentional modifications of functionality which have been attempted, those based on modification of the protein by physical restructuring have been most successful, and those based on modification of the protein by chemical methods have been least successful. The distinction between these two approaches is not as simple as it might sound; a physical restructuring such as extrusion processing certainly involves covalent chemical modification of the protein. Nevertheless, there is a very large potential for restructuring proteins by a redistribution of intra- and intermolecular interactions without damage to the nutritional value of the protein, whereas nutritional damage to the protein almost always accompanies chemical modification of the protein. Chemical modifications suitable for food protein modification should be reversible after exposure to the acidic conditions of the stomach and should release chemicals which can be shown to be nonhazardous.

Literature Cited

1. Maitland, G. C., Smith, E. B., *Chem. Soc. Rev.* (1973) **2,** 181.
2. Israelachvili, J. N., *Q. Rev. Biophys.* (1973) **6,** 341.
3. Huggins, M. L., *in* "Structural Chemistry and Molecular Biology," A. Rich and N. Davidson, Eds., W. H. Freeman, San Francisco, 1968, p 761.
4. Ewing, G. E., *Acc. Chem. Res.* (1975) **8,** 185.
5. Chothia, C., *Nature* (1975) **254,** 304.
6. Klapper, M. H., *Biochim. Biophys. Acta* (1971) **229,** 557.
7. Richards, F. M., *J. Mol. Biol.* (1974) **82,** 1.
8. Finney, J. L., *J. Mol. Biol.* (1975) **96,** 721.
9. Chothia, C., Janin, J., *Nature* (1975) **256,** 705.
10. Pauling, L., Corey, R., Branson, H., *Proc. Nat. Acad. Sci. USA* (1951) **37,** 207.
11. Pimentel, G. C., McClellan, A. L., "The Hydrogen Bond," Freeman, San Francisco, 1960.
12. Jencks, W. P., "Catalysis in Chemistry and Enzymology," McGraw-Hill, New York, 1969.
13. Kollman, P. A., Allen, L. C., *Chem. Rev.* (1972) **72,** 283.
14. Kresheck, G. C., Klotz, I. M., *Biochemistry* (1969) **8,** 8.
15. Klotz, I. M., Franzen, J. S., *J. Am. Chem. Soc.* (1962) **84,** 3461.
16. Roseman, M., Jencks, W. P., *J. Am. Chem. Soc.* (1975) **97,** 631.
17. Hunkapiller, M. W., Smallcombe, S. H., Whitaker, D. R., Richards, J. H., *Biochemistry* (1973) **12,** 4732.
18. Fersht, A. R., *Cold Spring Harbor Symp. Quant. Biol.* (1971) **36,** 71.
19. Fersht, A. R., Requena, Y., *J. Mol. Biol.* (1971) **60,** 279.
20. Flogel, M., Albert, A., Biltonen, R., *Biochemistry* (1975) **14,** 2616.
21. Richards, F. M., Wyckoff, H. W., *in* "The Enzymes," Vol. 4, P. D. Boyer, Ed., 3rd ed., Academic, New York, p. 647.
22. Tanford, C., "The Hydrophobic Effect," Wiley-Interscience, New York, 1973.
23. Brandts, J. F., Oliveira, R. J., Westort, C., *Biochemistry* (1970) **9,** 1038.
24. Tanford, C., *J. Am. Chem. Soc.* (1962) **84,** 4240.
25. Clothia, C., *Nature* (1974) **248,** 338.
26. Bohon, R. L., Claussen, W. F., *J. Am. Chem. Soc.* (1951) **73,** 1571.
27. Lee, B., Richards, F. M., *J. Mol. Biol.* (1971) **55,** 379.
28. Tanford, C., *Adv. Prot. Chem.* (1968) **23,** 121; and (1970) **24,** 1.
29. Brant, D. A., Miller, W. G., Flory, P. J., *J. Mol. Biol.* (1967) **23,** 47.
30. Page, M. I., Jencks, W. P., *Proc. Nat. Acad. Sci. USA* (1971) **68,** 1678.
31. Means, G. E., Ryan, D. S., Feeney, R. E., *Acc. Chem. Res.* (1974) **7,** 315.
32. Ruhlmann, A., Kukla, D., Schwager, P., Bartels, K., Huber, R., *J. Mol. Biol.* (1973) **77,** 417.

33. Fastrez, J., Fersht, A. R., *Biochemistry* (1973) **12,** 2025.
34. Deisenhofer, J., Steigemann, W., *in* "Bayer-Symposium V. Proteinase Inhibitors," H. Fritz, H. Tschesche, L. J. Greene, and E. Truscheit, Eds., Springer-Verlag, Heidelberg, 1974, p 464.
35. Winkler, F. K., Dunitz, J. D., *J. Mol. Biol.* (1971) **59,** 169.
36. Sweet, R. M., Wright, H. T., Janin, J., Clothia, C. H., Blow, D. M., *Biochemistry* (1974) **13,** 4212.
37. Robertus, J. D., Kraut, J., Alden, R. A., Birktoft, J. J., *Biochemistry* (1972) **11,** 4293.
38. Ramachandran, G. N., Ed., "Treatise on Collagen, Vol. 1, Academic, London, 1967.
39. Harrington, W. F., Rao, N. V., *Biochemistry* (1970) **9,** 3717.
40. Hauschka, P. V., Harrington, W. F., *Biochemistry* (1970) **9,** 3734, 3745, 3754.
41. Cooper, C., *J. Mol. Biol.* (1971) **55,** 123.
42. Piez, K. A., Torchia, D. A., *Nature* (1975) **258,** 87.
43. Karel, M., *J. Food Sci.* (1973) **38,** 756.
44. Pomeranz, Y., *Adv. Food Res.* (1973) **20,** 153.
45. Webb, B. H., Johnson, A. H., Alford, J. A., "Fundamentals of Dairy Chemistry," Avi, Westport, Conn., 1974.
46. Thompson, J. B., Buddemeyer, B. D., *Cereal Chem.* (1954) **31,** 296.
47. Tenney, R. J., Schmidt, D. M., *Baker's Dig.* (1968) **42** ,38.
48. Tsen, C. C., Hoover, W. J., Phillips, D., *Baker's Dig.* (1971) **45,** 20.
49. Fullington, J. G., *Cereal Chem.* (1974) **51,** 250.
50. Hoseney, R. C., Finney, K. F., Pomeranz, Y., *Cereal Chem.* (1970) **47,** 135.
51. Bernardin, J. E., Kasarda, D. D., *Cereal Chem.* (1973) **50,** 529.
52. Friedman, M., "The Chemistry and Biochemistry of the Sulfhydryl Group in Amino Acids, Peptides and Proteins," Pergamon, New York, 1973.
53. Joceyln, P. C., "Biochemistry of the SH Group," Academic, London, 1972.
54. Givol, D., deLorenzo, F., Goldberger, R. F., Anfinsen, C. B., *Proc. Nat. Acad. Sci. USA* (1965) **53,** 676.
55. Wetlaufer, D. B., Ristow, S., *Ann. Rev. Biochem.* (1973) **42,** 135.
56. Moss, R., *Cereal Sci. Today* (1974) **19,** 557.
57. Dahle, L. K., Hinz, R. S., *Cereal Chem.* (1966) **43,** 682.
58. Jones, I. K., Carnegie, P. R., *J. Sci. Food Agric.* (1969) **20,** 60.
59. Mecham, D. K., *Cereal Chem.* (1959) **36,** 134.
60. Tsen, C. C., *Cereal Chem.* (1969) **46,** 435.
61. Tsen, C. C., *Baker's Dig.* (1973) **47,** 44.
62. Kilborn, R. H., Tipples, K. H., *Cereal Chem.* (1973) **50,** 70.
63. Bloksma, A. H., *Cereal Chem.* (1975) **52,** 146.
64. Ewart, J. A. D., *J. Sci. Food Agric.* (1968) **19,** 617.
65. Jones, I. K., Carnegie, P. R., *J. Sci. Food Agric.* (1971) **22,** 358.
66. Bernardin, J. E., Kasarda, D. D., *Cereal Chem.* (1973) **50,** 735.
67. Means, G. E., Feeney, R. E., "Chemical Modification of Proteins," Holden-Day, San Francisco, 1971.
68. Gross, E., Adv. Chem. Ser., **160,** 37.
69. Van Beek, L., Feron, V. J., DeGroot, A. P., *J. Nutr.* (1974) **104,** 1630.
70. Richardson, T. R., Adv. Chem. Ser. **160,** 185.
71. Evans, M. T. A., Irons, L. I. (to Unilever N.V.), Patent No. **1,951,247,** Germany, May 27, 1970; *Chem Abstr.* (1970) **73,** 34013d.
72. Melnychyn, P., Stapley, R. B. (to Carnation Co.), So. African Patent No. **6,807,706,** So. Africa, June 27, 1969; *Chem. Abstr.* (1970) **72,** 65575x.
73. Patent Application No. **6,919,461,** Netherlands, 1970 (to Unilever N.V.); *Food Sci. Abstr.* (1971) **3,** 10P1707.
74. Patent Application No. **6,919,619,** Netherlands, 1970 (to Unilever N.V.); *Food Sci. Abstr.* (1971) **3,** 10P1708.
75. Chen, L., Richardson, T. R., Amundson, C. H., *J. Milk Food Technol.* (1975) **38,** 89.

76. McElwain, M. D., Richardson, T. R., Amundson, C. H., *J. Milk Food Technol.* (1975) **38**, 521.
77. Grant, D. R., *Cereal Chem.* (1973) **50**, 417.
78. Groninger, H. S., Jr., *J. Agric. Food Chem.* (1973) **21**, 978.
79. Nemethy, G., Leach, S. J., Scheraga, H. A., *J. Phys. Chem.* (1966) **70**, 998.

RECEIVED January 26, 1976. Work supported by the Research Division, College of Agricultural Life Sciences, University of Wisconsin—Madison.

Enzymatic Modification

5

Enzymatic Modification of Proteins Applicable to Foods

JOHN R. WHITAKER

Department of Food Science and Technology, University of California, Davis, Calif. 95616

Enzymes are presently used to modify proteins through hydrolysis of peptide bonds. Hydrolysis can lead to changes in solubility and functional properties of the proteins including taste changes and aggregation as in cheese and plastein formation. Enzymes in vivo carry out extensive post-translational modifications of proteins including cross-linking of polypeptide chains, phosphorylation, glycosylation, hydroxylation, and methylation. These enzymatically catalyzed chemical modifications of proteins will occur in in vitro systems, thus making it potentially feasible to use them to modify the functional and nutritional properties of proteins. This paper describes the types of post-translational modifications that occur in proteins, with particular emphasis on the enzymes involved and their substrate specificities.

In food manufacture, lipids and carbohydrates are often intentionally modified in order to impart certain desired characteristics to foods. Important examples are the hydrogenation of polyunsaturated lipids to increase the melting point, and the acetylation and carboxymethylation of carbohydrates to change their textural characteristics. Modification of food proteins to give different functional properties as well as to improve their nutritional values is not used to a large extent commercially. As requirements for high quality protein increase, it will become more and more necessary and desirable to change the properties of seed proteins, fish protein concentrate, single cell protein, leaf protein, etc. to meet the world's protein needs. Cereals and legumes genetically selected for high yields of protein generally have a lower nutritional value (*1*).

In order to satisfy the nutritional needs, we can (a) return to lower yielding, higher nutritive value varieties, (b) continue to look for genetically selected varieties that have both high yield and high nutritional value, or (c) modify chemically or enzymatically the proteins from the higher yielding varieties to give them the desired nutritional and functional properties. It is likely that both of the last two methods will be needed.

The functional and nutritional properties of food proteins may be improved by the use of specific enzymes to partially hydrolyze the proteins, to incorporate cross-linkages into the proteins, or to add specific functional groups to the proteins. These reactions can be carried out under mild conditions and, because of specificity of the enzyme-catalyzed reactions, are not likely to lead to toxic products. Through enzymatic modification it should be possible to increase the quality of some plant and animal proteins.

In this chapter we describe briefly present uses of proteolytic enzymes for modifying proteins through partial hydrolysis of the proteins, which in some cases leads to aggregation and peptide bond formation, as these reactions are described in more detail in Chapters 6 and 7. The major emphasis of this chapter will be an exploration of those enzymes which bring about aggregation of proteins, cross-link formation, and side chain modification through post-translational changes in the polypeptide chain. These enzymatically catalyzed reactions have only recently been described in the literature, and to our knowledge their application to food systems has not been considered previously. Some of the reactions are so demanding in specificity requirements that they do not at this time appear applicable to food systems; others appear to be awaiting only the ingenuity of the scientist and technologist to achieve importance.

It seems to us that the time is right and the field is wide open for definitive research on the further characterization and application of enzymes to food protein systems so as to change their solubility, spreadability, whippability, foamability, and nutritional properties, permitting one to tailor proteins for specific food uses. Associated areas of investigation include the development of sources (microbial?) of enzymes which will make them readily available at competitive prices and investigation of the changes in functional, nutritional, and digestibility properties of the chemically modified proteins. Perhaps it will prove more feasible to develop, through genetic engineering, crops which yield high quantities of relatively low grade protein and then to upgrade the quality of the protein through chemical and enzymatic processing than to build both high yield and nutritional quality into the same gene pool.

Table I. Proteolytic Enzymes in Protein Modification

Food	*Purpose or Action*
Baked goods	Softening action in doughs. Cut mixing time, increase extensibility of doughs. Improvement in texture, grain and loaf volume. Liberate β-amylase.
Brewing	Body, flavor, and nutrient development during fermentation. Aid in filtration and clarification. Chillproofing.
Cereals	Modify proteins to increase drying rate, improve product handling characteristics. Production of miso and tufu.
Cheese	Casein coagulation. Characteristic flavor development during aging.
Chocolate-cocoa	Action on beans during fermentation.
Egg, egg products	Improve drying properties.
Feeds	Waste product conversion to feeds. Digestive aids, particularly for pigs.
Fish	Solubilization of fish protein concentrate. Recovery of oil and proteins from inedible parts.
Legumes	Hydrolyzed protein products. Removal of flavor. Plastein formation.
Meats	Tenderization. Recovery of protein from bones.
Milk	Coagulation in rennet puddings. Preparation of soybean milk.
Protein hydrolysates	Condiments such as soy sauce and tamar sauce. Bouillon. Dehydrated soups. Gravy powders. Processed meats. Special diets.
Antinutrient factor removal	Specific protein inhibitors of proteolytic enzymes and amylases. Phytate.[a] Gossypol.[a] Nucleic acid.[a]
Wines	Clarification.[a]
in vivo processing[b]	Conversion of zymogens to enzymes. Fibrinogen to fibrin. Collagen biosynthesis. Proinsulin to insulin. Macromolecular assembly.

[a] In large part caused by other than proteolytic enzymes.
[b] Representative examples given.

Protein Hydrolysis

Of the many actual and potential uses of enzymes for chemical modification and improvement of proteins, that of hydrolysis of proteins is the most widely used. Hydrolysis involves the action of selected proteolytic enzymes to split specific peptide bonds in a protein. Along with a decrease in size of the protein there are changes in the solubility and functional properties of the product. Some of the uses of proteolytic enzymes in protein modification are shown in Table I.

Of the many functions of proteolytic enzymes listed in Table I, the most extensively used commercially are: chillproofing of beer, production of cheese, tenderization of meats, and production of protein hydrolysates. Two of the most active research areas at the moment include use of proteolytic enzymes for plastein formation (*see* "Proteolytic-Induced Aggregation of Proteins," p. 99) and the solubilization of fish protein concentrate.

Chillproofing. Proteolytic enzymes, primarily papain, have been used routinely in the chillproofing of beer since the process was patented in 1911 by Wallerstein (*2*). The process is rather simple in that papain is allowed to digest the proteins contributed by the cereal, malt, and yeast to the polypeptide stage. Control of the process is critical in that the proteins must be hydrolyzed to a stage where they do not precipitate when beer is chilled to 4°C in a refrigerator, yet the polypeptides left must be large enough to entrap the CO_2 required for head formation on the beer.

Meat Tenderization. Tenderization of meat may be effected by the action of the endogenous proteolytic enzymes, especially the catheptic enzymes found in the lysosomes of cells, and perhaps by a neutral protease found in or near the myofibrillar proteins (*3*). This process, possibly involving action of the enzyme(s) on the Z-line of the sarcosome of the muscle, requires 14–28 days of holding at room temperature. Meat can be tenderized also by the use of exogenous proteolytic enzymes such as papain, ficin, bromelain, fungal proteases, or a combination of these enzymes. Methods of application of the enzymes include: (a) injection of the live animal just prior to slaughter (Swift process; Ref. *4*), (b) injection of the carcass after slaughter (*5*), or (c) topical application of the enzyme as a solution or powder. Proper control of the process is needed to avoid localized proteolysis as in topical application or in injection of the live animal where the enzyme becomes concentrated in the organs, particularly the liver. Unlike the tenderization brought about by the endogenous enzymes, the exogenous enzymes act on the connective tissue proteins, elastin, and collagen (*6, 7*) as well as on the myofibrillar proteins (*8*).

Solubilization of Protein. Fish protein concentrate has high nutritional quality as determined both from its essential amino acid composition and from animal feeding experiments. Unfortunately, the concentrate is quite insoluble in water because of its denaturation by the solvent extraction method used in processing; thus it contributes no functional properties to a food and must be used in bakery products primarily. A potentially useful method of solubilizing the protein is by proteolysis (*9–12*). As is the case with protein hydrolysates of casein and soybean protein, bitter peptides are formed during the hydrolysis. Papain and ficin produce more of these bitter peptides than does Pronase, for example (*12*). Pronase was found to produce a more brothy taste (*13*). A possible method of removing the bitter peptides is to convert the concentrated protein hydrolysate to plastein by further proteolytic enzyme action (*14*) to remove the bitter peptides.

Proteolytic-Induced Aggregation of Proteins

It is well known that the specific scission of one or a few peptide bonds in a precursor protein is required for activation of several digestive enzymes (*15*), the enzymes of the blood clotting system (*16*), complement action (*17*), and insulin (*18*), as well as others. In some cases, conformational changes occur following peptide bond scission to form the specific active site.

Specific proteolytic scission may also result in an ordered alignment of the modified polypeptide chains into macromolecular structures and aggregates. Aggregation can precede proteolysis or it can be a direct result of proteolysis. Some typical examples are post-translational cleavage in collagen (*19*), κ-casein (*20*), fibrinogen (*21*), several viral systems (*22*), immunoglobulin light chains (*23, 24*), and plastein formation (*25*). We shall restrict this discussion to collagen, fibrinogen, the casein system, and plastein formation.

Collagen. The nascent polypeptide chains of procollagen undergo extensive modification while still attached to the ribosome (Figure 1). These modifications include hydroxylation of proline and lysine residues, glycosylation of hydroxylysine residues, and at least the initial stages of chain aggregation, generation of the triple helix, and disulfide bond formation. At this stage the pro-α1 chain (and pro-α2 chain, if appropriate for the particular collagen) has a molecular weight of 115,000–140,000 (*26*).

Either the pro-α1 chains or the assembled procollagen is secreted from the cell where conversion of procollagen to collagen takes place. In this conversion, the teleopeptides are removed (*27*) by procollagen peptidase to reduce the size of the α1 chains to 95,000. Procollagen

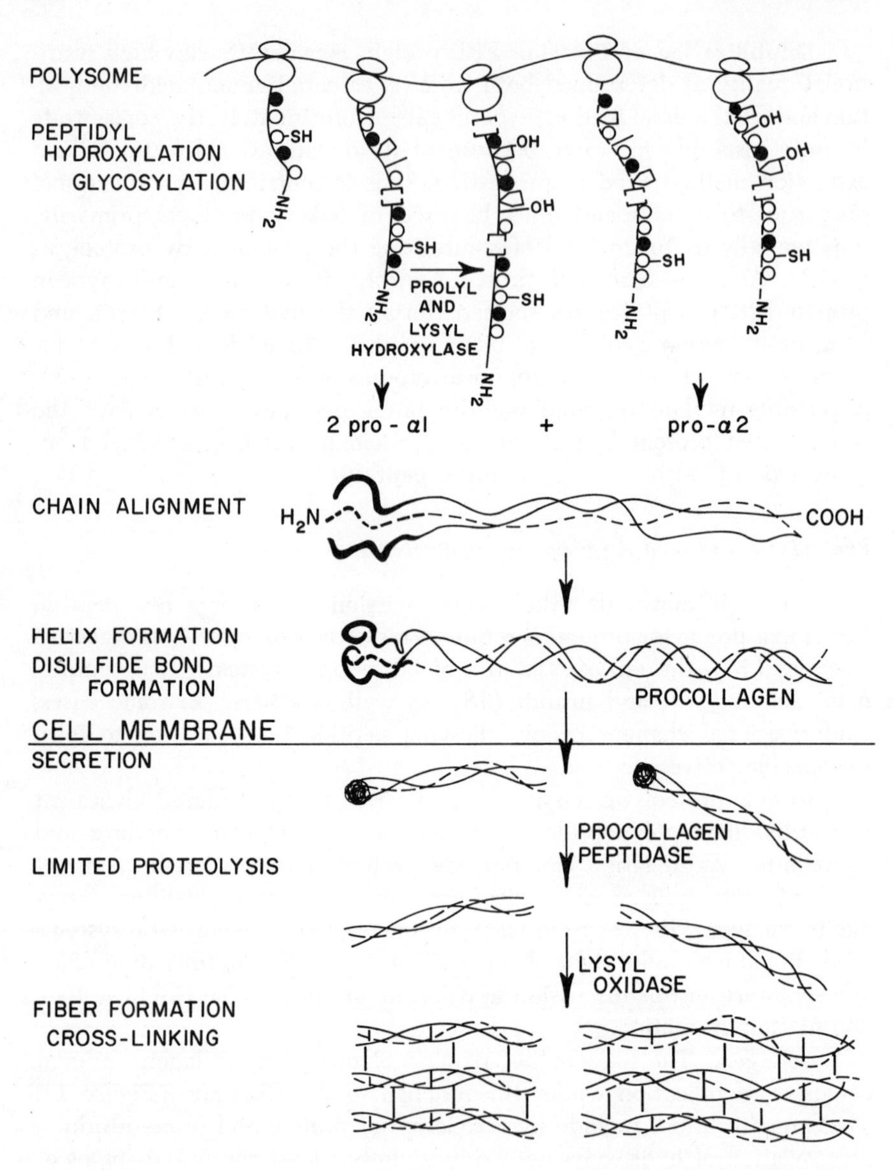

Figure 1. Proposed scheme for in vivo post-translational enzymatic modifications involved in collagen formation. (Adapted from Refs. 110 *and* 111.*)*

peptidase hydrolyzes the pro-α1 chain at a peptide bond where glutamic acid (or glutamine) is the amino group donor to the peptide bond. Subsequent to hydrolysis, the glutamic acid (or glutamine) residue cyclizes to a pyrrolidone carboxylic acid residue. Aggregation to the triple stranded procollagen molecule precedes proteolysis.

After proteolysis to form the collagen molecules of molecular weight ~ 285,000 the molecules are aligned both end-to-end in a staggered array and laterally to give cylindrical fibrils with diameters ranging from ca. 50Å to 2000Å, depending on the tissue and stage of development (*28*). These cylindrical fibrils then undergo further aggregation in a variety of ways depending on both the tissue and the biological function of collagen in that tissue (*29, 30*). The fibrils may align themselves in parallel array as in the fibers of tendon and the dermis of skin, as stacked sheets in the cornea of the eye, or as a more or less random crosshatch of fibers as seen in healing wounds. The aggregation of collagen fibrils

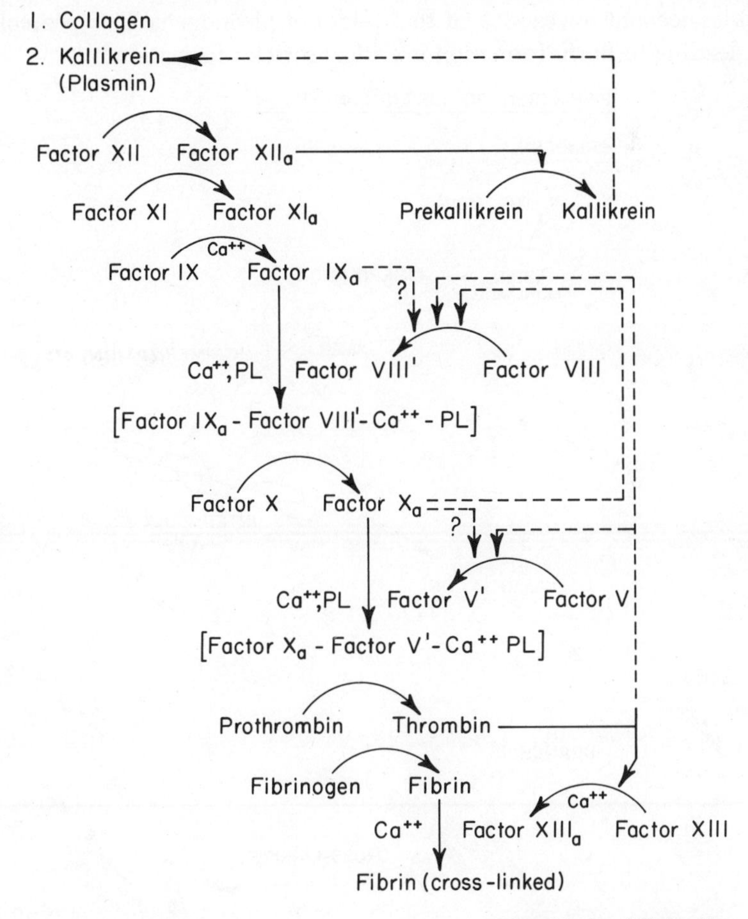

Figure 2. Proposed mechanism for blood clotting in mammalian plasma in the intrinsic system. The factor on the left side of reaction (zymogen) is converted to active enzyme by proteolysis. PL = phospholipids. (16a)

is stabilized by inter- and intramolecular covalent cross-links as described elsewhere in this chapter.

Fibrin. Bovine and human fibrinogens are glycoproteins of molecular weight ca. 340,000 (*21*). The molecule is composed of three nonidentical polypeptide chains—α, β, and γ—held together by anywhere from 21–22 to 32–34 disulfide bonds.

Fibrinogen is converted to fibrin by the action of thrombin on the α and β chains of fibrinogen. The proteolytic scission events leading up to thrombin formation are complex (Figure 2). Thrombin hydrolyzes each chain at two specific Arg–Gly bonds to release two fibrinopeptides A (from α-chain) and two fibrinopeptides B (from β-chain). The fibrinopeptides account for ca. 3% of the weight of fibrinogen. This proteolytic step, leading to fibrin formation is critical for clot formation.

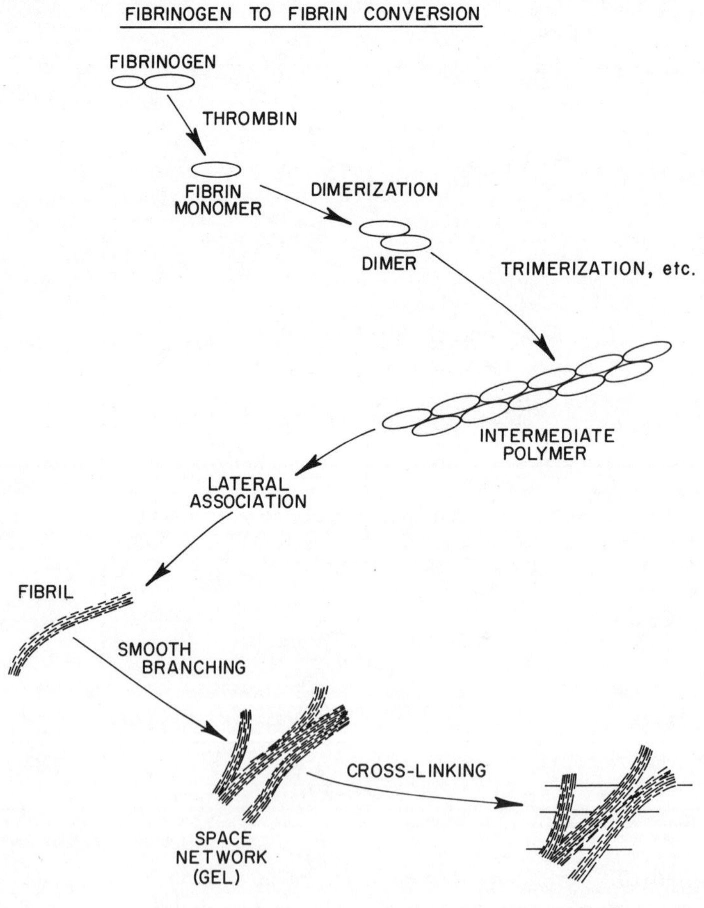

Figure 3. Association of fibrin monomers to form clot in the blood clotting process. (Adapted from Ref. 21).

After proteolysis, the fibrin molecules undergo both end-to-end and lateral aggregation followed by intertwining of the fibrils as shown schematically in Figure 3. The exact nature of the aggregation appears to depend on several environmental factors (*31*), and numerous models have been proposed in attempts to explain the details of the aggregation (*21*).

The clot formed by aggregation of fibrin is soluble in dilute acid and base and 6*M* urea, and does not have sufficient stability for permanent blood clotting. The aggregate is stabilized by an enzyme, a transglutaminase, which cross-links the chains by formation of ϵ-*N*-(γ-glutamyl)-lysine residues between a glutamine side chain of one fibrin monomer with a lysine side chain on another fibrin monomer (*32, 33, 34*). The enzyme is officially named Factor XIII, although numerous other names have been used (*see* section on Cross-Linking). There are eventually about six cross-links formed per fibrin monomer, two involving cross-linking of γ-γ chains (*35*) and four involving the cross-linking of α-α chains (*36*) of the fibrin. The crosslinked fibrin is very insoluble and resistant to lysis.

Factor XIII will slowly form ϵ-*N*-(γ-glutamyl)lysine cross-linkages in fibrinogen, in itself and in other proteins (*37*). Transaminases from other sources will also cross-link fibrin, as well as other proteins. In addition, they will covalently attach amino-containing substrates to the γ-carboxyl group of glutamic acid residues in proteins (*38*).

Casein. In milk, casein exists as large macromolecular micelles composed of α_s-, β-, and κ-caseins as well as Ca^{2+} and perhaps other proteins in minor amounts (*39*). There is still substantial disagreement on the exact composition and organization of the micelle, and more particularly on the location of the three caseins (*20, 40*). Regardless of the casein micelle model chosen, the specific proteolytic cleavage of a single peptide bond in κ-casein (*41*) destroys the integrity of the micelle, causing a reorganization of the caseins and micelles to form a clot. The casein micelles can also be destroyed and the caseins caused to aggregate by lowering the pH of milk to less than five, as for example in cottage cheese production by the fermentation method with lactic acid-forming bacteria.

At pH 5–6, rennin rapidly hydrolyzes a specific phenylalanylmethionine peptide bond in κ-casein (*41*). Splitting of this specific peptide bond is sufficient to destroy the ability of κ-casein to stabilize the casein micelles. The initial proteolytic step, with an activation energy of about 6 kcal/mol, is the only enzymatic step involved in clotting. The reorganization step(s), with an activation energy of about 40 kcal/mol, is not enzymatic. Further slow proteolysis of the aggregate occurs during ripening of the curd (*40*) and results in characteristic flavor development, including bitter taste with some proteolytic enzymes.

Jollès (*20*) has presented an intriguing discussion of the similarities between the clotting of casein and fibrin including amino acid sequence similarities between the key proteins, κ-casein and fibrinogen.

More specific details of milk clotting in relation to cheese production are given in Chapter 7 of this book.

Plastein Formation. The ability of some proteolytic enzymes to convert soluble proteins to an insoluble aggregate was noted as early as 1947 (*42, 43*) in the formation of plakalbumin from albumin by subtilisin. More recently, Fujimaki and co-workers have investigated this reaction for removal of flavor constituents from soybeans as well as a method of covalently attaching essential amino acids to the protein.

Isolated soybean protein still contains some of the characteristic soybean flavor, which makes it unacceptable for use in many bland foods. The flavor component is not removed by solvent extraction but can be removed by limited proteolysis of the protein with pepsin for example (*44*). However, the pepsin hydrolysate is quite bitter. The bitter taste can be removed by neutralizing the hydrolysate, concentrating it to 20% solids, and treating with α-chymotrypsin. At this high concentration of peptides, α-chymotrypsin is able to synthesize new peptide bonds leading to an increase from 5% to 75% in trichloroacetic acid insoluble material and to elimination of the bitter taste. The properties of the protein-like substance, plastein, are quite different from those of the original material; the plastein is quite digestible by pepsin and trypsin in vitro and has high nutritional quality based on its amino acid composition (*45*).

The nutritional quality of a protein can be increased by the plastein reaction (*46*). Following partial hydrolysis of a protein by pepsin, the ethyl ester of a limiting amino acid such as methionine or cystine, or a partial hydrolysate of another protein which is limiting in another amino acid residue, can be added to the hydrolysate and covalently linked through plastein formation.

Fish protein concentrate, although a good protein source nutritionally, is very insoluble; this limits its use in foods. The protein can be solubilized by proteolytic enzymes, but the hydrolysate is bitter (*12*). It would be useful to investigate the plastein reaction for removal of the bitter taste as well as for changing the solubility properties of the digest (*14*). The plastein reaction is described more fully in Chapter 6 of this monograph.

Cross-Linkages Formed in Proteins

Cross-linkages are incorporated into proteins after translation, and the data support the hypothesis that cross-linkages are incorporated into proteins after they are excreted from the cell. Types of cross-linkages found in proteins are indicated in Table II. In the following discussion we shall deal with the systems: disulfide bond formation, cross-linking in the connective tissue proteins, cross-linking in blood clotting, and lipoxygenase-catalyzed cross-linking of wheat proteins.

Disulfide Bonds. It is generally true that intracellular proteins do not contain disulfide bonds, while a larger number of extracellular proteins do. Is this because the greater reducing nature of the intracellular environment prevents disulfide bond formation? Are disulfide bonds in extracellular proteins the result of extracellularly catalyzed oxidation, and is the disulfide bond in extracellular proteins a major stabilizing factor? Could one change the functional properties of a food protein by rearrangement of the disulfide bonds? The answer to all these questions is probably yes.

There are at least two types of enzyme systems involved in the formation and breakage of disulfide bonds of cystine residues in proteins. A thiol-disulfide interchange enzyme (protein disulfide-isomerase; EC 5.3.4.1; other name, S-S-rearrangase) was first described in 1963 (*47*) and was subsequently purified from beef liver (*48, 49*). The molecular weight of the enzyme is 42,000. The enzyme contains three half-cystine residues, one of which must be cysteine in order for the enzyme to be active. The enzyme catalyzed the rearrangement of random "incorrect" pairs of half-cystine residues to the native disulfide bonds in several protein substrates. Low levels of mercaptoethanol were required for activity unless the enzyme was reduced prior to use. The efficiency of the enzyme in catalyzing the interconversion of disulfide bonds was found to be a function of the number of disulfide bonds in the substrate. Purification of a thiol-disulfide interchange enzyme from *Candida claussenii* has been described recently (*50*).

The reaction catalyzed by the thiol-disulfide interchange enzyme may be depicted as shown in Equation 1:

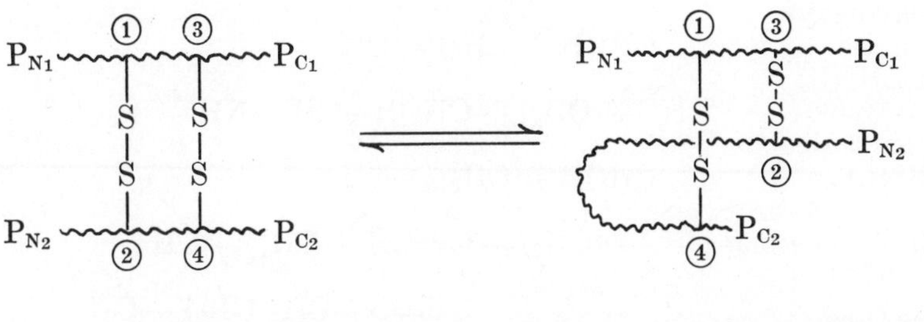

$$P_{N_1}\sim\overset{①}{\underset{|}{}}\sim\overset{③}{\underset{|}{}}\sim P_{C_1} \quad (S\text{—}S \text{ between } ① \text{ and } ②;\ S\text{—}S \text{ between } ③ \text{ and } ④) \quad P_{N_2}\sim ② \sim ④ \sim P_{C_2} \rightleftharpoons P_{N_1}\sim ① \sim ③ \sim P_{C_1};\ ①\text{—}S\text{—}S\text{—}④;\ ③\text{—}S\text{—}S\text{—}②;\ P_{N_2}\sim ② ;\ ④ \sim P_{C_2} \qquad (1)$$

A second type of enzyme system which reduces disulfide bonds in proteins has been described. Protein-disulfide reductase(NAD(P)H) [NAD(P)H:protein-disulfide oxidoreductase; EC 1.6.4.4], purified from pea seeds (*51*), catalyzes the Reaction 2, shown on p. 111.

Table II. Cross-

Name	*Structure*[a]
Disulfide	HC—CH_2—S—S—CH_2—CH
Dehydrolysinonorleucine (deLNL)	HC—$(CH_2)_3$—CH_2—N=CH—$(CH_2)_3$—CH
Lysinonorleucine (LNL)	HC—$(CH_2)_3$—CH_2—NH—CH_2—$(CH_2)_3$—CH
"Aldol" (ALAL)	HC—$(CH_2)_2$—CH(CHO)—CHOH—$(CH_2)_3$—CH and HC—$(CH_2)_2$—C(CHO)=CH—$(CH_2)_3$—CH
Dehydrohydroxylysinonorleucine ($deHLNL_a$ or $deHLNL_b$)	HC—$(CH_2)_2$—CHOH—CH_2—N=CH—$(CH_2)_3$—CH and HC—$(CH_2)_3$—CH_2—N=CH—CHOH—$(CH_2)_2$—CH
δ-Hydroxylysinonorleucine (HLHL)	HC—$(CH_2)_3$—CH_2—NH—CH_2—CHOH—$(CH_2)_2$—CH
Dehydrohydroxylysinohydroxynorleucine (deHLHNL)	HC—$(CH_2)_2$—CHOH—CH_2—N=CH—CHOH—$(CH_2)_2$—CH
Hydroxylysinohydroxynorleucine HLHNL)	HC—$(CH_2)_2$—CHOH—CH_2—NH—CH_2—CHOH—$(CH_2)_2$—CH
Dehydromerodesmosine (deM)	HC—$(CH_2)_2$—C(—CH=N—$CH_2(CH_2)_3$—CH)=CH—$(CH_2)_3$—CH
Mesodesmosine (M)	HC—$(CH_2)_2$—C(—CH_2—NH—CH_2—$(CH_2)_3$—CH)=CH—$(CH_2)_3$—CH

Linkages in Proteins

Comment

Possibly formed by reaction of lysine and α-amino adipic acid δ-semialdehyde

Reduction of dehydrolysinonorleucine

Formed from condensation of two α-amino adipic acid δ-semialdehyde residues

Condensation of α-amino adipic acid δ-semialdehyde and hydroxylysine

Condensation of δ-hydroxy-α-amino adipic acid δ-semialdehyde and lysine

Reduction of dehydrohydroxylysinonorleucine

Condensation of δ-hydroxy-α-amino adipic acid δ-semialdehyde and hydroxylysine

Reduction of dehydrohydroxylysinohydroxynorleucine

Condensation of "Aldol" (above) with lysine

Reduction of dehydromerodesmosine

Table II.

Name	*Structure*[a]
Dehydrohydroxymerodesmosine (deHM)	$\rangle HC-(CH_2)_2-C{=}CH-(CH_2)_3-CH\langle$; C bonded to $CH{=}N-CH_2CHOH(CH_2)_2-CH\langle$
Hydroxymerodesmosine (HM)	$\rangle HC-CH_2-C{=}CH-(CH_2)_3-CH\langle$; C bonded to $CH_2-NH-CH_2CHOH(CH_2)_2-CH\langle$
Dihydrodesmosines (dihydro D)	$\Delta^{3,4}$,$\Delta^{5,6}$-dehydrodesmopiperidines of ring types A and D[b]
Desmosines (D)	$\Delta^{1,2}$,$\Delta^{3,4}$,$\Delta^{5,6}$-dehydrodesmopiperdines of ring types A and D[b]
Tetrahydrodesmosines (tetrahydro D)	$\Delta^{3,4}$ or $\Delta^{5,6}$-dehydrodesmopiperidines of ring types A and D[b]
"Compound 285"	$\rangle HC-CH{=}CH-C-C{=}C(CH_3)-CH{=}CH-CH\langle$; upper C bonded to $CH{=}CH-CH_2-CH\langle$; $C{=}CH-NH-CH_2-(CH_2)_3-CH\langle$
Histidino-merodesmosine (HiM)	$\rangle HC-(CH_2)_2-CH-CH-(CH_2)_3-CH\langle$; second CH bonded to imidazole N (ring bearing $CH_2-CH\langle$) ; first CH bonded to $CH_2-NH-CH_2-(CH_2)_3-CH\langle$
Histidino-hydroxymerodesmosine (HiHM)	$\rangle HC-(CH_2)_2-CH-CH-(CH_2)_3-CH\langle$; second CH bonded to imidazole N (ring bearing $CH_2-CH\langle$) ; $CH_2-NH-CH_2-CHOH-(CH_2)_2-CH\langle$

Continued

Comment

Condensation of "Aldol" (above) with hydroxylysine

Reduction of dehydrohydroxymerodesmosine

Condensation of α-amino adipic acid δ-semialdehyde with dehydromerodesmosine

From dehydrogenation of dihydrodesmosines

From reduction of dihydrodesmosines

From oxidation of the desmosines?

Condensation of an imidazole ring of an histidine residue with merodesmosine

Condensation of an imidazole ring of an histidine residue with hydroxymerodesmosine

Table II.

Name	*Structure*[a]
Aldol-histidine (AHi)	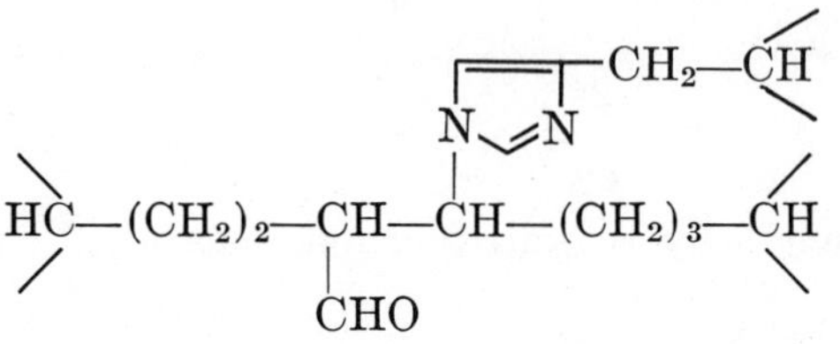 $>$HC—$(CH_2)_2$—CH(—CHO)—CH(—N-imidazole—CH_2—CH$<$)—$(CH_2)_3$—CH$<$
ε-N-(γ-Glutamyl)-lysine	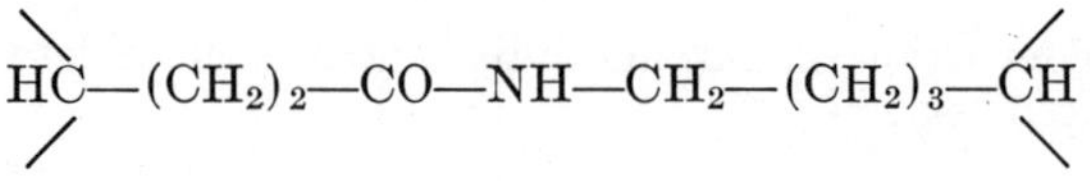 $>$HC—$(CH_2)_2$—CO—NH—CH_2—$(CH_2)_3$—CH$<$
ε-N-(β-Aspartyl)-lysine	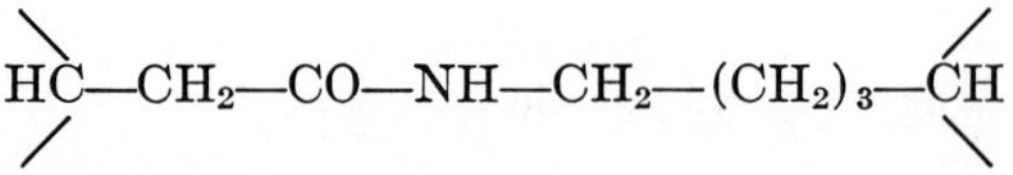 $>$HC—CH_2—CO—NH—CH_2—$(CH_2)_3$—CH$<$

[a] In each case the structure is drawn to the α-carbon atom.
[b] The structures of desmopiperidines A and D are:

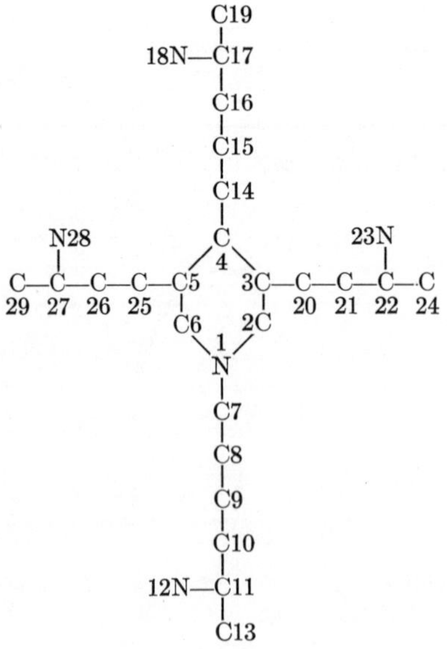

```
            C19
             |
      18N—C17
             |
            C16
             |
            C15
             |
            C14                                                        N23
             |                                                          |
  N28        C          23N           N28          C          C—C—C—C
   |        / 4 \         |             |          / 4 \      /20 21 22 24
C—C—C—C—C5      3C—C—C—C—C     C—C—C—C—C5      3C
29 27 26 25 |     |  20 21 22 24  29 27 26 25 |      |
           C6    2C                           C6     2C         17
             \ 1 /                              \ 1 /  \    C—C—C—C—C
              N                                  N        14 15 16 |  19
              |                                  |                 N18
             C7                                 C7
              |                                  |
             C8                                 C8
              |                                  |
             C9                                 C9
              |                                  |
             C10                                C10
              |                                  |
      12N—C11                            12N—C11
              |                                  |
             C13                                C13
             (A)                                (D)
```

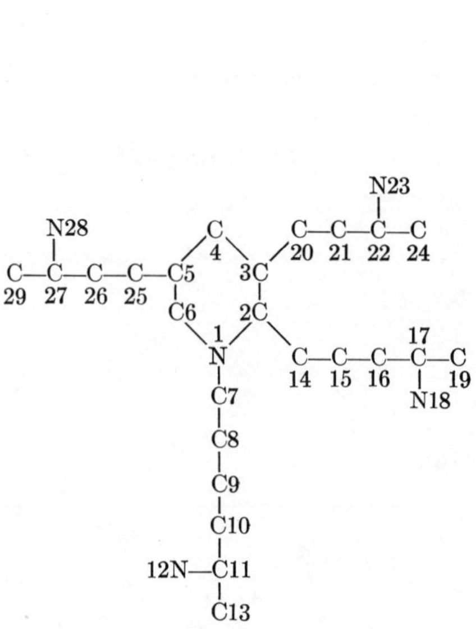

C-11, C-17, C-22, and C-27 are α-carbons initially of lysine residues.

Continued

Comment

Condensation of aldol with histidine

Condensation of glutamine and lysine residues

Condensation of asparagine and lysine residues

$$\text{NAD(P)H} + \text{protein disulfide} = \text{NAD(P)}^{+} + \text{protein dithiol} \quad (2)$$

The enzyme, found in a number of plant tissues, reduces the disulfide bonds in a number of proteins. Its activity is strongly inhibited by sulfhydryl reagents. The enzyme is different from cystine reductase and glutathione reductase which do not reduce protein disulfide bonds.

Protein-disulfide reductase(glutathione) [glutathione:protein-disulfide oxidoreductase; EC 1.8.4.2) has been reported in hepatic tissue (*52*). The enzyme rapidly cleaved the three disulfide bonds of insulin and the disulfide bonds of other proteins. The K_m for reduced glutathione was $8.9 \times 10^{-3}M$, a very high value. The enzyme catalyzed the reaction (Equation 3) from either direction.

$$\begin{aligned} &2\ \text{Glutathione} + \text{protein–disulfide} = \\ &\text{oxidized glutathione} + \text{protein–dithiol} \end{aligned} \quad (3)$$

Recently, a heat-labile compound in meat extract which decreases the heat stability of soybean trypsin inhibitors was reported (*53*). It is tempting to speculate that this is some type of protein-disulfide reductase, but further work is needed on this point.

Cross-Linkages in Connective Tissue. Collagen undergoes post-translational enzymatically-catalyzed chemical modification involving (a) hydroxylation of lysine and proline, (b) glycosylation of lysine, (c) peptide bond cleavage leading to *N*-terminal pyrrolidone carboxylic acid residues, and (d) cross-linking reactions.

The numerous cross-linkages found in collagen are the result of the oxidation of lysine and hydroxylysine to α-amino adipic acid δ-semialdehyde (allysine) and δ-hydroxy α-amino adipic acid δ-semialdehyde (hydroxyallysine), respectively, following the assemblage of the collagen chains into the macromolecular structure (Equations 4 and 5):

$$\underset{\text{lysine}}{>HC-(CH_2)_3-CH_2-NH_2} \xrightarrow{[O]} \underset{\alpha\text{-amino adipic acid } \delta\text{-semialdehyde}}{>HC-(CH_2)_3-CHO} + NH_3 \quad (4)$$

$$\underset{\text{Hydroxylysine}}{>HC-(CH_2)_2-CHOH-CH_2-NH_2} \xrightarrow{[O]}$$

$$\underset{\delta\text{-hydroxy } \alpha\text{-amino adipic acid } \delta\text{-semialdehyde}}{>HC-(CH_2)_2-CHOH-CHO} + NH_3 \quad (5)$$

The enzyme which catalyzes the oxidation of lysine and hydroxylysine residues in collagen and elastin is lysyl oxidase. The enzyme, first demonstrated in extracts of tissues of embryonic chick (*54*), has been partially purified and characterized (*55*). It has a molecular weight of 170,000, requires O_2, Cu^{2+}, and pyridoxal phosphate (*56*) and is inhibited

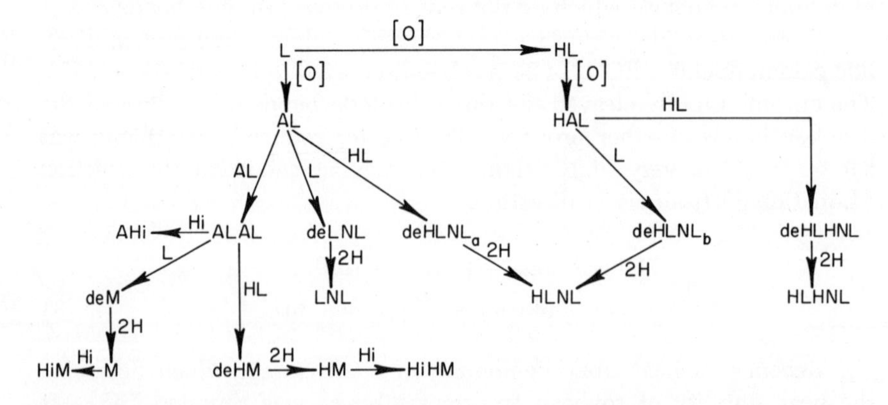

Figure 4. Possible routes of biosynthesis of cross-linkages in collagen. L, lysine; HL, hydroxylysine; Hi, histidine; definition of other symbols in Table II. (Based on data presented in Refs. 59 *and* 60*).*

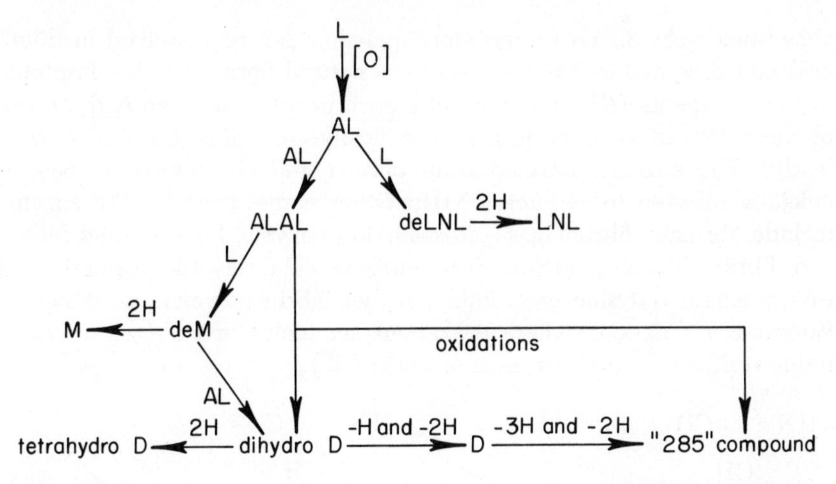

Figure 5. Possible routes of biosynthesis of cross-links in elastin. See *Figure 4 and Table II for definition of symbols used* (59).

by β-aminopropionitrile. Lysyl oxidase can catalyze the oxidation of lysine and hydroxylysine in vitro leading to cross-linking of collagen (*56–58*). An enzyme extracted from chick embryo bone produced identical cross-linkages to those found in natural collagen when it was allowed to act on bone collagen from lathyritic chicks (*58*). Similar cross-links were found in extracted porcine tendon collagen when it was treated with an extract from porcine bone (*56*).

The cross-linkages found in collagen and elastin arise from the nonenzymatic interaction of α-amino adipic acid δ-semialdehyde (AL) and δ-hydroxy α-amino adipic acid δ-semialdehyde (HAL) with another residue of the aldehyde or with lysine, hydroxylysine, or histidine residues located at specific positions in the collagen polypeptide chains. Possible relationships among the different compounds in Table II are shown in Figures 4 and 5 for collagen and elastin.

Cross-Linkages Following Fibrin Clotting. The molecular events leading to formation of a stable fibrin clot include a cascade of interrelated zymogen-enzyme transformations as described above in Figure 2. The final proteolytic step in this process is the action of thrombin on fibrinogen as shown in Equation 6:

$$\text{Fibrinogen} \xrightarrow[\text{H}_2\text{O}]{\text{thrombin}} \begin{array}{c}\text{fibrin monomer} \\ + \text{ fibrinopeptides}\end{array} \rightarrow \text{fibrin gel} \qquad (6)$$

Following fibrin monomer formation there is an aggregation of monomers to form fibrils which then become intertwined as depicted above in Figure 3. This intertwined polymer can be dissolved in dilute acid and base and in 8*M* urea, whereas natural fibrin clots are insoluble in these reagents (*61, 62*). The difference between the two is the result of the action of an enzyme that specifically cross-links the fibrils covalently. The enzyme, purified from plasma and characterized (*63*), is officially referred to as Factor XIII. Other names used for the enzyme include fibrinase, fibrinoligase, cross-linking enzyme, Laki–Lorand factor, and fibrin-stabilizing factor. The enzyme catalyzes the formation of ε-*N*-(γ-glutamyl)lysine cross-links between fibrin polymers as shown in Equation 7 (*32, 33*). There are about six moles of ε-*N*-(γ-glutamyl)-lysine residues formed per mole of fibrin (*35*).

$$\begin{array}{c} \sim HN \quad CO\sim \\ \diagdown \diagup \\ CH \\ | \\ (CH_2)_3 \\ | \\ CH_2 \\ | \\ NH_2 \\ O{=}C{-}NH_2 \\ | \\ CH_2 \\ | \\ CH_2 \\ | \\ CH \\ \diagup \diagdown \\ \sim HN \quad CO\sim \end{array} \xrightarrow[Ca^{2+}]{\text{Factor XIII}} \begin{array}{c} \sim HN \quad CO\sim \\ \diagdown \diagup \\ CH \\ | \\ (CH_2)_3 \\ | \\ CH_2 \\ | \\ NH \\ | \\ O{=}C \\ | \\ CH_2 \\ | \\ CH_2 \\ | \\ CH \\ \diagup \diagdown \\ \sim HN \quad CO\sim \end{array} + NH_3 \qquad (7)$$

As shown by the reaction in Equation 7, Factor XIII is a transaminase (transglutaminase) enzyme. Similar enzymes have been isolated from a variety of tissues (*64*). These enzymes are capable of incorporating a variety of amino compounds into proteins with the addition taking place exclusively at glutamine residues. Therefore, the enzyme appears to have specificity toward glutamine residues of proteins. What other amino acid residues around the glutamine are required is not shown. These transaminases are activated by calcium ion and sulfhydryl compounds.

It would be useful to look at the possibility of using these enzymes for formation of cross-links in proteins leading to textured products. Asquith et al. (*65*) have recently reviewed the occurrence of isopeptide bonds in proteins.

Lipoxygenase-Catalyzed Cross-Linkages in Proteins. In order to produce the right texture in bread and similar products, chemicals such as bromates, acetone peroxides, ascorbic acid, azodicarbonamide, calcium peroxide, chlorine, chlorine dioxide, iodates, and nitrosyl chloride have been used, or proposed for use, in flour or dough (*66*). These "conditioners" produce reactions which lead to cross-link formation among the protein molecules, particularly gluten, and thereby facilitate entrapment

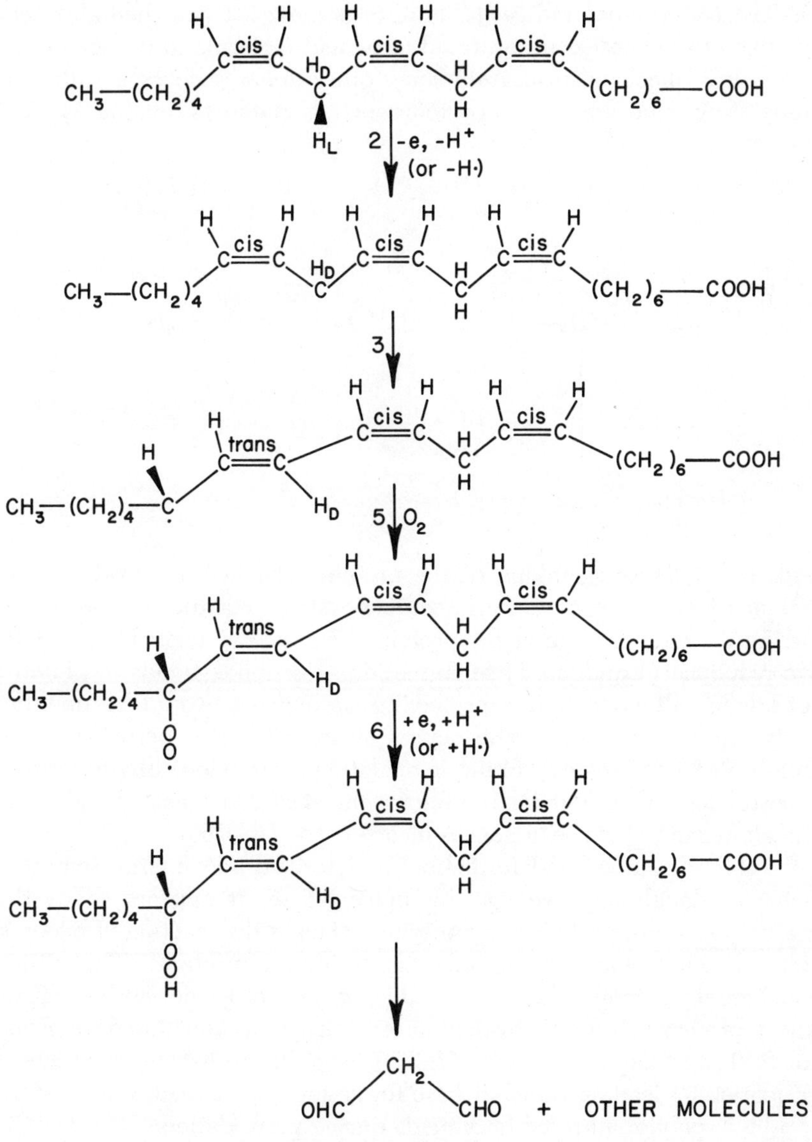

Figure 6. Reaction catalyzed by lipoxygenase. Malondialdehyde is formed from further oxidation of species V.

of carbon dioxide during the rising process. "Conditioning" of dough can be brought about also by lipoxygenase.

Lipoxygenase is added to dough by blending small amounts of soybean flour, rich in lipoxygenase, into wheat flour. Lipoxygenase acts on unsaturated fatty acids to produce initially fatty acid hydroperoxy free radicals which eventually lead to production of malondialdehyde (Figure 6). The hydroperoxy radical (Figure 6, formula D) will then abstract a hydrogen from another unsaturated fatty acid molecule to perpetuate the free radical mechanism or from some other donor such as a sulfhydryl group of cysteine residues on proteins such as gluten (Equation 8). This

$$2\,\text{—}(CH_2)_4\text{—}\underset{\substack{|\\O\\O\\\cdot}}{CH}\text{—}CH{=}CH\text{—} + 2\,{>}HC\text{—}CH_2SH \rightarrow$$

$$2HC\text{—}CH_2\text{—}S\cdot + 2\,\text{—}(CH_2)_4\text{—}\overset{H}{\underset{\substack{|\\O\\O\\H}}{C}}\text{—}CH{=}CH\text{—} \qquad (8)$$

$$\downarrow \qquad HC\text{—}CH_2SH \swarrow$$

$$>HC\text{—}CH_2\text{—}S\text{—}S\text{—}CH_2\text{—}CH<$$

leads to –S–S– cross-linking of the protein. The hydroperoxide formed here or by other means can also oxidize cysteine residues to cystine furthering the cross-linkage of the protein. The relative contribution of the two reactions (Equation 8) to formation of disulfide bonds in gluten is not known, although it is expected the second one would be quantitatively more important. This lipoxygenase-catalyzed peroxidation of unsaturated fatty acids resulting in oxidation of protein sulfhydryl groups is postulated to account for the improvement of soft wheat dough when a small amount of soybean flour is incorporated (*67–70*).

While disulfide bond formation as described above is an important factor in dough improvement by lipoxygenase, it appears to us that another reaction could also be important. This is the reaction of malondialdehyde with ϵ-amino groups of lysine residues of proteins. This bifunctional aldehyde might react with ϵ-amion groups of lysine residues of the wheat proteins to form covalent cross-linkages as illustrated in Equation 9. Tappel and co-workers (*71, 71_a*) have shown that the first reaction (Equation 9) leading to Schiff base formation occurs readily in mixtures of protein-polyunsaturated fatty acids during peroxidation.

It is tempting to speculate that reduction of the Schiff base might occur by oxidation of protein cysteine residues. Experimental work is

$$2\ \left[\sim\mathrm{HN{-}CH(CO{\sim}){-}(CH_2)_3{-}CH_2{-}NH_2}\right] + \mathrm{CH_2(CHO)_2} \longrightarrow \sim\mathrm{HN{-}CH(CO{\sim}){-}(CH_2)_3{-}CH_2{-}N{=}CH{-}CH_2{-}CH{=}N{-}CH_2{-}(CH_2)_3{-}CH(CO{\sim}){-}NH}\sim \xrightarrow{+4\mathrm{H}} \sim\mathrm{HN{-}CH(CO{\sim}){-}(CH_2)_3{-}CH_2{-}NH{-}CH_2{-}CH_2{-}CH_2{-}NH{-}CH_2{-}(CH_2)_3{-}CH(CO{\sim}){-}NH}\sim \qquad (9)$$

needed to determine if this type of reaction occurs in wheat dough, and its possible nutritional significance.

Protein Phosphorylation and Dephosphorylation

At least six different types of amino acid residues in proteins undergo enzymatic phosphorylation following translation of the protein. These include: the hydroxyl groups of serine, threonine, and hydroxylysine; the imidazole group of histidine; the guanidino group of arginine; and the ϵ-amino group of lysine as shown in Table III. The possibility of acyl phosphate linkages in proteins has not been unequivocally eliminated (*72*).

N^3-Phosphohistidine has been isolated from an alkaline hydrolysate of liver acid phosphatase (*73*) while N^1-phosphohistidine has been isolated from prostate, placenta, and wheat germ acid phosphatases (*74, 75*). N^3-Phosphohistidine is found in histone H4 (IV, F2al) (*76*). Phosphohistidines, substituted on either ring nitrogen atom, and ϵ-*N*-phospholysine were isolated from a purified citrate cleavage enzyme (*77*). There is a good possibility that certain acidic nuclear proteins may contain ϵ-*N*-phospholysine and ω-*N*-phosphoarginine (*78*). ϵ-*N*-Phospholysine is present in histone I (F1) (*78a*). Phosphothreonine is found in certain proteins of which phosvitin is an example (*79, 80*).

The major phosphorylated amino acid in proteins, however, is phosphoserine. It is found, in one or two residues, in enzymes such as alka-

Table III. Phosphorylated Amino Acid Derivatives in Proteins

Derivative	*Formula*[a]
Phosphoserine	$>HC—CH_2—OPO_3^{2-}$
Phosphothreonine	$>HC—HCOPO_3^{2-}—CH_3$
O-Phosphohydroxy-lysine	$>HC—CH_2—CH_2—HCOPO_3^{2-}—CH_2—NH_2$
ε-*N*-Phospholysine	$>HC—(CH_2)_3—CH_2—NHPO_3^{2-}$
ω-*N*-Phosphoarginine	$>HC—(CH_2)_2—CH_2—NH—C({=}NH)—NH—PO_3^{2-}$
N^1-Phosphohistidine	$>HC—CH_2$—(imidazole ring, $^{2-}O_3P—N$ at N¹)
N^3-Phosphohistidine	$>HC—CH_2$—(imidazole ring, $N—PO_3^{2-}$ at N³)

[a] The formulas show the α-carbon and the side chain of the amino acid residues.

line phosphatase, phosphorylase kinase, glycogen synthetase, troponin, RNA polymerase, polynucleotide, phosphorylase, and triglyceride lipase. Phosphoserine is also found in nonenzyme proteins which include histones, protamiens, ribosomal proteins, membrane proteins, ovalbumin, casein, and phosvitin often in substantial amounts. For example, there are 119 phosphoserines and only one phosphothreonine residue in phosvitin (*72*).

The remainder of our discussion on phosphoproteins will deal with the types of enzymes involved in phosphorylation of nonenzyme proteins, substrate specificity of those enzymes, and enzymatic dephosphorylation of phosphoproteins.

Phosphorylation of proteins probably occurs after translation of the protein (*81*). *t*-RNA specific for phosphoserine occurs in rat and rooster liver, but it has not been shown that the phosphorylated amino acid can be incorporated directly into proteins (*82*). In this case, it appears that serine is phosphorylated after it combines with the specific *t*-RNA (*83*, *84*). Sites of phosphorylation are probably determined by the amino acid sequence of the protein (*see below*), and extent of phosphorylation may be limited by formation of secondary, tertiary, and quaternary structure.

The enzymes involved in phosphorylation of serine and threonine residues of proteins are protein kinases (ATP:protein phosphotransferase; EC 2.7.1.37). Protein kinases with specificity toward protamines and histones are classified as protamine kinases (ATP:protamine *O*-phosphotransferase, EC 2.7.1.70) by the International Union of Pure and Applied Chemistry and the International Union of Biochemistry. Because of marked specificity differences between the two groups, it is preferable to separate the enzymes into the histone kinases and the protamine kinases.

$$\text{ATP} + \text{a protein} = \text{ADP} + \text{a phosphoprotein} \qquad (10)$$

The term protein kinase was first used to describe the enzymes of yeast, liver, and brain which phosphorylated casein and phosvitin (*85–87*). More recently it has been proposed that the term protein kinase be restricted to those enzymes which catalyze the transfer of the γ-phosphate of ATP to serine and threonine hydroxyls of proteins and are regulated by cyclic nucleotides (*88*). The other class of protein kinases, the phosphoprotein kinases, are not affected by cyclic nucleotides, and they can often utilize GTP as a source of phosphate. Wherever appropriate, we shall refer to these two groups of enzymes as cyclic AMP-dependent or cyclic AMP-independent protein kinases, respectively (*89*).

The cyclic AMP-dependent protein kinases are widely distributed in animal tissues but have not yet been found in higher plants (*89*). They are made up of regulatory (R) and catalytic (C) subunits. Cyclic AMP binds to the regulatory subunit causing dissociation of the RC complex to form the active subunit (C) (Equation 11):

$$\underset{\text{inactive}}{\text{RC}} \underset{-\,\text{cAMP}}{\overset{+\,\text{cAMP}}{\rightleftharpoons}} \underset{\text{active}}{\text{C} + \text{R}\cdot\text{cAMP}} \qquad (11)$$

Cyclic AMP-dependent protein kinases have the following general properties (*89*). The pH optima are in the range of 6–9, with molecular weights of 123,000 for skeletal muscle enzyme to $\sim$ 160,000 for anterior pituitary enzyme. The K_m for ATP is 7–20μM in the presence of saturating levels of Mg^{2+} (1–3mM). GTP, UTP, CTP, and dTTP have K_m values at least 15 times larger than for ATP. With some enzymes, ATP cannot be replaced by these other trinucleotides, while with other enzymes they are inhibitory. The K_a for cyclic AMP is $\sim 1 \times 10^{-8} M$.

Cyclic AMP-independent protein kinases are found in numerous animal tissues. Some are specific for basic proteins such as the histones and protamines. Others are specific for acidic proteins such as phosvitin and casein, while others may show both types of activity. Enzymes that catalyze the phosphorylation of phosvitin and casein have been purified from calf brain (*90, 91*), cytosol of liver of estrogen-treated roosters (*92*), the Golgi apparatus of the lactating mammary gland (*93*), and dogfish skeletal muscle (*94*). The calf brain and rooster liver enzymes do not phosphorylate histones or protamines. The rooster liver enzyme catalyzes the phosphorylation of phosvitin at ten times the rate of casein. The mammary gland enzyme readily phosphorylates unphosphorylated casein, phosphorylates β-lactoglobulin, α-lactalbumin, and native casein less readily, and does not appreciably phosphorylate histone, phosvitin, and lysozyme. The dogfish skeletal muscle enzyme phosphorylates protein at the relative rates of protamine, 100; phosvitin, 20; histone H4 (F2al), 10; and casein, 5. Protamines are extensively phosphorylated by a protamine kinase from testes of the rainbow trout (*95*) which has at least ten times more activity on protamine than on histone.

A cyclic AMP independent protein kinase which catalyzes the phosphorylation of histidine in histones has been found in nuclei from rat tissue and in Walker-256 carcinosarcoma cell nuclei (*96*). Two histone kinases (ATP:histone *N*-phosphotransferase) have been partially purified from the nuclei of the carcinosarcoma cells (*78a*). One of these enzymes preferentially phosphorylates histone H4 (IV, F2al) at an optimum pH of 9.5, while the other preferentially phosphorylates histone I (F1) at an optimum pH of 6.5. Both enzymes had an absolute requirement for Mg^{2+}, required similar levels of ATP, and were not stimulated by added cyclic AMP or cyclic GMP. The pH 9.5 kinase was strongly inhibited by relatively low concentrations of GTP and CTP, but the pH 6.5 kinase was not affected. The pH 9.5 kinase phosphorylates the only two histidines (His-18 and His-75) in histone H4 to form 3-phosphohistidine. The pH 6.5 kinase phosphorylates the ε-amino group of a lysine residue in histone I. Location of this lysine is not known. The phospho groups of both these derivatives are acid labile.

Specificity of the protein kinases for substrate is thought to be determined by the primary sequence of amino acids. In general, in in vitro experiments denatured proteins and proteolytic digests (*86*) are better substrates than are the native proteins—an important observation with respect to the possible use of enzymatic phosphorylation to chemically modify food proteins. We shall discuss substrate specificity in terms of what is known about the sites of phosphorylation in several proteins of importance in foods.

Ovalbumin can be isolated as the di-, mono-, and unphosphorylated protein. Although there are 35 residues of serine and 16 of threonine per mole of ovalbumin (MW 45,000), only two serine residues are phosphorylated. The amino acid sequences around these residues are (*72*):

Phosphopeptide I –Asp–*SerP*–Glu–Ile–Ala–Glu–Cys– (12)

Phosphopeptide II –Ala–Gly–Arg–Glu–Val–Val–Gly–*SerP*–Ala– (13)

Phosvitin is the principal phosphoprotein of egg yolk. Hen egg yolk phosvitin contains 10% phosphate by weight with 120 of the total 222 residues per mole (MW 35,500) phosphorylated. There are 119 residues of phosphoserine and one residue of phosphothreonine. About 5–7 of the total serine residues of phosvitin are not phosphorylated (*97*). A distinctive feature of the primary sequence of phosvitin is the repeating units of phosphoserine followed by a basic amino acid residue (*72*). This is illustrated by the following nine sequences of amino acids found in phosvitin:

$-(\mathrm{SerP})_6$–Arg; $-(\mathrm{SerP})_6$,Lys–; $-(\mathrm{SerP})_6$,Lys–; $-(\mathrm{SerP})_6$,His–;
$-(\mathrm{SerP})_6$–His–; $-(\mathrm{SerP})_3$,Glu–; $-(\mathrm{SerP})_3$,Leu–;
$-(\mathrm{SerP})_6$–Arg–; $-(\mathrm{SerP})_8$–Arg–

In contrast to phosvitin, the phosphate groups of bovine α_{s1}- and β-caseins are located in acidic parts of the proteins. The locations of the eight phosphates of α_{s1}-casein, variant B, have been determined (*98*). They are located in peptides with the following sequences.

–Asp–Ile–Gly–*SerP*–Glu–*SerP*–Thr–Glu– (14)
45 50

–Glu–Ala–Glu–*SerP*–Ile–*SerP*–*SerP*–*SerP*–Glu–Glu–Ile (15)
65 70

–Val–Pro–Asn–*SerP*–Val–Gln–Glu– (16)
75

–Val–Pro–Asn–*SerP*–Ala–Glu–Glu– (17)
115

Seven of the eight phosphate groups of α_{s1}-casein are contained in a 30-amino acid sequence between residues 46 and 75. This same region contains 10 of the total 26 Asp and Glu residues of the molecule and only one basic amino acid residue (Lys-58) and three hydrophobic residues (Ile-65, Ile-71, and Val-72). Thus, the region of the molecule is highly negatively charged and hydrophilic in character. There is a total of 16 serine and 5 threonine residues in α_{s1}-casein, variant B.

β-Casein contains five phosphate residues all as phosphoserine. The amino acid sequences containing these phosphoserine residues in β-casein, variant A^2 (*99*), are:

–Ile–Val–Glu–*SerP*–Leu–*SerP*–*SerP*–*SerP*–Glu–Glu–Ser– (18)
15 20

–Lys–Phe–Gln–*SerP*–Glu–Glu– (19)
35

Positions 15–20 contain four of the five phosphoserine residues of β-casein with the remaining residue at 35. Seven of the total of 20 Asp and Glu residues are in the first 21 residues of β-casein.

There are great similarities between α_{s1}- and β-caseins in the location of the phosphoserine residues. They are located in the N-terminal parts of the molecules which are also quite negative and hydrophilic in nature. There is a near identical phosphoserine cluster in the two proteins.

–Glu–*SerP* [Ile 65 / Leu 16]$_{\alpha / \beta}$ –*SerP*–*SerP*–*SerP*–Glu–Glu– (20)

The single phosphoserine residue of κ-casein is also in a hydrophilic region of the molecule (*100*), adjacent to the carbohydrate moiety of the molecule.

CHO
|
–Thr–Glx–Pro–Thr–*SerP*–Pro–Thr– (21)

The serine phosphorylated in histone I (F1) is one amino acid removed from a basic residue and in proximity to two proline residues (*101*):

–Lys–Ala–*Ser*P–Gly$(Pro)_2$–Val–Ser–Glu–Leu–Ile–Thr–Lys– (22)

Protamine is phosphorylated on serine residues adjacent to arginine residues (*102*):

–Pro–$(Arg)_2$–*SerP*–*SerP*–*SerP*–Arg–Pro–Val–$(Arg)_3$–
Pro–$(Arg)_2$–Val–*Ser*P–$(Arg)_2$– (23)

Trout testis histone H2A is phosphorylated on the N-terminal acetylserine residue (*103*):

N–Ac*SerP*–Gly–Arg–Gly– (24)
1

Calf thymus histone H4 (IV, F2a1) is phosphorylated by a specific enzyme on His-18 and His-75. The sequences around these residues are (*76*):

$$\underset{16}{\text{–Lys}}(\text{Ac})_{0,1}\text{–Arg–}\mathit{HisP}\text{–Arg–}\underset{20}{\text{Lys}}(\text{CH}_3)_{1,2}\text{–Val–} \quad (25)$$

$$\text{–Thr–Glu–}\underset{75}{\mathit{HisP}}\text{–Ala–Lys–Arg–Lys–} \quad (26)$$

On the basis of data presently available, one would conclude that phosphorylation of serine in proteins probably occurs when the serine residues are adjacent to or in close proximity (sequence-wise) to basic or acidic amino acid residues. In the tertiary structure it would be expected that these residues would be near the surface of the protein. Recently, it has been shown for a wide variety of *O*-phosphoryl proteins that protein substrates of intracellular protein kinases all have a common feature when the caseins are excluded; in general, all phosphorylated sites are separated from either lysine or arginine (in two cases, histidine) by no more than two amino acid residues (*103a*).

Dephosphorylation. In vivo, phosphoprotein phosphatases (phosphoprotein phosphohydrolase; EC 3.1.3.16) participate in the regulation of phosphate turnover and in rapid reversal of the phosphorylating reactions, yet there is very little information on this important group of enzymes. Alkaline (*104*) and acid (*105*) phosphatases of milk can hydrolyze the phosphate groups of phosphoserine residues in the caseins. *E. coli* alkaline phosphatase dephosphorylates many phosphoproteins (*106, 107*). A phosphoprotein phosphatase of liver dephosphorylates phosphorylated histones and protamines but has little or no activity on casein or phosvitin (*108*). The 60-fold purified enzyme had an apparent K_m for dephosphorylation of histone I (F1) of $2 \times 10^{-5}M$ and a pH optimum of 7–8. Histone phosphatase activity was detected in all eukaryotic cells examined, but it was not found in the extracts of several prokaryotes.

A great deal more needs to be known about these important phosphoprotein phosphatases and the in vivo role they may play in the regulation of cellular activity and phosphate turnover. An immobilized enzyme system for the phosphorylation of proteins has been described (*108a*).

Hydroxylation

In some proteins, proline and lysine are hydroxylated following translation of the polypeptide chain to give hydroxyproline and 5-hydrox-

Table IV. Hydroxylated Amino Acid Residues Found in Proteins

Derivative	*Formula*[a]
5-Hydroxylysine	$>HC-(CH_2)_2-CHOH-CH_2NH_2$
4-Hydroxyproline	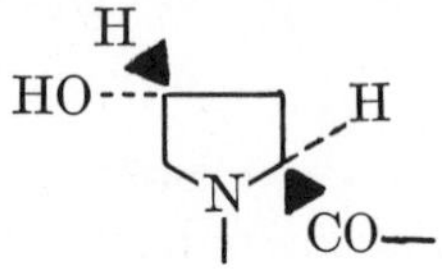 [structure: pyrrolidine ring with HO and H at C-4, H and CO— at C-2, N—]
3-Hydroxyproline	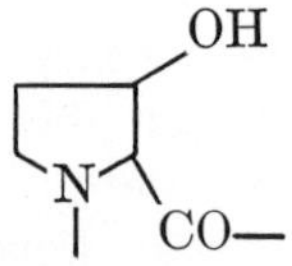 [structure: pyrrolidine ring with OH at C-3, CO— at C-2, N—]

[a] The formula for the 5-hydroxylysine residue depicts the α-carbon and the side chain.

ylysine, respectively. 4-Hydroxyproline is the major derivative of proline in collagen; however, there is one residue of 3-hydroxyproline in both the α_1 and α_2 chains near the C-terminal end (*59*). Enzyme preparations from earthworm cuticle have been reported to hydroxylate both the 3 and 4 positions of proline (*109*). The structure of these hydroxylated amino acid residues are shown in Table IV.

Hydroxylation of proline and lysine occurs following translation of the polypeptide chain, but probably begins before the nascent chain is released from the ribosomes (*59, 110, 111*). Hydroxylation must take place while the chain is still in the random coil configuration since native collagen is resistant to hydroxylation (*see below*). In fact, it appears that hydroxylation is essential for normal secretion of the pro α-chain from the cell (*112, 113*). Present data indicate it takes ca. 6 min to synthesize a pro α-chain (*26*) and that hydroxyproline appears in the chain after about 2 min from start of synthesis (*114*). Prolyl hydroxylase is present within the cell and is distributed throughout the lumen of the cisternae (*115*). Lysyl hydroxylase probably follows a similar pattern. Figure 1 shows a postulated scheme for the hydroxylation of proline and lysine.

Hydroxylation of collagen represents a major post-translational chemical modification. Collagen is the most abundant protein in vertebrates with about one-third of the total amino acid residues being either proline or hydroxyproline. On the average, hydroxyproline is one-tenth of the total amino acid residues of collagen. This is about 100 residues per

α-chain ($\sim$ 1015 amino acid residues). Hydroxylation of proline in the nascent pro α-chain is completed essentially within the 6 min needed for biosynthesis of the chain (*116*). The number of hydroxylysine residues in collagen ($\sim$ 5) is much lower than the number of hydroxyproline residues, which agrees with the lower level of activity of this enzyme in cells (*111*).

Because of its importance, much effort has been devoted to research on prolyl hydroxylase (prolyl-glycyl-peptide,2-oxoglutarate:oxygen oxidoreductase; EC 1.14.11.2). Other names that have been used for this enzyme include collagen proline hydroxylase, protocollagen hydroxylase, protocollagen proline hydroxylase, peptidyl proline hydroxylase, and proline,2-oxoglutarate dioxygenase (recommended trivial name). Prolyl hydroxylase has been found in every animal cell line tested (*117, 118*) and has been detected in plant tissues (*119*) and certain microorganisms that produce hydroxyproline-containing actinomycins (*120*).

Prolyl hydroxylase has been extensively purified from rat skin (*121*) and from chick embryo (*122*). The rat skin enzyme was found to have a molecular weight of 130,000 and to contain two polypeptide chains, while the enzyme from chick embryo had a molecular weight of 248,000 and contained four polypeptide chains. More recent work on the mouse- and rat-skin enzymes indicate molecular weights of 300,000–400,000 (*111*).

Molecular O_2 (*123*), Fe^{2+} (*124*), ascorbate (*125*), and α-ketoglutarate (*126*) are required for activity of prolyl hydroxylase. A possible mechanism for prolyl hydrolase is shown in Equation 27 (*111*):

$$\underset{\alpha\text{-Ketoglutarate}}{{}^{\ominus}OOC{-}CH_2{-}CH_2{-}C(=O){-}COO^{\ominus}} \xrightarrow[-CO_2]{O_2,\ \text{Enzyme}} \underset{\text{Peroxysuccinic acid}}{{}^{\ominus}OOC{-}CH_2{-}CH_2{-}C(=O){-}O{-}O^{\ominus}} \xrightarrow{\text{prolyl residue}} \underset{\text{Succinic acid}}{{}^{\ominus}OOC{-}CH_2{-}CH_2{-}COO^{\ominus}} + \underset{\text{Hydroxyproline residue}}{\text{HO-Pro-CO-}} \qquad (27)$$

The substrate specificity of prolyl hydroxylase has been well studied. Native collagen is not hydroxylated to its fullest extent and can serve as a substrate for the enzyme. Collagen cannot undergo hydroxylation in the triple-helix conformation but must first be denatured to the random-coil configuration (*127*). Protocollagen (unhydroxylated collagen) prepared in the presence of an Fe^{2+} chelator is an excellent substrate. Sequential polymers of the type (Pro-Gly-Pro)$_n$ serve as substrates (*128*) with the hexapeptide ($n = 2$) having detectable activity and the nonapeptide having appreciable activity (*129*). As n increased, K_m decreased but V_{max} remained the same. Even the tripeptide Pro-Pro-Gly can be hydroxylated (*130*).

The peptide bradykinin and its derivatives have been very useful in elucidating the specificity of prolyl hydroxylase (*131*). The primary sequence of bradykinin is: H–Arg–Pro–Pro–Gly–Phe–Ser–Pro–Phe–Arg–OH (with Pro-3 marked "3"). Bradykinin is hydroxylated by prolyl hydroxylase only at Pro-3. This work on the substituted bradykinins, as well as other data, permitted the conclusion that the minimum specificity requirement of the enzyme is the –X–Pro–Gly– triplet. X can be almost any amino acid except glycine.

Prolyl hydroxylase from the earthworm requires the same co-factors as the animal enzymes, but it has a different substrate specificity. It requires that the glycine residue precede the proline rather than follow it. Thus, the peptide (Gly–Pro–Ala)$_n$ is a substrate for earthworm prolyl hydroxylase but not for mammalian and bird enzymes (*132*). Prolyl hydroxylase from carrots appears to be very similar in co-factor requirements and specificity to that of the animal enzymes (*119*). The plant enzyme could hydroxylate collagen prepared from chick-embryo tibias and the animal enzyme could hydroxylate the unhydroxylated plant substrate dehydroxyextensin.

Lysyl hydroxylase (peptidyllysine,2-oxoglutarate:oxygen 5-oxidoreductase; EC 1.14.11.4) has not been studied as extensively as prolyl hydroxylase, but it appears to have very similar properties, including requirement for the same co-factors as prolyl hydroxylase (*111*). Chick-embryo lysyl hydroxylase was separated into two active fractions of molecular weights ca. 200,000 and 550,000 (*133*). Naturally occurring compounds and synthetic substrates containing the sequence –X–Lys–Gly– are readily hydroxylated by lysyl hydroxylase (*134*) . The K_m values decreased as the length of chain of polymers (–X–Lys–Gly–)$_n$ increased but V_{max} remained constant.

Hydroxylation of proline and lysine is an irreversible post-translational enzymatically catalyzed chemical modification of proteins.

Glycosylation

The addition of sugar residues to a polypeptide chain represents a major post-translational enzymatic chemical modification of proteins.

Many proteins contain carbohydrate, including those found in plasma, muscle, milk, eggs, plants, bacteria, and fungi. Glycoproteins are a major part of mucus, various other secretions, connective tissue, and cell membranes. Glycoproteins range in carbohydrate content from less than 1% to more than 80% by weight. Despite the large number of glycoproteins and the tremendous variation in carbohydrate content, there are only six amino acid residues which are involved in carbohydrate–peptide attachment in proteins. These are: asparagine, serine, threonine, cysteine, hydroxylysine, and hydroxyproline. So far, carbohydrate–peptide attachments involving aspartic acid, glutamic acid, glutamine, lysine, arginine, and histidine have not been found. Identified carbohydrate–peptide bonds in glycoproteins are shown in Table V.

Table V. Structures of Carbohydrate-peptide Linkages in Glycoproteins

Name	*Structure*[a]	*Reference*
Asparagine 4-*N*-(2-Acetamido-2-deoxy)-β-D-glucopyranosyl-L-asparaginyl-	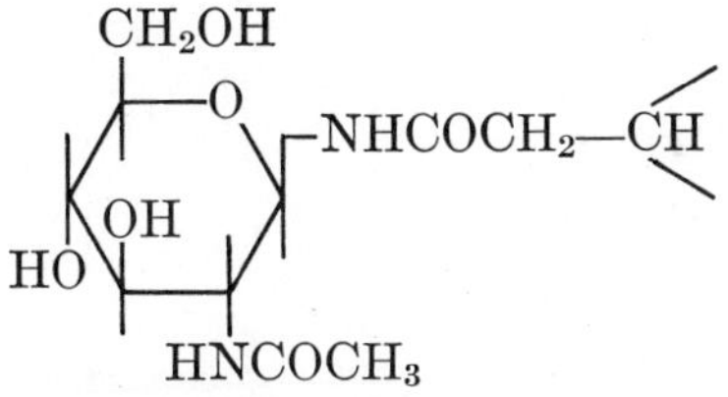 CH_2OH; O; $-NHCOCH_2-CH$; OH; HO; $HNCOCH_3$	*135*
Serine and threonine *O*-2-Acetamido-2-deoxy-α-D-galactopyranosyl-L-seryl- (or -L-threonyl-)	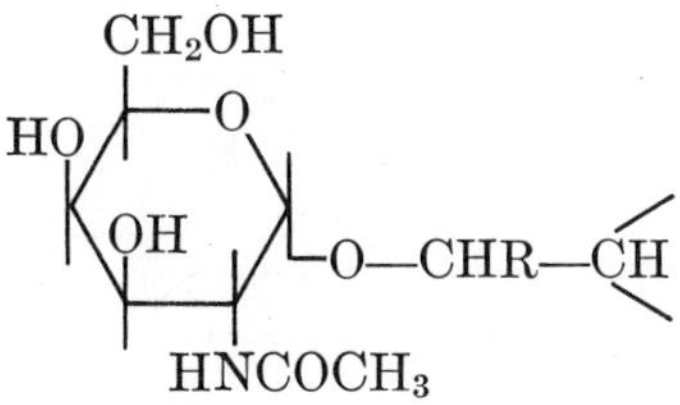 CH_2OH; HO; O; OH; $-O-CHR-CH$; $HNCOCH_3$	*135*
O-2-Acetamido-2-deoxy-D-glucopyranosyl-L-seryl-	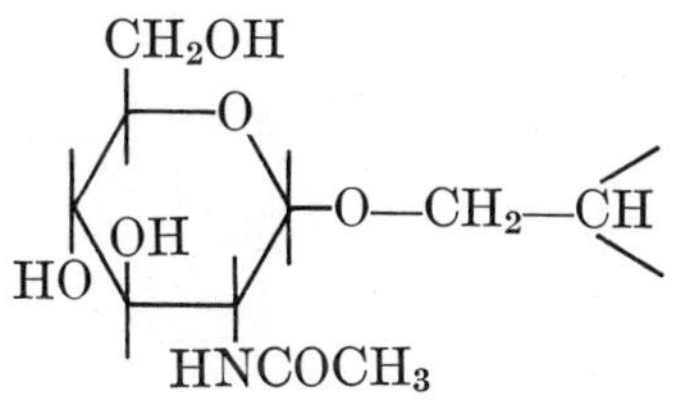 CH_2OH; O; $-O-CH_2-CH$; OH; HO; $HNCOCH_3$	*136*

Table V. Continued

Name	*Structure*[a]	*Reference*
O-D-Galactosyl-L-seryl- (or -L-threonyl-)	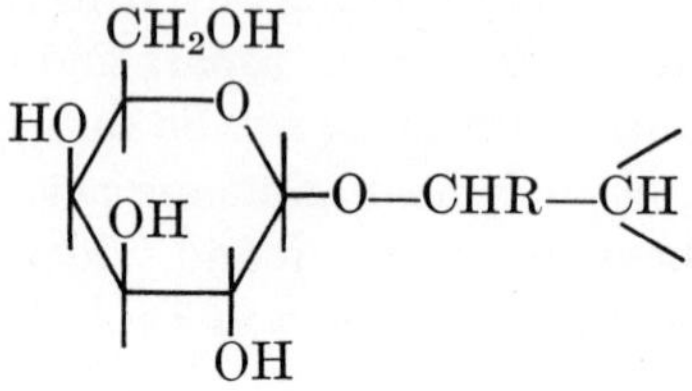 CH_2OH; O; HO; OH; OH; —O—CHR—CH	*137*
O-D-Mannosyl-L-seryl- (or -L-threonyl-)	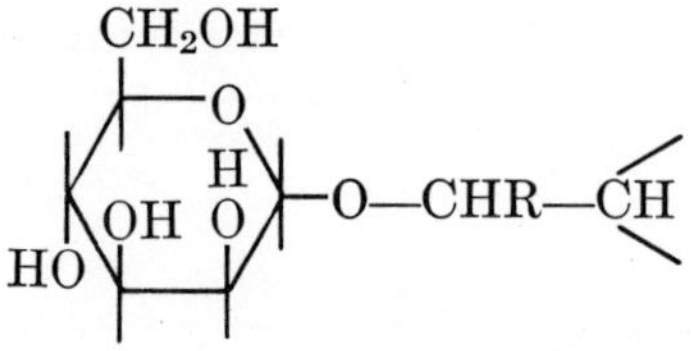 CH_2OH; O; H; OH O; HO; —O—CHR—CH	*138*
O-β-D-Xylopyranosyl-L-seryl-	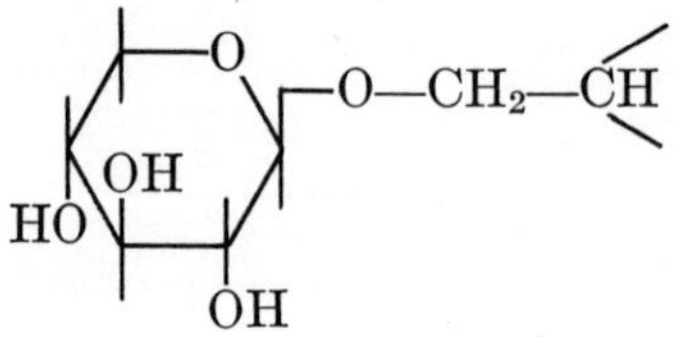 O; OH; HO; OH; —O—CH_2—CH	*139*
O-α-L-Fucopyranosyl-L-threonyl-	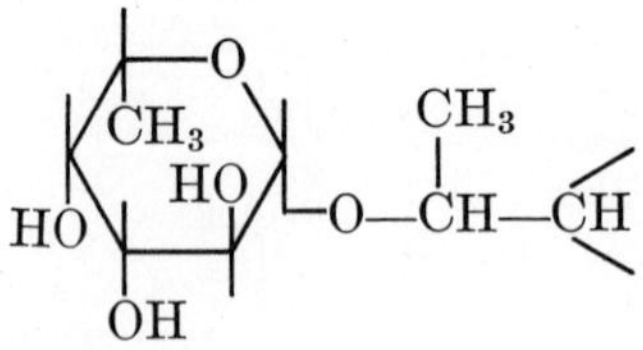 O; CH_3; HO; HO; OH; —O—CH(CH_3)—CH	*140*
Hydroxylysine 5-*O*-β-D-Galactopyranosyl-L-hydroxylysyl-	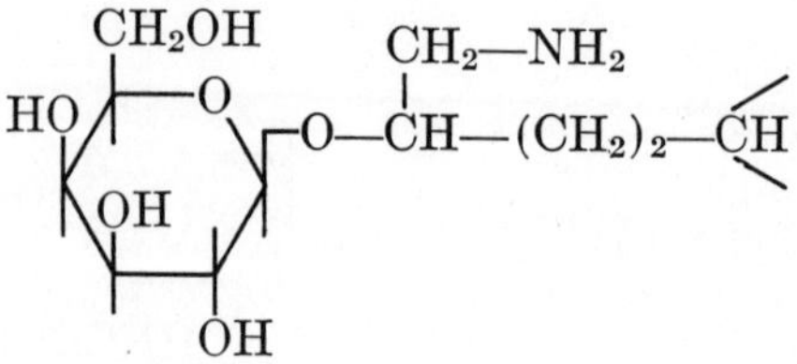 CH_2OH; O; HO; OH; OH; —O—CH(CH_2—NH_2)—$(CH_2)_2$—CH	*141*

Table V. Continued

Name	*Structure*[a]	*Reference*
Hydroxyproline O-D-Galactosyl-4-*trans*-L-hydroxy-prolyl-	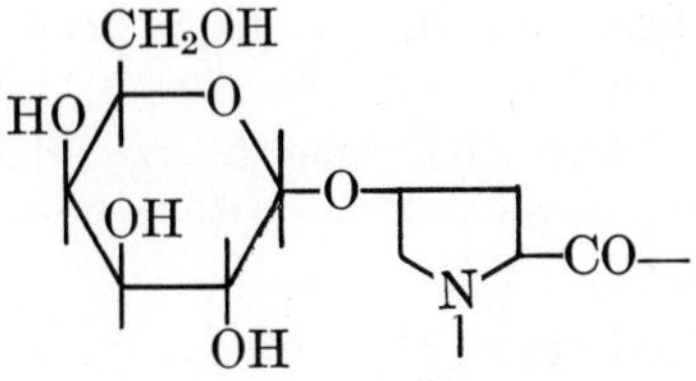 CH_2OH; HO; O; OH; OH; —O—; N; —CO—	*142*
O-L-Arabinosyl-4-*trans*-L-hydroxy-prolyl-	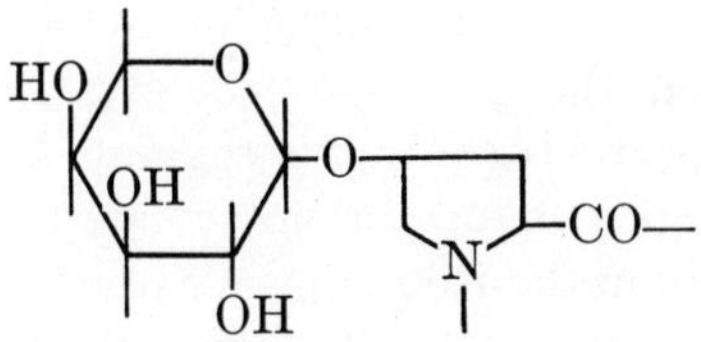 HO; O; OH; OH; —O—; N; —CO—	*143*
Cysteine S-D-Galactosyl-L-cysteinyl-	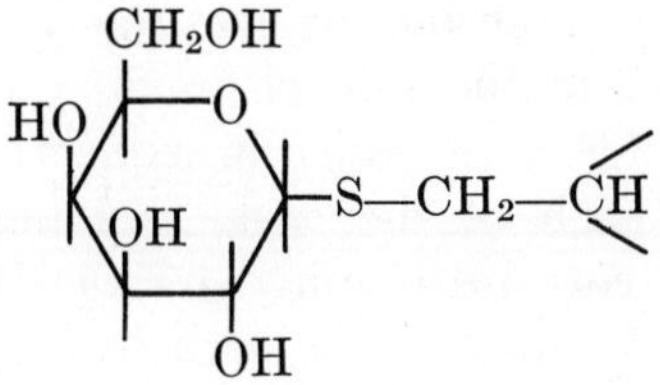 CH_2OH; HO; O; OH; OH; —S—CH_2—CH	*144*
S-D-Glucosyl-L-cysteinyl-	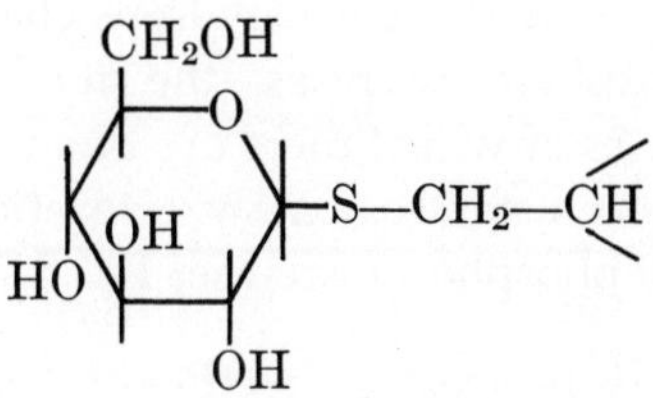 CH_2OH; O; OH; HO; OH; —S—CH_2—CH	*145*

[a] The structures are shown up to the α-carbon of the amino acid residue. Other carbohydrate residues may be attached to the carbohydrate residue shown.

As shown in Table V the residue attached to the amide bond of asparagine is always an *N*-acetylglucosamine. *N*-Acetylgalactosamine, *N*-acetylglucosamine, galactose, mannose, and xylose are found attached to the hydroxyl group of serine (or threonine), while only galactose residues have been found on 5-hydroxylysine. Both arabinose and galactose are found attached to the hydroxyl group of 4-*trans*-hydroxyproline in plants, but such a linkage has not been found in animals. Quite recently, galactose and glucose have been found attached to cysteine residues.

The carbohydrate chain may be extended by addition of other sugar residues. For the nature of these the reader is referred to several excellent reviews (*146–150*).

The carbohydrate residues of glycoproteins are attached to the polypeptide backbone following its translation on the ribosome. In the case of 5-hydroxylysine and 4-hydroxyproline, hydroxylation of lysine and proline must occur prior to attachment of carbohydrate. The enzymes for glycosylation, the glycosyltransferases, are found in high amounts in the Golgi apparatus (*151*); this suggests that addition of the carbohydrate chain begins before or immediately after release of the nascent peptide chain from the ribosome. Complete synthesis of the glycoprotein probably occurs prior to secretion from the cell; in fact, glycosylation may be a prerequisite for secretion. Soluble glycosyltransferases from goat colostrum have also been studied (*152*).

The first enzymatic step involves the covalent attachment of the first carbohydrate unit to the side chain of a specific amino acid residue. Further sugar residues are added in sequential order by specific glycosyltransferases for each of the sugar residues. It appears that specific multienzyme systems are required for the biosynthesis of each type of polymer (*153*). Frequently the carbohydrate chain terminates with an *N*-acetylneuraminic acid residue.

Phosphotransferases. There are several types of enzymes involved in the biosynthesis of the carbohydrate chains of glycoproteins. These include the phosphotransferases, the nucleotidyltransferases, and the glycosyltransferases of which there are two types.

The phosphotransferases catalyze reactions as shown for hexokinase (ATP:D-hexose 6-phosphotransferase; EC 2.7.1.1) in Equation 28:

$$\text{ATP} + \text{D-hexose} = \text{ADP} + \text{D-hexose 6-phosphate} \qquad (28)$$

Hexokinase has activity with D-glucose, D-mannose, and D-glucosamine as acceptors. There are phosphotransferases specific for all of the monosaccharides.

Hexose 1-phosphates may be synthesized directly by some of the phosphotransferases (i.e., ketohexokinase or galactokinase), or the hexose

6-phosphates can be isomerized to hexose 1-phosphates by another type of phosphotransferase which catalyzes intermolecular phosphate transfer as shown for phosphoglucomutase (α-D-glucose-1,6-bisphosphate:α-D-glucose-1-phosphate phosphotransferase; EC 2.7.5.1) in Equation 29:

α-D-Glucose 1,6-bisphosphate + α-D-glucose 1-phosphate =
α-D-glucose 6-phosphate + α-D-glucose 1,6-bisphosphate (29)

Glucose 1-phosphate can also be formed directly by an hexosyltransferase, phosphorylase (1,4-α-D-glucan:orthophosphate α-glucosyltransferase; EC 2.4.1.1)

$(1,4\text{-}\alpha\text{-D-Glucosyl})_n$ + orthophosphate = $(1,4\text{-}\alpha\text{-D-glucosyl})_{n-1}$ +
α-D-glucose 1-phosphate (30)

Nucleotidyltransferases. The nucleotidyltransferases catalyze the reaction of nucleotide triphosphates with the hexose 1-phosphates as shown for the specific enzyme glucose 1-phosphate uridyltransferase (UTP:α-D-glucose-1-phosphate uridyltransferase; EC 2.7.7.9) in Equation 31:

UTP + α-D-glucose 1-phosphate = pyrophosphate + UDP-glucose (31)

Glycosyltransferases. One type of glycosyltransferase catalyzes the addition of the UDP monosaccharide to the growing carbohydrate chain on the glycoprotein. This enzyme is specific for at least the UDP-sugar and the other sugar residue on the glycoprotein. Equation 32 shows a typical reaction for glycoprotein β-galactosyltransferase (UDPgalactose: 2-acetamido-2-deoxy-D-glycopeptide galactosyltransferase; EC 2.4.1.38):

UDPGalactose + 2-acetamido-2-deoxy-D-glucosyl-glycopeptide =
UDP + 4-0-β-D-galctosyl-2-acetamido-2-deoxy-
D-glucosyl-glycopeptide (32)

Some examples of glycosyltransferases that have been partially characterized are shown in Table VI.

A second type of glycosyltransferase adds a monosaccharide directly to the side chain of a specific amino acid residue. A glycosyltransferase capable of synthesizing the *N*-acetylgalactosaminyl-*O*-threonine bond found in submaxillary glycoproteins has been highly purified (*160*). The enzyme was highly specific for the transfer of *N*-acetylgalactosamine to a specific site on receptor proteins prepared from bovine submaxillary glycoprotein. The pH optimum was 6.5–7.5, the enzyme required Mn^{2+} (6m*M* maximal activity) and had K_m values of $6 \times 10^{-5} M$ and $9.5 \times$

Table VI. Some Glycosyltransferases Involved in Glycoprotein Biosynthesis

Type[a]	*Trivial Name*	*Systematic Name*	*Reference*
Protein			
	UDPacetylgalactosamine-protein acetylgalactosaminyltransferase	UDP-2-acetamido-2-deoxy-D-galactose:protein acetamidodeoxygalactosyltransferase (EC 2.4.1.41)	*154*
	UDPgalactose-collagen galactosyltransferase	UDPgalactose:5-hydroxylysine-collagen galactosyltransferase (EC 2.4.1.50)	*155*
	UDPglucose-collagen glucosyltransferase[b]	UDPglucose:5-hydroxylysine-collagen glucosyltransferase (EC 2.4.1.66)	*141, 155*
Glycoprotein			
	Glycoprotein β-galactosyltransferase	UDPgalactose:2-acetamido-2-deoxy-D-glucosyl-glycopeptide galactosyltransferase (EC 2.4.1.38)	*156*
	UDP-*N*-acetylglucosamine-glycoprotein *N*-acetylglucosaminyltransferase	UDP-2-acetamido-2-deoxy-D-glucose: glycoprotein 2-acetamido-2-deoxy-D-glycosyltransferase (EC 2.4.1.51)	*157*
	GDPfucose-glycoprotein fucosyltransferase	GDPfucose:glycoprotein fucosyltransferase (EC 2.4.1.68)	*158*
	Sialyltransferase	CMP-*N*-acetylneuraminate: D-galactosyl-glycoprotein *N*-acetylneuraminyltransferase (EC 2.4.99.1)	*159*

[a] Type is based on the acceptor molecule and nature of the bond formed.
[b] There is some doubt about this enzyme since galactose is stated to be the only monosaccharide found to be attached directly to 5-hydroxylysine in collagen (*150*).

$10^{-4}M$ for *N*-acetylgalactosamine and protein receptor, respectively. The enzyme did not catalyze transfer to a Pronase digest of the receptor protein but did so with heat and urea denatured receptor proteins (*161*).

Substantial data are available on the specificity requirements of the glycosyltransferases which add sugar units directly to amino acid residues of the polypeptide chain. It appears that substitution of the amide group of an asparagine residue by a sugar occurs only when the asparagine is in the amino acid sequence –*Asn*–X–Thr(or Ser)– (*162, 163, 164, 165*). (Amino acid residue italicized indicates site of the glycopeptide bond.) This has been found to be true of all glycoproteins studied so far except hen egg yolk phosvitin (*166*) which has the sequence –Ser–*Asn*–Ser–Gly–(SerP)$_8$– around the Asn residue to which the carbohydrate is attached. It has been suggested that the nature of X determines the complexity of the carbohydrate side chain (*167*). When the side chain of the amino acid X is polar, the carbohydrate chain is more complex than when the side chain of the amino acid X is apolar. There are some exceptions to this, however.

Not all sequences –Asn–X–Thr(or Ser)– are glycosylated in proteins, however. Hunt and Dayhoff (*168*) searched the amino acid sequences of 264 proteins. In only 101 cases did they find the occurrence of the sequence –Asn–X–Thr(or Ser)–, and in only 20 cases did the sequence occur in glycosylated form. They did conclude that this sequence is probably the key recognition sequence of the glycosyltransferase. Glycosylation may not occur because of the absence of one or more of the required enzymes, because the protein becomes folded into tertiary structure before glycosylation can occur or the carbohydrate groups have been removed prior to isolation.

The glycosyltransferase which glycosylates the hydroxyl group of 5-hydroxylysine in collagens and similar proteins appears to recognize the amino acid sequence –Gly–X–*Hyl*–Gly–Y–Arg–. Whereas the amino acids X and Y are quite variable, the positions of Gly, Hyl, and Arg are invariant in the glycopeptides of ten vertebrate and invertebrate collagens (*169*).

Less information is available about the possible recognition site of the glycosyltransferases which add sugar residues to serine or threonine in proteins. In some cases there appears to be a correlation between the occurrence of proline residues and glycosylation. In bovine myelin Al protein, the carbohydrate chain occurs in the sequence –Pro–Arg–*Thr*$_{98}$–Pro–Pro–Pro–Ser– (*170*). In the hinge region of rabbit IgG immunoglobulin the sequence is –Ser–Lys–Pro–*Thr*–Cys–Pro–Pro–Pro–Glu– (*171*), while in the β-subunit of human chorionic gonadotropin the sequence is –Pro–Pro–Pro–*Ser*–Leu–Pro–*Ser*–Pro–*Ser*–Arg– (*172*). Fetuin also contains three proline residues in the immediate vicinity of the glycopeptide bond (*150*). On the other hand, the sequence around the glycopeptide bond in freezing point-depressing glycoprotein is –*Thr*–Ala–Ala–*Thr*–Ala–Ala–*Thr*– (*173*) and in chondroitin sulfate proteoglycan is –Glu–Gly–*Ser*–Ala–Gly– (*174*).

The sequence of amino acids around the glycosylated cysteine in human erythrocyte membrane glycopeptide is (Glc)$_3$–*Cys*–Glu–His–Ser–His–Asp–His–Gly–Ala–COOH (*145*), while in a glycopeptide isolated from urine the sequence is (Gal)$_2$–*Cys*–Glu–His–Ser–His–Asp–Gly–Ala–COOH (*144*). Thus these glycosylated cysteine peptides are very similar.

The enzyme systems described above can function in vitro to glycosylate proteins with the requisite amino acid sequence. The hydroxylysine residues of the collagen-like proteins from the silk of the sawfly *Nematus ribesii* will accept D-galactose residues in vitro from UDP-galactose in the presence of hydroxylysine galactosyltransferase from rat kidney cortex (*175*). The basic protein from bovine myelin can accept *N*-acetyl-D-galactosamine in vitro (*170*). A specific threonine residue in bovine submaxillary glycoprotein is glycosylated in vitro (*160*) at an even greater rate when the protein is denatured (*161*). Therefore, it is the primary structure and not the tertiary structure of the polypeptide that provides the specificity requirements of this group of glycosyltransferases. UDP-galactose-collagen galactosyltransferase is not specific for the source of collagen in catalyzing the addition of a galactose unit to an hydroxylysine residue (*176*). However, it cannot catalyze the transfer of galactose to free hydroxylysine or to small peptides containing hydroxylysine. On the other hand, UDPglucose:galactosyl-hydroxylysine-collagen glucosyltransferase can transfer a glucose unit to the galactosylhydroxylysine glycopeptide regardless of the size of the substrate (*177*).

Removal of Glycosyl Groups. Removal of the carbohydrate chains from the glycoproteins probably occurs in a sequential manner by the combined action of several glycosidases. The sequential action of sialidase, β-galactosidase, and β-*N*-acetylglucosaminidase has been shown to release monosaccharides from fetuin, fetuin glycopeptides, and modified fetuin (*178*). These enzymes occur in rat kidney lysosomes. It is not clear whether in vivo removal of the carbohydrate chains occurs other than in the lysosomes. Data support the hypothesis that the catabolism of some of the serum glycoproteins and hormones may be regulated through the carbohydrate moieties. Studies with a number of glycoproteins indicate that removal of the terminal sialic acid residue leads to increased removal of the modified protein from circulation in rats (*179, 180*).

Methylation and Demethylation

Since the discovery in 1959 of a methylated lysine derivative in flagella protein of *Salmonella typhimurium* (*181*), many methylated amino acid derivatives have been found in a large number of proteins (Table VII). Three excellent recent reviews are available on this subject (*182, 183, 184*).

The various methylated amino acid residues are widely distributed in nature, and the lysine and arginine derivatives were present in every tissue examined except *Eubacterium*, which did not contain any methylated arginine (*195*). These derivatives are found in highly specialized proteins such as histones, myosin, actin, flagella protein, opsin, and ribosomal proteins. In addition, ε-*N*-trimethyllysine has been found in cytochrome c from wheat germ, *Neurospora* (*196*), yeast (*197*), and protozoa (*198*), but does not occur in animal cytochrome c's. ε-*N*-Trimethyl-5-L-hydroxylysine has been found in the diatom cell wall (*187*). The three methylated arginine derivatives are found mainly in histones, myosin, and myelin. The reader is referred to the review by Paik and Kim (*183*) for a comprehensive listing of the occurrence of the various methylated amino acid derivatives.

The number of residues of methylated amino acids in proteins is generally low and runs from one to four residues per molecule protein. Matsuoka (*195*) has made a comprehensive survey of the methylated lysines and arginines in a large number of proteins (*see* Ref. *183* for English translation). Methylation of the amino acid residues occurs after polypeptide formation (*199*).

Protein Methylases. Because of the work of Paik and Kim primarily, much is known about the enzymes involved in the methylation of proteins. They have isolated three protein methylases to various stages of purity and have shown that each has unique specificity properties. All of the protein methylases studied require S-adenosyl-L-methionine as co-factor, although there have been reports of methionine serving as the methyl donor (*200, 201*).

PROTEIN(ARGININE) METHYLTRANSFERASE. The enzyme(s) which methylates arginine residues of proteins was named protein methylase I by Paik and Kim (*202*). The recommended name is protein(arginine) methyltransferase (*S*-adenosyl-L-methionine:protein *N*-methylttransferase, EC 2.1.1.23). The enzyme(s) catalyzes the reaction

$$S\text{-adenosyl-L-methionine} + \text{protein} = S\text{-adenosyl-L-homocysteine} + \text{protein-containing } N\text{-methylarginine} \quad (33)$$

The question of whether one or more enzymes is involved in the biosynthesis of N^G-mono-, N^G,N^G-di- and N^G,N'^G-dimethylarginine residues cannot be answered at the moment. (The superscript G indicates methylation on nitrogens of the guanido group; *see* Table VII for specific structures.) Different patterns of methylation argue for more than one enzyme. Only N^G,N^G-dimethylarginine, up to four residues per 5×10^5 g protein, was found in myosin prepared from developing leg muscle of chick embryos (*203*). On the other hand, N^G,N'^G-dimethylarginine and N^G-monomethylarginine were found in the ratio of 1.6:10 in the encephalitogenic basic protein of myelin (*204*). Nakajima et al. (*205*) found 5–10 times more N^G,N^G-dimethylarginine than N^G,N'^G-dimethylarginine in various tissues.

Table VII. Methylated Amino Acid Residues Occurring in Proteins

Amino Acid Residue	*Formula*	*Reference*[a]
ε-*N* Monomethyl-lysine	$\rangle HC-(CH_2)_3-CH_2-NHCH_3$	*181*
ε-*N*-Dimethyl-lysine	$\rangle HC-(CH_2)_3-CH_2-N(CH_3)_2$	*185*
ε-*N*-Trimethyl-lysine	$\rangle HC-(CH_2)_3-CH_2-\overset{+}{N}(CH_3)_3$	*186*
ε-*N*-Trimethyl-5-hydroxylysine	$\rangle HC-(CH_2)_2-CHOH-CH_2-\overset{+}{N}-(CH_3)_3$	*187*
N^G-Monomethyl-arginine	$\rangle HC-CH_2)_3-NH-C(=NH)-NHCH_3$	*188*
N^G,N^G-Dimethyl-arginine	$\rangle HC-(CH_2)_3-NH-C(=NH)-N(CH_3)_2$	*189, 190*
N^G,N'^G-Dimethyl-arginine	$\rangle HC-(CH_2)_3-NH-C(=NCH_3)-NHCH_3$	*189, 190*
3-*N*-Methyl-histidine	$\rangle HC-CH_2$–(imidazole ring: N, N–CH_3)	*191*
1-*N*-Methyl-histidine[b]	$\rangle HC-CH_2$–(imidazole ring: CH_3N, N)	*191*
Carboxyl methylglutamate	$\rangle HC-CH_2CH_2-\overset{O}{\overset{\Vert}{C}}-OCH_3$	*192, 193*
Carboxyl methylaspartate	$\rangle HC-CH_2-\overset{O}{\overset{\Vert}{C}}-OCH_3$	*192, 193*
O-Methylserine	$\rangle HC-CH_2-OCH_3$[c]	*194*
O-Methylthreonine	$\rangle HC-CHOCH_3-CH_3$[c]	*194*

[a] One of the earliest reports.
[b] Found in very low amounts in organ proteins.
[c] Postulated derivative.

Protein(arginine) methyltransferase has been purified approximately 34-fold from calf thymus (*202*). The enzyme has a pH optimum of 7.4 and a K_m of $2.1 \times 10^{-6} M$ for S-adenosyl-L-methionine, which is at least 100 times lower than the in vivo level of this co-factor. The enzyme may have a molecular weight as high as 1–2 million (*206*). Although myelin protein is the best substrate for the enzyme (*207*), it has ben shown to methylate various histones, bovine serum γ-globulin, trypsin inhibitor, and pancreatic ribonuclease (*208*).

Little is known about the specific requirements of the enzyme. Guinea pig brain enzyme specifically methylates only Arg-107 of the 19 arginine residues in an encephalitogenic basic protein of human myelin (*204*). The same residue is also methylated in a similar protein from monkey, bovine, rabbit, guinea pig, rat, chicken, and turtle (*209*). The amino acid sequence around Arg-107 of the bovine basic protein is

$$\text{–Gly–}\underset{107}{\text{Arg}}(CH_3)_{1,2}\text{–Gly–Leu–Ser–Leu–Ser–Arg–} \quad (34)$$

PROTEIN(LYSINE) METHYLTRANSFERASE. The enzyme which methylates lysine residues of proteins was named protein methylase III by Paik and Kim (*210*). Its recommended trivial name is protein(lysine) methyltransferase (S-adenosyl-L-methionine:protein-lysine methyltransferase; EC 2.1.1.25). Protein(lysine) methyltransferase was found in all rat organs examined and was localized exclusively in the nuclei. Paik and Kim (*210*) solubilized the enzyme from an acetone powder of calf thymus and purified it 1.3-fold. The enzyme was difficult to work with in the solubilized state, since its activity was lost on overnight storage at either − 10° or 3°C. The enzyme was most effective in methylating histones, especially arginine-rich histone. Denaturation of histone by heating at 100°C for 30 min had no effect on the rate at which protein-(lysine) methyltransferase methylated it. Polylysine and protamine were methylated at slower rates, but horse heart cytochrome c did not serve as substrate. K_m for S-adenosyl-L-methionine was $3.0 \times 10^{-6} M$.

Just as with protein(arginine) methyltransferase, it is likely that there are several protein(lysine) methyltransferases. *Neurospora* and wheat germ cytochrome c's contain only ε-*N*-trimethyllysine (*196*), while pea embryo histone III and bovine retina opsin contain either ε-*N*-mono- or ε-*N*-dimethyllysine, but not ε-*N*-trimethyllysine (*211*). Flagella protein from *Salmonella serpens* contains only ε-*N*-monomethyllysine (*212*).

Some information is available on the sites of methylation in histones and cytochrome c's. Wheat germ cytochrome c contains 11 residues of lysine and two residues of ε-*N*-trimethylysine (*196*). The ε-*N*-trimethyllysine residues are located on residues 72 and 86 as shown in the following peptide sequences of wheat germ cytochrome c:

$$\underset{70}{\text{–Asn}}\text{–Pro–Lys}(CH_3)_3\text{–Lys–}\underset{74}{\text{Tyr–}} \quad (35)$$

$$\underset{86}{\text{–Lys}}(CH_3)_3\text{–Lys–Pro–Gln–}\underset{90}{\text{Asp–}} \quad (36)$$

Residue 72 is in a region which is invariant in plant and microbial cytochrome c's and residue 86 is in a region which is characterized in almost all cytochromes as being either Lys–Lys or Lys–Lys–Lys. *Neurospora* cytochrome c contains ε-*N*-trimethyllysine at residue 72 only, and animal and most plant cytochrome c's are not methylated.

The sequence of amino acids around residue 20 methylated in calf thymus histone H4 (other names, IV, F2a1) is (*76*):

$$\underset{17}{\text{–Arg}}\text{–His–Arg–}\underset{20}{\text{Lys}}(CH_3)_{1,2}\text{–Val–Leu–Arg–} \quad (37)$$

In calf thymus histone H3(III, F3) the sequences are (*76*):

$$\text{–Ala–Arg–}\underset{9}{\text{Lys}}(CH_3)_{0\text{-}3}\text{–Ser–Thr–} \quad (38)$$

$$\text{–Ala–Arg–}\underset{27}{\text{Lys}}(CH_3)_{0\text{-}3}\text{–Ser–Ala–} \quad (39)$$

These limited data indicate that methylation occurs in a location where there are two or more basic amino acid residues. Methylation occurs near the N-terminal end of the histones but not in the methylated cytochrome c's.

PROTEIN *O*-METHYLTRANSFERASE. Protein methylase II, which methylates the γ or β carboxyl group of glutamic and/or aspartic acid residues has been purified from calf thymus (*193*) and rat and human blood (*213*). The recommended trivial name of this enzyme is protein *O*-methyltransferase (*S*-adenosyl-L-methionine:protein *O*-methyltransferase; EC 2.1.1.24). The enzyme catalyzes the reaction:

$$\begin{gathered}S\text{-Adenosyl-L-methionine} + \text{protein} = \\ S\text{-adenosyl-L-homocysteine} + O\text{-methylprotein}\end{gathered} \quad (40)$$

where the enzyme forms a methyl ester in a number of proteins. The methyl group(s) incorporated becomes volatile following either acid or alkaline hydrolysis of the protein and is recovered as methanol.

The enzyme from calf thymus has been purified some 2500-fold (*214*). It exists in several molecular forms with molecular weights ranging from 16,000 to 48,000 (*215*). Isoelectric focusing of the purified enzyme gave four enzyme peaks with pI of 4.95, 5.50, 6.00, and 6.20. The pH optimum of the enzyme depended on the substrate. It was 6.0 with calf thymus protein, histone, and fibrinogen, 6.1 for blood plasma, 6.5 for γ-globulin, and 7.4 for gelatin (*214*). The K_m for S-adenosyl-L-methionine ranged from 1–3 $\times 10^{-6} M$ for various protein substrates. The protein *O*-methyltransferase isolated from rat and human erythrocytes had similar properties to that of the calf thymus enzyme (*214*).

In vitro, the activity of calf thymus protein *O*-methyltransferase on proteins decreased in the order: gelatin, denatured calf thymus cytosol proteins (F-P-100), fibrinogen, γ-globulin, ovalbumin, pancreatic ribonuclease, and histones (*193*). Denatured ribonuclease, oxidized or reduced to cleave the four disulfide bonds, gave a much higher rate of methylation than did the native protein, although the final degree of methylation was the same (*216*). Tryptic or chymotryptic digests of ribonuclease were still methylated (*216*). However, free aspartic or glutamic acid, dipeptides or homopolymers of these dicarboxylic acids, or glutathione were not methylated (*217*). In glutathione, the γ-carboxyl group is involved in peptide linkage. A protein *O*-methyltransferase from rat skin and spleen had the best activities on ovalbumin and sericin with lower rates of activity on human albumin, bovine fibrinogen, and various γ-globulin preparations (*192*).

The best evidence that protein *O*-methyltransferase methylates the carboxyl groups of aspartic and glutamic acid residues in proteins comes from treating the enzyme methylation product with lithium borohydride followed by hydrolysis in 6*N* HCl. Two new hydroxy amino acids, γ-hydroxy-α-aminobutyric acid and δ-hydroxy-α-aminovaleric acid, were obtained (*183*).

PROTEIN(HISTIDINE) METHYLTRANSFERASE. An enzyme which methylates histidine in proteins to give primarily 3-methylhistidine residues has been observed in myofibrillar protein and in the sarcoplasmic fraction of muscle homogenates (*218*). S-Adenosyl-L-methionine serves as the methyl donor for the enzyme. The enzyme has not been solubilized and purified. Very little is known about the substrate specificity of protein-(histidine) methyltransferase. Actins from a wide variety of species consistently contain one 3-*N*-methylhistidine residue per molecule (*191, 219*). It appears that myosin from white muscle contains two residues of 3-*N*-methylhistidine (one residue per heavy chain), whereas myosin from red muscle contains no 3-*N*-methylhistidine (*220*). The amino acid sequence around the methylated residue of rabbit skeletal muscle is (*221*):

$$-\text{Asp–Val–Asp–His}(\text{CH}_3)\text{–Gln–Thr–Tyr–} \qquad (41)$$

The amino acid sequence in rabbit cardiac muscle myosin is homologous with that around the 3-*N*-methylhistidine residue in rabbit skeletal muscle myosin, yet it does not contain a 3-*N*-methylhistidine residue.

A methylase in bull seminal plasma which incorporates the methyl group of *S*-adenosyl-L-methionine into endogenous seminal plasma protein has been described (*194*). The protein methylase required a heterologous DNA source, had a pH optimum of 8.1, and its activity was enhanced by Mg^{2+}, NH_4^+, and reduced glutathione. The authors suggest, on rather tentative data, that the enzyme methylates the hydroxyl group of serine and/or threonine residues in proteins.

Protein Demethylases. Removal of methyl groups from proteins proceeds very slowly and there is controversy among workers as to whether they are removed at all. In synchronized HeLa S-3 cell cultures about 2% removal of the methyl groups per hour was observed (*222*). The methyl group of ε-*N*-methyllysine residues of calf thymus histone is removed by an enzyme found in several rat organs (*223*), with kidney having the highest activity. The product of demethylation is formaldehyde. The demethylase activity in rat kidney is located primarily in the nuclear and mitochondrial fractions and appears to be identical to ε-alkyllysinase (ε-alkyl-L-lysine:oxygen oxidoreductase; EC 1.5.3.a) which dealkylates free ε-*N*-mono- and ε-*N*-dimethyllysine with an oxygen consumption of 0.5 mole to give an equivalent amount of lysine and formaldehyde (*224*).

The slow rate of removal of methyl groups from specifically methylated proteins appears to eliminate the possibility that methylation is a biological control process. It also raises questions of whether proteins chemically methylated to prevent the Maillard reaction during processing will prove to be commercially important. A great deal more work is needed on this point.

The protein methylases which add methyl groups to lysine, arginine, and histidine residues of proteins may be too specific to have potential application in food protein modification. Protein *O*-methyltransferase appears to methylate the carboxyl groups of numerous proteins and could have broad enough specificity to be important for increasing the hydrophobic properties of proteins through esterification of carboxyl groups.

Acetylation and Deacetylation

Two types of acetylated amino acid residues occur in proteins. The N-terminal amino acid residue in a protein may be acetylated following proteolytic removal of the terminal methionine (eukaryotes) or α-*N*-formyl methionine (prokaryotes) residue subsequent to translation. Data are not readily available on these reactions. The second type involves the acetylation of lysine residues in proteins, specifically the histones, following completion of the translation.

The reaction catalyzed by protein(lysine) transacetylase is

$$\text{protein} + \text{acetyl CoA} = \text{CoA} + \text{acetylated protein} \qquad (42)$$

leading to formation of ϵ-*N*-acetyllysine ($\rangle$HC–$(CH_2)_3$–CH_2–NH–$COCH_3$).

Three histone-specific acetyltransferases have been partially purified and characterized from rat thymus nuclei (*225*). The enzymes were extracted from rat thymus nuclei by sonication in the presence of 1*M* ammonium sulfate and separated into two active fractions (A and B) by DEAE-cellulose chromatography. Fraction B was further separated into two active fractions (B_1 and B_2) by gel filtration on Sephadex G-200. Each fraction was then purified further by chromatography on hydroxyapatite. The molecular weights, determined by Sephadex G-200 and by sucrose density gradient centrifugation, were 99,000, 110,000, and 92,000 for enzymes A, B_1, and B_2, respectively. All three enzymes required acetyl CoA as acetate donor, and the activity of the enzymes was inhibited by *p*-chloromercuribenzoate. Acetyltransferase A preferentially acetylated histone I (F1) and also poly-L-lysine. Acetyltransferase B_1 and B_2 preferred histone H4 (other names IV, F2a1) and did not acetylate poly-L-lysine and histone H3 (III, F3). In addition to ϵ-*N*-acetyllysine, two other unidentified amino acid derivatives were obtained from a digest of histone H4 acetylated by the two B enzymes.

The amino sequence around the ϵ-*N*-acetyllysine residues is known in several histones. In calf thymus histone H4 the sequence is (*76*):

$$\text{–Gly–Gly–Ala–}\underset{16}{\text{Lys}}(\text{Ac})_{0,1}\text{–Arg–His–Arg–} \qquad (43)$$

while in calf thymus histone H3 the sequences are:

$$\text{–Gly–Gly–}\underset{14}{\text{Lys}}(\text{Ac})_{0\text{-}1}\text{–Ala–Pro–} \qquad (44)$$

$$\text{–Ala–Thr–}\underset{23}{\text{Lys}}(\text{Ac})_{0,1}\text{–Ala–Ala–} \qquad (45)$$

Acetylation occurs near the N-terminal end of the protein and near the sites of methylation (*see* section on Methylation and Demethylation) of these histones. Residue 16 is the major site of acetylation in both pea and calf thymus histones H4, but other sites may also be acetylated in some cases, particularly residue 8 in pea histone (*226*):

$$-\text{Gly–Gly–}\underset{8}{\text{Lys(Ac)}_{1,0}}\text{–Gly–Leu–} \quad (46)$$

DeLange et al. (*227*) suggested that the recognition site for acetylation of calf thymus histones H3 and H4 by one of the acetylation enzymes might be –Lys–X–Y–Arg–Lys–. Other enzymes with different recognition sites were also indicated by the data.

Dixon et al. (*103*) have suggested there are two distinct types of sites for acetylation on histones. Type A site is characterized by an ε-*N*-acetyllysine residue bordered on either side by a small neutral amino acid, while type B site is characterized by an ε-*N*-acetyllysine residue present as a member of Lys–Arg–, Arg–Lys, or Lys–Lys pairs. Perhaps two distinct protein(lysine) acetyltransferases are involved. The sequences around ε-*N*-acetyllysine residues in histones from rainbow trout testis are shown in Table VIII. This hypothesis holds also for the pea and calf thymus histone sequences listed above.

Acetylation of the histones occurs near the N-terminal end and in the region where these proteins undergo methylation and phosphorylation reactions.

Table VIII. Classification of Acetylation Sites in Trout Testis Histones (*103*)

Histone	*Type A*	*Type B*
H2A (IIb1,F2a2)	–Gly–Lys(Ac)–Thr– 5	
H2B (IIb2,F2b)	–Ala–Lys(Ac)–Ser– 5	–Lys–Lys(Ac)–Gly– 10
	–Thr–Lys(Ac)–Ser– 18	–(Ser)–Lys(Ac)–Lys– 13
H3 (III,F3)	–Gly–Lys(Ac)–Ala– 14	–Arg–Lys(Ac)–Ser– 9
	–Thr–Lys(Ac)–Ala– 23	–Arg–Lys(Ac)–Gln– 18
H4 (IV,F2a1)	–Gly–Lys(Ac)–Gly– 5	–Ala–Lys(Ac)–Arg– 16
	–Gly–Lys(Ac)–Gly– 8	
	–Gly–Lys(Ac)–Gly– 12	

It is thought that the histones undergo rapid acetylation and deacetylation depending on the stage of maturation of cells. Histone deacetylases have been shown to be present in calf thymus (*227a, b*) as well as other animal and plant tissues (*227c, d*). Calf thymus histone deacetylase acts specifically on the ε-*N*-acetylated lysine residues in arginine-rich histones H3 and H4 (*227e*). One enzyme has been purified some 500-fold and shown to have a molecular weight of 150,000–160,000 (*227f*).

Other Post-Translational Modifications of Proteins

Several other enzymatic modifications of proteins which probably occur after translation of a polypeptide chain are listed in Table IX. We shall describe each briefly since none of the modifications would appear at this time to be generally useful for modification of proteins in foods.

γ-Carboxyglutamic acid was recently identified as a component of the vitamin K-dependent region of bovine prothrombin (*228*) and has subsequently been shown to be present in rat prothrombin and bovine factor X as well (*237*). There appear to be at least ten residues of γ-carboxyglutamic acid in the nonthrombin-generating region of bovine prothrombin. These are in close proximity, located at positions 7, 8, 15, 17, 20, 21, 26, 27, 30, and 33 in prothrombin (*238*). γ-Carboxyglutamic acid is essential for calcium binding by prothrombin and by blood clotting factor X (*239*). Vitamin K probably plays a direct or indirect role in the carboxylation of glutamic acid (*240*). The amino acid sequence around the γ-carboxyglutamic acid residues in prothrombin is (*238, 241*):

–Phe–Leu–*Gla*–*Gla*–Val–Arg - - - Asn–Leu–*Gla*–Arg–*Gla*–Cys–Leu–
5 10 13 15 19

–*Gla*–*Gla*–Pro–Cys–Ser–Arg–*Gla*–*Gla*–Ala–Phe– (47)
20 25 26 27

Gla–Ala–Leu–*Gla*–Ser–Leu–
30 33

Pyrrolidone carboxylic acid is the N-terminal amino acid residue of the β-chain of fibrinogen from several species (*229, 230, 242*). The amino acid sequence containing pyrrolidone carboxylic acid (Glu) is (*243*):

[Glu–Gly–Val–Asn–Asp–Asn–Glu–Glu–Gly– (48)

The N-terminal residue of the α1 and α2 chains of collagen is pyrrolidone carboxylic acid (*231*). Action of procollagen peptidase on

Table IX. Some Other Post-Transitional Enzymatic Modifications of Proteins

Name	*Structure*[a]	*Reference*
Glutamic acid		
γ-Carboxyglutamic acid	HC—CH_2—CH(COOH)(COOH)	*228*
Pyrrolidone carboxylic acid	O=C—CH_2—CH_2—CH(—CO—)—NH (ring)	*229–231*
Tyrosine		
Iodinated tyrosine	HC—CH_2—C_6H_4(nI)—OH	*232*
Tyrosine-*O*-sulfate	HC—CH_2—C_6H_4—OSO_3H	*233*
Dihydroxyphenylalanine	HC—CH_2—C_6H_3(OH)—OH	*234*
AMP-Tyrosine	HC—CH_2—C_6H_4—O—PO_2H—O—CH_2—(ribose, OH, OH)—adenine (H_2N)	*235*

Table IX. Continued

Name	*Structure*[a]	*Reference*
Peptidyl tyrosine	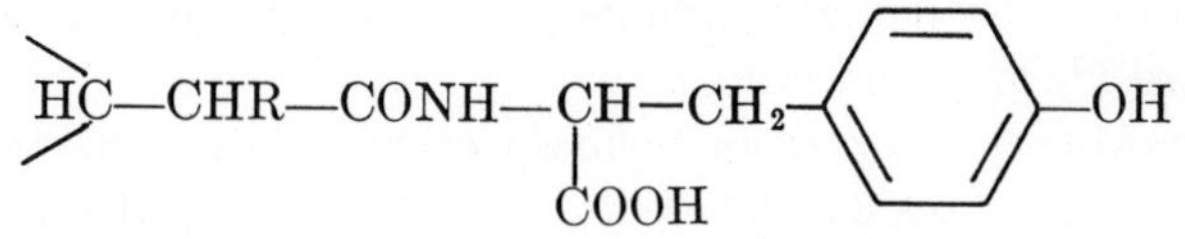 >HC—CHR—CONH—CH(COOH)—CH_2—C_6H_4—OH	*236*

[a] Structures drawn to the α-carbon of the amino acid residue.

procollagen gives cleavage at –X–Gln– or –X–Glu– bonds with subsequent cyclization of the N-terminal glutamic acid (or glutamine) residue to pyrrolidone carboxylic acid.

Tyrosine-*O*-sulfate is present in many fibrinopeptides B (*233*). It is present in fish, frog, and a large number of mammalian fibrinogens but not in that of higher primates and the rat (*242*).

In vitro studies on the ability of horseradish peroxidase or lactoperoxidase to catalyze iodination of proteins at tyrosine residues are becoming quite frequent (*244-248*). The studies are designed either to provide better understanding of the incorporation of iodine into thyroid proteins or to develop better ways of producing radioactively labeled proteins for tracer studies. An immobilized enzyme system has been developed to achieve this second purpose (*248*). A large number of proteins incorporate iodine by this method which is superior to iodination with chloroamine-T or ICl. These iodinated proteins should prove invaluable in nutritional studies.

The possibility of using tyrosinase to oxidize selected tyrosines in or near the active sites of enzymes has been considered for a number of years. The difficulty has been to achieve oxidation in native proteins. Recently, glyceraldehyde-3-phosphate dehydrogenase and α-glycerophosphate dehydrogenase have been shown to be oxidized by tyrosinase (*234*). Tyrosinase-catalyzed oxidation of glyceraldehyde-3-phosphate dehydrogenase gave two residues of oxidized tyrosine per subunit, while there was an oxidation of three to four tyrosine residues in α-glycerophosphate dehydrogenase.

Recent primary structure sequence studies on *Escherichia coli* glutamine synthetase demonstrated that the covalently bound active site adenylic acid residue is attached to a tyrosine residue (*235*). The sequence of amino acids around the derivated tyrosine residue is:

–Asp–Asn–Leu–Tyr(AMP)–Asp–Leu–Pro–Pro– (49)

This is an example of several enzymes in which an essential co-factor is covalently bound to the protein. For example, lipoic acid and biotin are covalently linked to the ε-amino group of a specific lysine residue in certain enzymes. In some cases, pyridoxal phosphate is bound to the protein through the formation of a Schiff base involving the carbonyl group of the co-factor and an ε-amino group of a lysine residue. In cytochrome c, the heme is attached through two thiol ether linkages to cysteine residues of the protein.

A tyrosine ligase present in brain extracts has been shown to add a tyrosine residue specifically to the carboxyl terminal end of α-tubulin (*236, 249*). The function of this enzyme in the assembly of the tubulin system is not clear since both tyrosylated and untyrosylated α-tubulin dimers can polymerize.

Effect of Protein Modification on Catabolic Rate

Before modified proteins can be used as human food, the effect of modification on the rate and extent of digestion of the protein must be known. To date, very little has been done on the digestibility of deliberately modified proteins.

It is generally accepted that the rate of proteolysis of a denatured protein is much higher than that of the same protein in the native state. Unfolding of proteins can increase proteolytic sensitivity by two or three orders of magnitude (*250, 251, 252*). Susceptibility to proteolysis is used to distinguish between native and denatured protein (*253*). The presence of ligands in a protein can greatly increase its resistance to proteolysis, presumably by maintaining the protein in the native state (*254, 255*). Therefore, one would expect an effect on the rate of proteolysis depending on whether the stability of the protein to denaturation has been changed by chemical modification.

In vivo, the extent of glycosylation of proteins may be an important controlling factor in their rate of turnover. Removal of the terminal sialic acid residue from such glycoproteins as orosomucoid, fetuin, ceruloplasmin, and thyroglobulin by neuraminidase prior to injection into rats led to increased rates of removal of the modified protein from circulation (*256*). Removal of both the sialic and galactose residues from ceruloplasmin gave rates similar to those in which the unmodified protein was injected (*179, 257*).

Recent experiments indicate that the lifetime of a protein in vivo may be a function of the rate of deamidation of its glutaminyl and asparaginyl residues (*258, 259*). The authors do not argue for a specific enzyme system for deamidation. Rather, they suggest that the proteins slowly undergo deamidation nonenzymatically with a half-time of approximately eight days. Correlation between in vitro rates of deamidation of model peptides and aldolase and cytochrome c were found (*259, 260*). However, many proteins undergo much faster rates of in vivo turnover, with ornithine transcarbamylase having a half-time of only 11 min (*261*).

Substitution of one amino acid residue with another may greatly affect the rate of digestibility of a protein. The incorporation of a valine or lysine analog into hemoglobin by reticulocytes led to rapid turnover of the protein (*262, 263*). Incorporation of canavanine into *Escherichia coli* proteins also led to rapid in vivo turnover of the proteins (*264*). Human hemoglobin Ann Arbor differs from normal hemoglobin by one amino acid replacement; however, it undergoes catabolism in vivo at much faster rates than normal hemoglobin (*265*). These factors should be considered not only in terms of chemical modification of proteins but also in genetic engineering designed to increase yield or nutritional values.

Feeding of casein treated with oxidized ethyl linoleate, at 9% protein level, caused depressed growth rate and enlargement of the liver (*266*). Such effects could not be the result of the oxidation of sulfhydryl groups only, but may be caused by additional reactions as discussed under the section on Cross-Linking in this chapter.

Modification of specific amino acid residues should affect the rate of proteolysis by those enzymes which have specificity for the unmodified residues. For example, modification of the lysine residues of a protein restricts the action of trypsin. Dimethylated casein undergoes a slower rate of hydrolysis by trypsin in vitro, and the extent of proteolysis is lower than that of casein (*267*). An unexpected finding was that the rate and extent of hydrolysis of dimethylated casein by α-chymotrypsin are also adversely affected, while hydrolysis by subtilisin is not. Peptides derived from dimethylated casein were inhibitory of α-chymotrypsin.

Partial removal of the phosphate groups of phosvitin by β-elimination in alkaline solution results in a decreased in vitro initial rate of hydrolysis by trypsin (*268*). The decreased rate of hydrolysis might be a result of (a) a change in conformation of the protein on removal of the phosphate groups, (b) cross-linking by the reaction of the dehydroalanine residues with lysine residues (to form lysinoalanine), or (c) racemization of some of the residues by the alkaline treatment.

The in vitro digestibility of heat-treated soybean protein by trypsin was increased by the breakage of disulfide bonds (*269*). This was caused primarily by increased digestibility of the trypsin inhibitors which are about 6% of the total protein. Even though the trypsin inhibitory activity had been destroyed by heat, the trypsin inhibitors were not readily hydrolyzed by trypsin until the disulfide bonds were broken by reduction or oxidation.

A great deal more work is needed on the effect of chemical modification on the in vitro and in vivo digestibility of proteins as well as on any effect modification might have on toxicity of the derivatives.

Acknowledgment

The assistance of Linda Geiger in compiling and checking references and Clara Robison in organization and typing the manuscript is gratefully acknowledged. Ideas for this paper and symposium and some experimental work were the result of research grant 5RO1 FD 00568.

Literature Cited

1. Foreman, M. J., Horstein, I., AGRO 18, The First Chemical Congress of the North American Continent, Mexico City, Nov. 30–Dec. 5, 1975.
2. Wallerstein, L., U. S. Patents 995,824–5 (1911).
3. Busch, W. A., Stromer, M. H., Goll, D. E., Suzuki, A., *J. Cell Biol.* (1972) **52**, 367.
4. Beuk, J. F., Savich, A. L., Goeser, P. A., Hogan, J. M., U. S. Patent 2,903,362 (1959).
5. Silberstein, O. O., U. S. Patent 3,276,879 (1966).
6. Hinricks, J. R., Whitaker, J. R., *J. Food Sci.* (1962) **27**, 250.
7. Yatco-Manzo, E., Whitaker, J. R., *Arch. Biochem. Biophys.* (1962) **97**, 122.
8. El-Gharbawi, M., Whitaker, J. R., *J. Food Sci.* (1963) **28**, 168.
9. Cheftel, C., *Ann. Technol. Agric.* (1972) **21**, 423.
10. Archer, M. C., Ragnarsson, J. O., Tannenbaum, S. R., Wang, D. I. C., *Biotechnol. Bioeng.* (1973) **15**, 181.
11. Faith, W. T., Jr., Steigerwalt, R. B., Robbins, E. A., U. S. Patent 3,697,285 (1972).
12. Hevia, P., Whitaker, J. R., Olcott, H. S., *J. Agric. Food Chem.* (1976) **24**, 383.
13. Fujimaki, M., Arai, S., Yamashita, M., Kato, H., Noguchi, M., *Agric. Biol. Chem. (Tokyo)* (1973) **37**, 2891.
14. Onoue, Y., Riddle, V. M., *J. Fish. Res. Board Can.* (1973) **30**, 1745.
15. Neurath, H., Walsh, K. A., White, W. P., *Science* (1967) **158**, 1638.
16. Davie, E. W., Kirby, E. P., *Curr. Top. Cell. Regul.* (1973) **7**, 51.
16a. Davie, E. W., Fujikawa, K., *Annu. Rev. Biochem.* (1975) **44**, 799.
17. Müller-Eberhard, H. J., *Ann. Rev. Biochem.* (1975) **44**, 697.
18. Steiner, D. F., Clark, J. L., Nolan, C., Rubenstein, A. H., Margoliash, E., Aten, B., Oyer, P. E., *Recent Progr. Horm. Res.* (1969) **25**, 207.
19. Bellamy, G., Bornstein, P., *Proc. Nat. Acad. Sci. USA* (1971) **68**, 1138.
20. Jollès, P., *Mol. Cell. Biochem.* (1975) **7**, 73.
21. Doolittle, R. F., *Adv. Protein Chem.* (1973) **27**, 1.
22. Hershko, A., Fry, M., *Ann. Rev. Biochem.* (1975) **44**, 775.
23. Milstein, C., Brownlee, G. G., Harrison, T. M., Mathews, M. B., *Nature New Biol.* (1972) **239**, 117.
24. Mach, B., Faust, C., Vassalli, P., *Proc. Nat. Acad. Sci. USA* (1973) **70**, 451.
25. Fukimaki, M., Yamashita, M., Arai, S., Kato, H., *Agric. Biol. Chem. (Tokyo)* (1970) **34**, 483.
26. Vuust, J., Piez, K. A., *J. Biol. Chem.* (1972) **247**, 856.
27. Speakman, P. T., *Nature* (1971) **229**, 241.
28. Traub, W., Piez, K. A., *Adv. Protein Chem.* (1971) **25**, 243.

29. Harkness, R. D., in "Treatise on Collagen," B. S. Gould, ed., Vol. 2, Part A, Academic, New York (1968), 247.
30. Fitton Jackson, S., in "Treatise on Collagen," B. S. Gould, ed., Vol. 2, Part B, Academic, New York (1968), 1.
31. Steiner, R. F., Laki, K., *Arch. Biochem. Biophys.* (1951) **34,** 24.
32. Lorand, L., Downey, J., Gotoh, T., Jacobsen, A., Tokura, S., *Biochem. Biophys. Res. Commun.* (1968) **31,** 222.
33. Matačić, S., Loewy, A. G., *Biochem. Biophys. Res. Commun.* (1968) **30,** 356.
34. Pisano, J. J., Finlayson, J. S., Peyton, M. P., *Science* (1968) **160,** 892.
35. Pisano, J. J., Finlayson, J. S., Peyton, M. P., Nagai, Y., *Proc. Nat. Acad. Sci. USA* (1971) **68,** 770.
36. Ball, A. P., Hill, R. L., McKee, P. A., *Abstracts, 3rd Congr. Int. Soc. Thromb. Haemostatis* (1972), 62.
37. Takagi, T., Doolittle, R. F., *Biochem. Biophys. Res. Commun.* (1973) **51,** 186.
38. Chung, S. I., *Ann. N. Y. Acad. Sci.* (1972) **202,** 240.
39. McKenzie, H. A., in "Milk Proteins—Chemistry and Molecular Biology," Vol. II, H. A. McKenzie, ed., Academic, New York, 1971, p. 87.
40. Waugh, D. F., in "Milk Proteins—Chemistry and Molecular Biology," Vol. II, H. A. McKenzie, ed., Academic, New York, 1971, p. 3.
41. Jollès, J., Alais, C., Jollès, P., *Biochim. Biophys. Acta* (1968) **168,** 591.
42. Linderstrøm-Lang, K., Ottesen, M., *Nature* (1947) **159,** 807.
43. Linderstrøm-Lang, K., Ottesen, M., *Compt. Rend. Trav. Lab., Carlsberg. Sèr Chim.* (1948) **26,** 404.
44. Fujimaki, M., Kato, H., Arai, S., Tamaki, E., *Food Technol.* (1968) **22,** 889.
45. Yamashita, M., Arai, S., Gonda, M., Kato, H., Fujimaki, M., *Agric. Biol. Chem. (Tokyo)* (1970) **34,** 1333.
46. Yamashita, M., Arai, S., Tsai, S.-J., Fujimaki, M., *Agric. Biol. Chem. (Tokyo)* (1970) **34,** 1593.
47. Epstein, C. J., Goldberger, R. F., Anfinsen, C. B., *Cold Spring Harbor Symp. Quant. Biol.* (1963) **28,** 439.
48. De Lorenzo, F., Goldberger, R. F., Steers, E., Givol, D., Anfinsen, C. B., *J. Biol. Chem.* (1966) **241,** 1562.
49. Fuchs, S., De Lorenzo, F., Anfinsen, C. B., *J. Biol. Chem.* (1967) **242,** 398.
50. Kurane, R., Minoda, Y., *Agric. Biol. Chem.* (1975) **39,** 1417.
51. Hatch, M. D., Turner, J. F., *Biochem. J.* (1960) **76,** 556.
52. Katzen, H. M., Tietze, F., Stetten, D., *J. Biol. Chem.* (1963) **238,** 1006.
53. Nordal, J., Fossum, K., *Z. Lebensm.-Unters. Forsch.* (1974) **154,** 144.
54. Pinnell, S. R., Martin, G. R., *Proc. Nat. Acad. Sci. USA* (1968) **61,** 708.
55. Siegel, R. C., Pinnell, S. R., Martin, G. R., *Biochemistry* (1970) **9,** 4486.
56. Fowler, L. J., Peach, C. M., Bailey, A. J., *Biochem. Biophys. Res. Commun.* (1970) **41,** 251.
57. Siegel, R. C., Martin, G. R., *J. Biol. Chem.* (1970) **245,** 1653.
58. Bailey, A. J., Fowler, L. J., *Biochem. Biophys. Res. Commun.* (1969) **35,** 672.
59. Gallop, P. M., Blumenfeld, O. O., Seifter, S., *Annu. Rev. Biochem.* (1972) **41,** 617.
60. Tanzer, M. L., *Science* (1973) **180,** 561.
61. Robbins, K. C., *Am. J. Physiol.* (1944) **142,** 581.
62. Laki, K., Lorand, L., *Science* (1948) **108,** 280.
63. Takagi, T., Konishi, K., *Biochim. Biophys. Acta* (1972) **271,** 363.
64. Chung, S. I., Folk, J. E., *Proc. Nat. Acad. Sci. USA* (1972) **69,** 303.
65. Asquith, R. S., Otterburn, M. S., Sinclair, W. J., *Angew. Chem.* (1974) **86,** 580.

66. Hlynka, I., Ed., "Wheat—Chemistry and Technology," Am. Assoc. Cereal Chemists, Inc., St. Paul, Minn., 1964.
67. Auerman, L. Ya., Ponomareva, A. N., Polandova, R. D., Kilmova, G. S., *Tr. Vses Nauch-Issled. Inst. Khlebopek Prom.* (1971) **12**, 95; *Chem. Abstr.* (1972) **77**, 46871c.
68. Graveland, A., *Getreide, Mehl Brot.* (1973) **27**, 316.
69. Frazier, P. J., Leigh-Dugmore, F. A., Daniels, N. W. R., Russell Eggitt, P. W., Coppock, J. B. M., *J. Sci. Food Agric.* (1973) **24**, 421.
70. Axelrod, B., ADV. CHEM. SER. (1974) **136**, 324.
71. Roubal, W. T., Tappel, A. L., *Arch. Biochem. Biophys.* (1966) **113**, 150.
71a. Chio, K. S., Tappel, A. L., *Biochemistry* (1969) **8**, 2827.
72. Taborsky, G., *Adv. Protein Chem.* (1974) **28**, 1.
73. Igarashi, M., Takahashi, H., Tsuyama, N., *Biochim. Biophys. Acta* (1970) **220**, 85.
74. VanEtten, R. L., Hickey, M. E., *Fed Proc. Fed. Am. Soc. Exp. Biol.* (1972) **31**, Abstr. No. 451.
75. VanEtten, R. L., McTigue, J. J., "Abstracts of Papers," 164th National Meeeting, ACS, 1972, BIOL 93.
76. DeLange, R. J., Smith, E. L., *Ciba Found. Symp.* (1975) **28**, 59.
77. Walinder, O., Zetterqvist, Ö., Engström, L., *J. Biol. Chem.* (1968) **243**, 2793.
78. Schiltz, E., Sekeris, C., *Hoppe-Seyler's Z. Physiol. Chem.* (1969) **350**, 317.
78a. Chen, C.-C., Smith, D. L., Bruegger, B. B., Halpern, R. M., Smith, E. A., *Biochemistry* (1974) **13**, 3785.
79. Allerton, S. E., Perlmann, G. E., *J. Biol. Chem.* (1965) **240**, 3892.
80. Rask, L., Walinder, Ö., Zetterqvist, O., Engström, L., *Biochim. Biophys. Acta* (1970) **221**, 107.
81. Schirm, J., Gruber, M., Ab, G., FEBS Lett. (1973) **30**, 167.
82. Mäenpää, P. H., Bernfield, M. R., *Proc. Nat. Acad. Sci. USA* (1970) **67**, 688.
83. Bernfield, M. R., Mäenpää, P. H., *Cancer Res.* (1971) **31**, 684.
84. Hatfield, D., Portugal, F. H., Caicuts, M., *Cancer Res.* (1971) **31**, 697.
85. Burnett, G., Kennedy, E. P., *J. Biol. Chem.* (1954) **211**, 969.
86. Sundararajan, T. A., Sampath-Kumar, K. S. V., Sarma, P. S., *Biochim. Biophys. Acta* (1958) **29**, 449.
87. Rabinowitz, M., Lipmann, F., *J. Biol. Chem.* (1960) **235**, 1043.
88. Rubin, C. J., Rosen, O. M., *Annu. Rev. Biochem.* (1975) **44**, 831.
89. Walsh, D. A., Krebs, E. G., *The Enzymes* (1973) **8**, 555.
90. Wålinder, O., *Biochim. Biophys. Acta* (1972) **258**, 411.
91. Wålinder, O., *Biochim. Biophys.* (1973) **293**, 140.
92. Goldstein, J. L., Hasty, M. A., *J. Biol. Chem.* (1973) **248**, 6300.
93. Bingham, E. W., Farrell, H. M., Jr., *J. Biol. Chem.* (1974) **249**, 3647.
94. Blum, H. E., Pocinwong, S., Fischer, E. H., *Proc. Nat. Acad. Sci. USA* (1974) **71**, 2198.
95. Jergil, B., Dixon, G. H., *J. Biol. Chem.* (1970) **245**, 425.
96. Smith, D. L., Bruegger, B. B., Halpern, R. M., Smith, R. A., *Nature* (1973) **246**, 103.
97. Rosenstein, R. W., Taborsky, G., *Biochemistry* (1970) **9**, 658.
98. Mercier, J. C., Grosclaude, F., Ribadeau-Dumas, B., *Eur. J. Biochem.* (1971) **23**, 41.
99. Ribadeau-Dumas, B., Brignon, G., Grosclaude, F., Mercier, J. C., *Eur. J. Biochem.* (1972) **25**, 505.
100. Fiat, A.-M., Alais, C., Jollès, P., *Eur. J. Biochem.* (1972) **27**, 408.
101. Langan, T., *Ann. N. Y. Acad. Sci.* (1971) **185**, 160.
102. Sanders, M. M., Dixon, G. H., *J. Biol. Chem.* (1972) **247**, 851.
103. Dixon, G. H., Candido, E. P. M., Honda, B. M., Louie, A. J., MacLeod, A. R., Sung, M. T., *Ciba Found. Symp.* (1975) **28**, 229.

103a. Williams, R. E., *Science* (1976) **192**, 473.
104. Zittle, C. A., Bingham, E. W., *J. Dairy Sci.* (1959) **42**, 1772.
105. Bingham, E. W., Zittle, C. A., *Arch. Biochem. Biophys.* (1963) **101**, 471.
106. Kabat, D., *Biochemistry* (1970) **9**, 4160.
107. Kabat, D., *Biochemistry* (1971) **10**, 197.
108. Meisler, M. H., Langan, T. A., *J. Biol. Chem.* (1969) **244**, 4961.
108a. Kinzel, V., Alonso, A., Kübler, D., *Eur. J. Biochem.* (1975) **55**, 361.
109. Nordwig, A., Pfab, F. K., *Biochim. Biophys. Acta* (1969) **181**, 52.
110. Bornstein, P., *Annu. Rev. Biochem.* (1974) **43**, 567.
111. Cardinale, G. J., Udenfriend, S., *Adv. Enzymol.* (1974) **41**, 245.
112. Ramaley, P. B., Rosenbloom, J., *FEBS Letters* (1971) **15**, 59.
113. Jimenez, S. A., Dehm, P., Olsen, B. R., Prockop, D. J., *J. Biol. Chem.* (1973) **248**, 720.
114. Blumenkrantz, N., Rosenbloom, J., Prockop, D. J., *Biochim. Biophys. Acta* (1969) **192**, 81.
115. Olsen, B. R., Berg, R. A., Kishida, Y., Prockop, D. J., *Science* (1973) **182**, 825.
116. Bellamy, G., Bornstein, P., *Proc. Nat. Acad. Sci. USA* (1971) **68**, 1138.
117. Green, H., Goldberg, B., *Proc. Nat. Acad. Sci. USA* (1965) **53**, 1360.
118. Langness, U., Udenfriend, S., *Proc. Nat. Acad. Sci. USA* (1974) **71**, 50.
119. Sadava, D., Chrispeels, M. J., *Biochim. Biophys. Acta* (1971) **227**, 278.
120. Katz, E., Prockop, D. J., Udenfriend, S., *J. Biol. Chem.* (1962) **237**, 1585.
121. Rhoads, R. E., Udenfriend, S., *Arch. Biochem. Biophys.* (1970) **139**, 329.
122. Berg, R. A., Prockop, D. J., *J. Biol. Chem.* (1973) **248**, 1175.
123. Prockop, D. J., Kaplan, A., Udenfriend, S., *Biochem. Biophys. Res. Commun.* (1976) **9**, 162.
124. Prockop, D. J., Juva, K., *Proc. Nat. Acad. Sci. USA* (1965) **53**, 661.
125. Peterkofsky, B., Udenfriend, S., *Proc. Nat. Acad. Sci. USA* (1965) **53**, 335.
126. Hutton, J. J., Tappel, A. L., Udenfriend, S., *Biochem. Biophys. Res. Commun.* (1966) **24**, 179.
127. Berg, R. A., Prockop, D. J., *Biochemistry* (1973) **12**, 3395.
128. Kivirikko, K. I., Prockop, D. J., *J. Biol. Chem.* (1967) **242**, 4007.
129. Hutton, J. J., Marglin, A., Witkop, B., Kurtz, J., Berger, A., Udenfriend, S., *Arch. Biochem. Biophys.* (1968) **125**, 779.
130. Kikuchi, Y., Fujimoto, D., Tamiya, N., *FEBS Letters* (1969) **2**, 221.
131. McGee, J. O'D., Rhoads, R. E., Udenfriend, S., *Arch. Biochem. Biophys.* (1971) **144**, 343.
132. Adams, E. (1973) as reported in: Cardinale, G. J., Udenfriend, S., *Adv. Enzymol.* (1974) **41**, 245.
133. Kivirikko, K. I., Prockop, D. J., *Biochim. Biophys. Acta* (1972) **258**, 366.
134. Kivirikko, K. I., Shudo, K., Sakakibara, S., Prockop, D. J., *Biochemistry* (1972) **11**, 122.
135. Marshall, R. D., Neuberger, A., in "Glycoproteins," A. Gottschalk, ed., Elsevier, Amsterdam, 1972, p. 322.
136. Marshall, R. D., Neuberger, A., *Adv. Carbohydr. Chem. Biochem.* (1970) **25**, 407.
137. Lee, Y. C., Lang, D., *J. Biol. Chem.* (1968) **243**, 677.
138. Sentandreu, R., Northcote, D. H., *Biochem. J.* (1968) **109**, 419.
139. Lindhal, U., Rodén, L., in "Glycoproteins," A. Gottschalk, Ed., Elsevier, Amsterdam, 1972, p. 491.
140. Hallgren, P., Lundblad, A., Svensson, S., *J. Biol. Chem.* (1975) **250**, 5312.
141. Butler, W. T., Cunningham, L. W., *J. Biol. Chem.* (1965) **241**, 3882.
142. Miller, D. H., Lamport, D. T. A., Miller, M., *Science* (1972) **1776**, 918.
143. Lamport, D. T. A., *Biochemistry* (1969) **8**, 1155.
144. Lote, C. J., Weiss, J. B., *FEBS Lett.* (1971) **16**, 81.

145. Weiss, J. B., Lote, C. J., Bobinski, H., *Nature (London) New Biol.* (1971) **234**, 25.
146. Spiro, R. G., *Annu. Rev. Biochem.* (1970) **39**, 599.
147. Pazur, J. H., Aronson, N. N., Jr., *Adv. Carbohydr. Chem. Biochem.* (1972) **27**, 301.
148. Gottschalk, A., Ed., "Glycoproteins," Elsevier, Amsterdam, 1972.
149. Marshall, R. D., *Annu. Rev. Biochem.* (1972) **41**, 673.
150. Spiro, R. G., *Adv. Protein Chem.* (1973) **27**, 349.
151. Schachter, H., Jabbal, I., Hudgin, R. L., Pinteric, L., *J. Biol. Chem.* (1970) **245**, 1090.
152. Johnston, I. R., McGuire, E. J., Jourdian, G. W., Roseman, S., *J. Biol. Chem.* (1966) **241**, 5735.
153. Roseman, S., in "Biochemistry of Glycoproteins and Related Substances," E. Rossi and E. Stoll, Eds., Proc. 4th Intern'l. Conference on Cystic Fibrosis of the Pancreas, S. Karger, Basel, 1968, 244.
154. McGuire, E. J. Roseman, S., *J. Biol. Chem.* (1967) **242**, 3745.
155. Bosmann, H. B., Eylar, E. H., *Biochem. Biophys. Res. Commun.* (1968) **33**, 340.
156. Spiro, M. J., Spiro, R. G., *J. Biol. Chem.* (1968) **243**, 6529.
157. Den, H., Kaufman, B., Roseman, S., *J. Biol. Chem.* (1970) **245**, 6607.
158. Bosmann, H. B., Hagopian, A., Eylar, E. H., *Arch. Biochem. Biophys.* (1968) **128**, 470.
159. Hickman, J., Ashwell, G., Morell, A. G., van den Hamer, C. J. A., Scheinberg, I. H., *J. Biol. Chem.* (1970) **245**, 759.
160. Hagopian, A., Eylar, E. H., *Arch. Biochem. Biophys.* (1969) **129**, 515.
161. Hagopian, A., Eylar, E. H., *Arch. Biochem. Biophys.* (1968) **126**, 785.
162. Eylar, E. H., *J. Theor. Biol.* (1965) **10**, 89.
163. Gottschalk, A., *Nature (London)* (1969) **222**, 452.
164. Spiro, R. G., *New Engl. J. Med.* (1969) **281**, 1043.
165. Neuberger, A., Marshall, R. D., in "Carbohydrates and Their Roles," H. W. Schultz, R. F. Cain, and R. W. Wrolstad, Eds., Avi, Westport, Conn., 1968, p. 115.
166. Shainkin, R., Perlmann, G. E., *J. Biol. Chem.* (1971) **246**, 2278.
167. Jackson, R. L., Hirs, C. H. W., *J. Biol. Chem.* (1970) **245**, 624.
168. Hunt, L. T., Dayhoff, M. O., *Biochem. Biophys. Res. Commun.* (1970) **39**, 757.
169. Isemura, M., Ikenaka, T., Matsushima, Y., *Biochem. Biophys. Res. Commun.* (1972) **46**, 457.
170. Hagopian, A., Westall, F. C., Whitehead, J. S., Eylar, E. H., *J. Biol. Chem.* (1971) **246**, 2519.
171. Symth, D. S., Utsumi, S., *Nature* (1967) **216**, 332.
172. Bahl, O. P., Carlsen, R. B., Bellisario, R., Swaminathan, N., *Biochem. Biophys. Res. Commun.* (1972) **48**, 416.
173. DeVries, A. L., Vandenheede, J., Feeney, R. E., *J. Biol. Chem.* (1971) **246**, 305.
174. Katsura, N., Davidson, E. A., *Biochim. Biophys. Acta* (1966) **121**, 120.
175. Spiro, R. G., Lucas, F., Rudall, K. M., *Nature New Biol.* (1971) **231**, 54.
176. Spiro, M. J., Spiro, R. G., *J. Biol. Chem.* (1971) **246**, 4910.
177. Spiro, R. G., Spiro, M. J., *J. Biol. Chem.* (1971) **246**, 4899.
178. Mahadevan, S., Dillard, C. J., Tappel, A. L., *Arch. Biochem. Biophys.* (1969) **129**, 525.
179. van den Hamer, C. J. A., Morell, A. G., Scheinberg, I. H., Hickman, J., Ashwell, G., *J. Biol. Chem.* (1970) **245**, 4397.
180. Morell, A. G., Gregoriadis, G., Scheinberg, I. H., Hickman, J., Ashwell, G., *J. Biol. Chem.* (1971) **246**, 1461.
181. Ambler, R. P., Rees, M. W., *Nature* (1959) **184**, 56.
182. Vickery, H. B., *Adv. Protein Chem.* (1972) **26**, 81.
183. Paik, W. K., Kim, S., *Adv. Enzymol.* (1975) **42**, 227.

184. Cantoni, G. L., *Annu. Rev. Biochem.* (1975) **44**, 435.
185. Paik, W. K., Kim, S., *Biochem. Biophys. Res. Commun.* (1967) **27**, 479.
186. Hempel, N. K., Lange, H. W., Birkofer, L., *Naturwissenschaften* (1968) **55**, 37.
187. Nakajima, T., Volcani, B. E., *Biochem. Biophys. Res. Commun.* (1970) **39**, 28.
188. Paik, W. K., Kim, S., *Biochem. Biophys. Res. Commun.* (1967) **29**, 14.
189. Paik, W. K., Kim, S., *J. Biol. Chem.* (1970) **245**, 88.
190. Kakimoto, Y., Akazawa, S., *J. Biol. Chem.* (1970) **245**, 5751.
191. Asatoor, A. M., Armstrong, M. D., *Biochem. Biophys. Res. Commun.* (1967) **26**, 168.
192. Liss, M., Edelstein, L. M., *Biochem. Biophys. Res. Commun.* (1967) **26**, 497.
193. Kim, S., Paik, W. K., *J. Biol. Chem.* (1970) **245**, 1806.
194. Sheid, B., Pedrinan, L., *Biochemistry* (1975) **14**, 4357.
195. Matsuoka, M., *Seikagaku* (1972) **44**, 364.
196. DeLange, R. J., Glazer, A. N., Smith, E. L., *J. Biol. Chem.* (1969) **244**, 1385.
197. DeLange, R. J., Glazer, A. N., Smith, E. L., *J. Biol. Chem.* (1970) **245**, 3325.
198. Hill, G. C., Chan, S. K., Smith, L., *Biochim. Biophys.* Acta (1971) **253**, 78.
199. Kim, S., Paik, W. K., *J. Biol. Chem.* (1965) **240**, 4629.
200. Murray, K., *Biochemistry* (1964) **3**, 10.
201. Gershey, E. L., Haslett, G. W., Vidali, G., Allfrey, V. G., *J. Biol. Chem.* (1969) **244**, 4871.
202. Paik, W. K., Kim, S., *J. Biol. Chem.* (1968) **243**, 2108.
203. Reporter, M., Corbin, J. L., *Biochem. Biophys. Res. Commun.* (1971) **43**, 644.
204. Baldwin, G. S., Carnegie, P. R., *Biochem. J.* (1971) **123**, 69.
205. Nakajima, T., Matsuoka, Y., Kakimoto, Y., *Biochim. Biophys. Acta* (1971) **230**, 212.
206. Lee, H. W., Paik, W. K., (1975). Reported in Paik, W. K., Kim, S., *Adv. Enzymol.* (1975) **42**, 227.
207. Sundarraj, N., Pfeiffer, S. E., *Biochem. Biophys. Res. Commun.* (1973) **52**, 1039.
208. Paik, W. K., Kim, S., *Arch. Biochem. Biophys.* (1969) **134**, 632.
209. Brostoff, S., Eylar, E. H., *Proc. Nat. Acad. Sci. USA* (1971) **68**, 765.
210. Paik, W. K., Kim, S., *J. Biol. Chem.* (1970) **245**, 6010.
211. Patthy, L., Smith, E. L., *J. Biol. Chem.* (1973) **248**, 6834.
212. Glazer, A. N., DeLange, R. J., Martinez, R. J., *Biochim. Biophys. Acta* (1969) **188**, 164.
213. Kim, S., *Arch. Biochem. Biophys.* (1973) **157**, 476.
214. Kim, S., *Arch. Biochem. Biophys.* (1974) **161**, 652.
215. Kim, S., Litwack, G., Paik, W. K.; as quoted in Paik, W. K., Kim, S., *Adv. Enzymol.* (1975) **42**, 227.
216. Kim, S., Paik, W. K., *Biochemistry* (1971) **10**, 3141.
217. Kim, S., Paik, W. K., *Anal. Biochem.* (1971) **42**, 255.
218. Krzysik, B., Vergnes, J. P., McManus, I. R., *Arch. Biochem. Biophys.* (1971) **146**, 34.
219. Weihing, R. R., Korn, E. D., *Biochemistry* (1971) **10**, 590.
220. Kuehl, W. M., Adelstein, R. S., *Biochem. Biophys. Res. Commun.* (1970) **39**, 956.
221. Huszar, G., Elzinga, M., *J. Biol. Chem.* (1972) **247**, 745.
222. Borun, T. W., Pearson, D. N., Paik, W. K., *J. Biol. Chem.* (1972) **247**, 4288.
223. Paik, W. K., Kim, S., *Biochem. Biophys. Res. Commun.* (1973) **51**, 781.
224. Kim, S., Benoiton, L., Paik, W. K., *J. Biol. Chem.* (1964) **239**, 3790.

225. Gallwitz, D., Sures, I., *Biochim. Biophys. Acta* (1972) **263,** 315.
226. DeLange, R. J., Fambrough, D. M., Smith, E. L., Bonner, J., *J. Biol. Chem.* (1969) **244,** 5669.
227. DeLange, R. J., Hooper, J. A., Smith, E. L., *Proc. Nat. Acad. Sci. USA* (1972) **69,** 882.
227a. Inoue, A., Fujimoto, D., *Biochem. Biophys. Res. Commun.* (1969) **36,** 146.
227b. Inoue, A., Fujimoto, D., *Biochim. Biophys. Acta* (1970) **220,** 307.
227c. Libby, P. R., *Biochim. Biophys. Acta* (1970) **213,** 234.
227d. Fujimoto, D., *J. Biochem. (Tokyo)* (1972) **72,** 1269.
227e. Inoue, A., Fujimoto, D., *J. Biochem. (Tokyo)* (1972) **72,** 427.
227f. Vidali, G., Boffa, L. C., Allfrey, V. G., *J. Biol. Chem.* (1972) **247,** 7365.
228. Nelsestuen, G. L., Zythovicz, T. H., Howard, J. B., *Mayo Clin. Proc.* (1974) **49,** 941.
229. Blombäck, B., Doolittle, R. F., *Acta Chem. Scand.* (1963) **17,** 1816.
230. Blombäck, B., Doolittle, R. F., *Acta Chem. Scand.* (1963) **17,** 1819.
231. Kang, A. H., Bornstein, P., Piez, K. A., *Biochemistry* (1967) **6,** 788.
232. Pommier, J., Nunez, J., Sokoloff, L., in "Further Advan. Thyroid Res.," Trans. 6th Int. Thyroid Conf., 1970 (1971) **1,** 433.
233. Bettelheim, F. R., *J. Am. Chem. Soc.* (1954) **76,** 2838.
234. Solti, M., Telegdi, M., *Acta Biochim. Biophys.* (1972) **7,** 227.
235. Heinrikson, R. L., Kingdon, H. S., *J. Biol. Chem.* (1971) **246,** 1099.
236. Barra, H. S., Arce, C. A., Rodríguez, J. A., Caputto, R., *Biochem. Biophys. Res. Commun.* (1974) **60,** 1384.
237. Zytkovicz, T. H., Nelsestuen, G. L., *J. Biol. Chem.* (1975) **250,** 2968.
238. Magnusson, S., Sottrup-Jensen, L., Petersen, T. E., Morris, H. R., Dell, A., *FEBS Lett.* (1974) **44,** 189.
239. Nelsestuen, G. L., Broderius, M., Zytkovicz, T. H., Howard, J. B., *Biochem. Biophys. Res. Commun.* (1975) **65,** 233.
240. Davie, E. W., Fujikawa, K., *Annu. Rev. Biochem.* (1975) **44,** 799.
241. Morris, H. R., *Biochem. Soc. Trans.* (1975) **3,** 465.
242. Wooding, G. L., Doolittle, R. F., *J. Hum. Evol.* (1972) **1,** 553.
243. Blombäck, B., in "Molecular Evolution 2. Biochemical Evolution and the Origin of Life," E. Schoffeniels, ed., North-Holland Publ., Amsterdam, 1971, p 112.
244. Pommier, J., Sokoloff, L., Nunez, J., *Eur. J. Biochem.* (1973) **38,** 497.
245. Krohn, K. A., Welch, M. J., *Int. J. Appl. Radiat. Isot.* (1974) **25,** 315.
246. Fritz, R. B., *J. Virol.* (1974) **13,** 42.
247. David, G. S., Reisfeld, R. A., *Biochemistry* (1974) **13,** 1014.
248. Karonen, S.-L., Mörsky, P., Siren, M., Seuderling, U., *Anal. Biochem.* (1975) **67,** 1.
249. Raybin, D., Flavin, M., *Biochem. Biophys. Res. Commun.* (1975) **65,** 1088.
250. Green, N. H., Neurath, H., *Proteins* (1954) **B2,** 1059.
251. Bennett, J. C., *Methods Enzymol.* (1967) **11,** 211.
252. Rupley, J. A., *Methods Enzymol.* (1967) **11,** 905.
253. Mihályi, E., "Application of Proteolytic Enzymes to Protein Structure Studies," Chemical Rubber Co., Cleveland (1972), p 179.
254. Fischer, E. H., Stein, E. A., *The Enzymes,* 1st ed. (1960) **4,** 313.
255. Azari, P. R., Feeney, R. E., *J. Biol. Chem.* (1958) **232,** 293.
256. Morell, A. G., Gregoriadis, G., Scheinberg, I. H., Hickman, J., Ashwell, G., *J. Biol. Chem.* (1971) **246,** 1461.
257. Morell, A. G., Irvine, R. A., Sternlieb, I., Scheinberg, I. H., Ashwell, G., *J. Biol. Chem.* (1968) **243,** 155.
258. Robinson, A. B., McKerrow, J. H., Cary, P., *Proc. Nat. Acad. Sci. USA* (1970) **66,** 753.
259. McKerrow, J. H., Robinson, A. B., *Science* (1974) **183,** 85.

260. Robinson, A. B., McKerrow, J. H., Legaz, M., *Int. J. Pept. Protein Res.* (1974) **6**, 31.
261. Russell, D. H., Snyder, S. H., *Mol. Pharmacol.* (1969) **5**, 253.
262. Rabinovitz, M., Fisher, J. M., *Biochem. Biophys. Res. Commun.* (1961) **6**, 449.
263. Rabinovitz, M., Fisher, J. M., *Biochim. Biophys. Acta* (1964) **91**, 313.
264. Prouty, W. F., Goldberg, A. L., *J. Biol. Chem.* (1972) **247**, 3341.
265. Adams, J. G., Winter, W. P., Rucknagel, D. L., Spencer, H. H., *Science* (1972) **176**, 1427.
266. Yanagita, T., Sugano, M., *Agric. Biol. Chem.* (1975) **39**, 63.
267. Galembeck, F., Ryan, D. S., Whitaker, J. R., Feeney, R. E., (1976) submitted for publication.
268. Gonzalez-Flores, E., Whitaker, J. R., (1975), unpublished data.
269. Boonvisut, S., M.S. Thesis, University of California, Davis, Calif., 1975.

RECEIVED January 26, 1976.

6

Enzymatic Protein Degradation and Resynthesis for Protein Improvement

MASAO FUJIMAKI, SOICHI ARAI, and MICHIKO YAMASHITA

Department of Agricultural Chemistry, University of Tokyo,
Bunkyo-ku, Tokyo 113, Japan

Enzymatic protein degradation and resynthesis processes and their application to improving the functional quality and nutritive values of food proteins are described. Although the mechanism of the resynthesis process, commonly known as the "plastein reaction," is complicated, data accumulated during the past three decades indicate that both condensation and transpeptidation reactions are involved; these reactions change a protein hydrolysate or an oligopeptide mixture into higher molecular substances and/or less soluble substances called "plastein." Controlled amounts of essential amino acids used in ester form can be incorporated during the resynthesis reaction; therefore, plasteins whose essential amino acid compositions have been improved can be prepared. Data are given from studies carried out with the purpose of maximizing the nutritive values of food proteins by enzymatic processes. These processes can also be used for improving the solubility of relatively insoluble proteins as well as removing impurities which may contribute to undesirable flavor, odor, color, or toxicity of high protein foods. While reaction feasibility has been demonstrated under laboratory conditions, the commercial feasibility of the processes, especially from the economic standpoint, remains to be demonstrated.

Edible substances are not always acceptable. Although a new edible protein is developed, it will not be useful as human food until it has achieved acceptability. Soybean protein, for example, is fairly nutritious and has some popularity as a food material. However, consumers often complain of its undesirable beany flavor. Except for a few unique examples (*1, 2*), pure proteins have no flavor. If a flavor is perceived in a protein, it is most likely produced by impurities. In some cases, preparation of proteins acceptable for human food requires that the impurities responsible for the undesirable flavor be removed; furthermore, the procedures used must not produce toxic compounds or other undesirable effects. For this purpose the present authors have investigated the use of proteases for removing miscellaneous impurities from crude food proteins through partial degradation of the proteins (*3, 4*). This process could have application on an industrial scale (*5*).

Degradation alone, however, is not always suitable because protein hydrolysates may have an objectionable bitter taste which prevents their use as food materials without some further treatment. As discussed later, several low-molecular-weight peptides are responsible for the bitterness. A possible solution to the bitter taste problem is its elimination through an enzymatic resynthesis process. Debittering can be achieved by means of the so-called plastein reaction (*6*) which has been studied previously in relation to protein biosynthesis (*7, 8, 9*). The product, generally called "plastein," formed through this reaction is usually characterized by its insolubility in aqueous trichloroacetic acid (*8*). Plasteins prepared from various food proteins through enzymatic degradation and resynthesis are generally bland in flavor, as discussed in detail in the present paper.

Besides acceptability, the nutritional value is a matter of great concern. In 1973 a Joint FAO/WHO Ad Hoc Expert Committee established guidelines for the amounts of essential amino acids required for human infants, children, and adults (*10*). The essential amino acid composition of available food proteins often differs from the recommended amounts, especially with respect to methionine, cystine, lysine, and tryptophan levels. Supplementing certain food proteins with necessary amounts of essential amino acids is important in improving their nutritional quality. Adding free amino acids, although simple and convenient, is not necessarily the best method. Covalent attachment of amino acids, by peptide bonding, may sometimes be more favorable because the amino acids are stable during food processing, storage, and cooking and are not lost through washing or discarding of the juices. Arai (*11*) has found a difference in stability between free methionine and methionine in peptides. Lis et al. (*12*) have reported differences in transport between oligopeptide methionine and free methionine (*12*) and between oligopeptide lysine and free lysine (*13*) which may have nutritional implications. Altschul (*14*) has recently reviewed the value of amino acids in proteins as natural aliments. Chemical methods could be used to incorporate an amino acid into a protein. At present these methods appear to have little feasibility. Through use of the enzymatic resynthesis reaction a particular L-amino acid can be incorporated covalently into a protein hydrolysate without difficulty. This aspect is discussed in the present chapter.

Therefore, a combined process of enzymatic protein degradation and resynthesis might be useful in improving both the acceptability and the nutritive values of food proteins at the same time. To carry out the process we have been using in most cases the well-defined proteases, pepsin (EC 3.4.4.1), α-chymotrypsin (EC 3.4.4.5), and papain (EC 3.4.4.10). Our subsequent discussion will therefore center on the three enzymes and their applications to protein degradation and resynthesis. Our use of the word resynthesis refers only to formation of peptide bonds without reference to molecular size of the final product.

Enzymatic Protein Degradation

A number of factors affect the rate and extent of protein degradation by proteolytic enzymes. These include the specificity of the enzyme, the extent of denaturation of the protein used as substrate as well as substrate and enzyme concentrations, pH, temperature and ionic strength of the reaction medium, presence of inhibitory substances, etc. It is beyond the scope of this review to discuss all these factors in detail. However, we do want to call attention to some of the more important factors and to refer the reader to articles such as those of Blow (*15*), Keil (*16*), and Killheffer and Bender (*17*).

Specificity. Pepsin, α-chymotrypsin, and papain have been investigated in detail as examples of acid-, serine-, and sulfhydryl proteases, respectively (*15, 16, 17, 18*). Pepsin generally shows specificity for peptide bonds of adjacent aromatic amino acid residues (*19*). In protein substrates, however, some peptide bonds, whose susceptibility would not have been predicted on the basis of enzyme activity on synthetic substrates are hydrolyzed. Pepsin action has been studied on poly-α-amino acids to evaluate the effect of chain length (*20, 21*). Fruton's group, using synthetic peptide substrates, has investigated the effects of elongated chains on the rates of their hydrolysis by pepsin (*22, 23*). α-Chymotrypsin has specificity for peptide bonds in which the carbonyl group is contributed by a tryptophan, tyrosine, or phenylalanine residue, although peptide bonds involving other amino acid residues may be hydrolyzed at a much lower rate. Papain hydrolyzes peptide bonds in which the carbonyl group is contributed by an arginine or lysine residue; again, hydrolysis is not restricted to bonds involving those two amino acids.

Size of the peptide can be of importance in determining relative specificity as shown by Morihara and co-workers (*24, 25*) for several different serine proteinases. Some typical data (*24*) are: α-chymotrypsin, $k_{cat}/K_m = 55$ ($mM^{-1} \cdot sec^{-1}$) for Ac-Phe-OMe and 2,000 ($mM^{-1} \cdot sec^{-1}$) for

Ac-Ala-Ala-Phe-OMe [35-fold increase]; subtilisin BPN′, $k_{cat}/K_m = 3.6$ ($mM^{-1} \cdot sec^{-1}$) for Ac-Phe-OMe and 930 ($mM^{-1} \cdot sec^{-1}$) for Ac-Ala-Ala-Phe-OMe [240-fold increase]; trypsin, $k_{cat}/K_m = 470$ ($mM^{-1} \cdot sec^{-1}$) for Z-Lys-OMe and 1,400 ($mM^{-1} \cdot sec^{-1}$) for Z-Ala-Ala-Lys-OMe [3-fold increase]; subtilisin BPN′, $k_{cat}/K_m = 2.3$ ($mM^{-1} \cdot sec^{-1}$) for Z-Lys-OMe and 2,800 ($mM^{-1} \cdot sec^{-1}$) for Z-Ala-Ala-Lys-OMe [1,200-fold increase]. (Z is the abbreviation for the benzyloxycarbonyl group.) Since k_{cat}/K_m can be used as a measure of the specificity of an enzyme (*26*), one would conclude from these data, as well as a great deal more in the literature, that specificity of an enzyme for substrate is modified by amino acid residues adjacent to the two amino acids contributing the scissile peptide bond. This concept is particularly important in the hydrolysis of proteins.

The concept of enzyme subsites for binding of the several amino acid residues on either side of the scissile bond to the enzyme was introduced first by Schechter and Berger (*27, 28*) from their studies on the substrate specificity of papain. This concept of subsites will have special relevance to the specificity of amino acid incorporation during the resynthesis reaction to be discussed later.

Substrate Denaturation. Another important factor related to the rate and extent of enzymatic degradation of a protein is the degree of the denaturation of the protein. Native proteins are generally not susceptible to degradation by proteases because of their compact conformation (*29*). In a completely unfolded protein molecule the amino acid residue side chains can rotate freely; this is called "complete denaturation" in polymer chemistry (*30*). In food technology, however, "denaturation" is measured in a different way. For example, the extent of denaturation of a protein might be determined by a decrease in its water-solubility. Protein molecules are readily denatured (unfolded) during food processing and storage. In many cases the unfolded molecules aggregate to a complex insoluble in water. It is unlikely that such an insolubilized protein is degraded as fast enzymatically as a completely denatured protein which is still in monomeric form and still soluble in water. However, insoluble food proteins can be degraded by means of proteases and can be solubilized efficiently. Although such an enzymatic process is quite complicated kinetically, Archer et al. (*31*) have analyzed such a process mathematically in the case of the solubilization of a fish protein concentrate (FPC) with a microbial protease (Monzyme). They found that adsorption of the enzyme onto an FPC particle occurs first; therefore, the rate of solubilization depends on the average size of the particles suspended in the reaction medium. Using the Mihalyi–Harrington equation (*32*), Archer et al. (*31*) analyzed the solubilization process following enzyme adsorption and found that both fast and slow reactions were involved. The fast reactions were ascribed to enzymatic dissociation of the insoluble macromolecules into soluble polypeptide fragments and the slow reactions to hydrolysis of these fragments into lower-molecular-weight peptides. Enzymatic protein degradation in solution is less complicated kinetically since rate of complex formation between the substrate and enzyme is diffusion-controlled and bond splitting becomes the rate-determining step. Thus, Michaelis–Menten kinetics can be applied in this case.

Optimum Conditions for Protein Degradation. Maximum rates of protein degradation theoretically occur when the substrate concentration is high enough to combine with all the enzyme molecules in the solution. A measure of this is given by knowing K_m which is equal to the substrate concentration at which the observed rate of the reaction is one-half that of the maximum rate. Theoretically the maximum rate will be achieved at substrate concentrations equal to or greater than 20 times K_m. Some typical values of K_m observed when proteases attack soluble proteins are: $K_m = 2$ (wt %) for pepsin acting on ovalbumin (*33*); $K_m = 0.5$ (wt %) for trypsin acting on gelatin (*34*); and $K_m = 0.42$ (wt %) for papain acting on hemoglobin (*35*).

Along with substrate concentration, the pH value is an important factor influencing the rate of reaction. Interestingly enough, a protease may have a different pH optimum for degradation and for resynthesis of

Table I. Optimum pH Values for Degradation of Soybean Protein with Proteases and Those for Resynthesis with the Same Proteases from a Peptic Hydrolysate of Soybean Protein (*37*)

		Optimum pH Value	
Enzyme	*Source*	*Degradation*	*Resynthesis*
Acid protease			
pepsin	Swine	1.6	4.5
Molsin	*Aspergillus saitoi*	2.8	5.5
Rapidase	*Trametes sanguinea*	3.0	4.0
Serine protease			
α-chymotrypsin	Bovine	7.8	5.0–6.0
trypsin	Bovine	8.0	5.5–6.5
subtilisin BPN′	*Bacillus subtilis*	7.5–8.5	6.0–7.0
Pronase	*Streptomyces griceus*	8.0–9.0	6.0
Bioprase	*Bacillus subtilis*	10.0–11.0	6.0
Coronase	*Rhizopus* WR-35	7.0	4.0–5.0
Prozyme	*Streptomyces* No. 1033	8.0–9.0	5.5–6.5
Esperase	*Bacillus subtilis*	9.0–10.0	6.5
Sulfhydryl protease			
papain	*Papaya carica*	5.0–6.0	5.0–6.0
bromelain	*Ananas sativus*	5.0–6.0	5.0–6.0
ficin	*Ficus carica*	5.0–6.0	5.0–6.0

Helvetica Chimica Acta

peptide bonds; this in an advantage in the plastein reaction (to be discussed later). A typical example is pepsin which catalyzes protein degradation at pH ~1.6 and transamidation (synthesis) of some synthetic substrates at a much higher pH (*36*). Yamashita et al. (*37*) have determined optimum pH values for degradation of denatured soybean protein and resynthesis of peptide bonds in the hydrolysate by various proteases at 37°C (Table I). These optimum pH values are experimental parameters which may change with different experimental conditions.

Enzymatic Protein Resynthesis

Conditions Required. The resynthesis process, commonly known as the plastein reaction, requires at least three conditions different from those required for maximum rates of degradation. First, the substrate must be a low-molecular-weight peptide mixture, preferably prepared by enzymatic protein degradation. Virtanen (*38*) found that, for the most efficient plastein formation from a zein hydrolysate, the hydrolysate should contain mostly tetra-, penta-, and hexapeptides. Determann et al. (*39*), after investigating plastein formation from synthetic oligopeptides with different chain lengths, concluded that a peptide size of five or six amino acid units is preferable. Tsai et al. (*40*), using fractions of a soybean protein hydrolysate separated on the basis of molecular size, found that the second and the third fractions, having average molecular weights of 1043 and 685, respectively, produced plasteins (i.e., 10% trichloroacetic acid-insoluble substances) much more effectively than did the lower and higher molecular weight fractions. Recently, Hofsten and Lalasidis (*41*) reported that a whey protein hydrolysate fraction with an average molecular weight of < 1000 can grow to form a plastein product with a molecular weight of 2000–3000. Therefore, the molecular size of substrate is one of the most important factors influencing the plastein formation.

Secondly, the substrate concentration of the reaction medium is an important factor in peptide bond resynthesis and, as seen from Figure 1, should be in the range of 20–40% (w/v). When a substrate is incubated at a concentration of about 7.5% (Figure 1), no reaction appears to occur as measured by trichloroacetic acid-solubility/insolubility. Lower substrate concentrations are more favorable for the degradation reaction.

Of special interest is the pH, the third factor influencing plastein formation. The optima for most proteases for plastein formation generally lie in a narrow range of pH 4–7 (Table I) measured with soybean globulin hydrolysates. Pepsin, for example, is an acid protease, and one of its characteristics is high degradative activity at low pH. Even at pH 1 it is active and able to degrade proteins. As Yamashita et al. (*37*) have demonstrated, no appreciable amount of plastein is formed at pH 1–2 even if all the other conditions are satisfied. The optimum for pepsin to produce plastein is pH ~ 4.5 (Table I). On the other hand, α-chymotrypsin, a serine protease, has maximum activity for protein degradation near pH 8. The optimum is at pH 7.8 for hydrolysis of soybean globulin with this enzyme (Table I). However, effective plastein formation from soybean globulin hydrolysate using α-chymotrypsin occurs at pH 5.3. In the case of papain pH 4–7 is best both for protein degradation and for plastein formation.

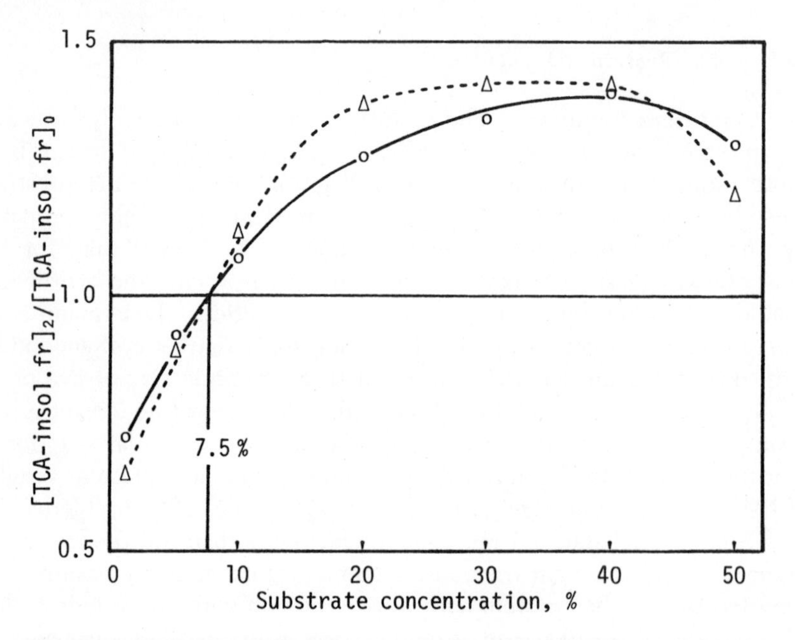

Figure 1. Effect of substrate concentration on the yield of 10% trichloroacetic acid (TCA)-insoluble fraction measured after a soybean protein hydrolysate (substrate) has been incubated with α-chymotrypsin (dotted curve) (42) or with immobilized α-chymotrypsin (solid curve) (43) at 37°C for 2 hr. The scale on the ordinate shows TCA-insolubility after incubation relative to that before incubation.

In summary, proper consideration of the three conditions of substrate size, substrate concentration, and pH is required for effective resynthesis.

Mechanism of Peptide Bond Resynthesis. Peptide bond resynthesis can occur by either transpeptidation or condensation. It is likely that in plastein reactions both reactions occur to variable extents depending on the experimental conditions and the nature of the peptides available. Below, we examine some of the features of each of these reactions.

Several workers have considered plastein formation to be primarily a type of transpeptidation which, as Horowitz and Haurowitz (*44*) have postulated, can be written schematically as:

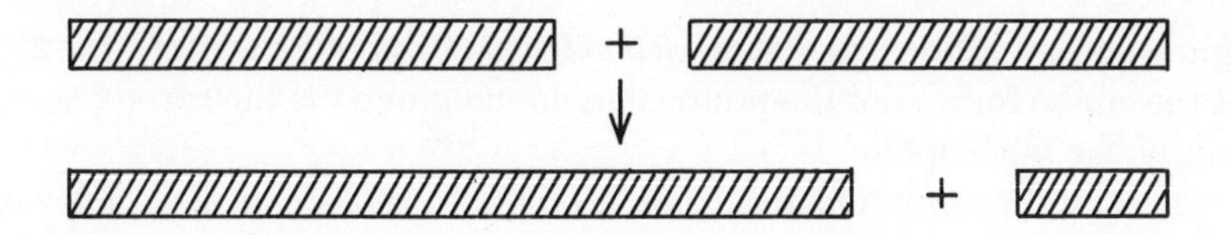

where the rectangles indicate peptides with various chain lengths. Several of the proteases used in plastein formation are known to form transitory acyl-enzyme intermediates in hydrolysis of substrates including proteins (*17*, *26*). The acyl-enzyme intermediate is then broken down through transfer of the acyl group to a nucleophile in the solution, generally water. However, the acyl group can be transferred to other nucleophiles (*26*) such as the amino groups of amino acids and peptides; this results in the synthesis of peptide bonds as shown in Equation 1. Whether or not there is an increase in size in the synthesis will be determined by the relative size of the peptide leaving vs. that of the peptide serving as the nucleophile.

$$\text{E–OH} + \text{–HN–}\underset{\text{H}}{\overset{\text{R}}{\text{C}}}\text{–}\overset{\text{O}}{\overset{\|}{\text{C}}}\text{–}\overset{\text{H}}{\text{N}}\text{–}\underset{\text{R}'}{\overset{\text{H}}{\text{C}}}\text{–}\overset{\text{O}}{\overset{\|}{\text{C}}}\text{–} \rightleftharpoons \underset{\text{Michaelis complex}}{\text{E}\cdot\text{S}} \rightleftharpoons \text{E–O–}\overset{\text{O}}{\overset{\|}{\text{C}}}\text{–}\underset{\text{H}}{\overset{\text{R}}{\text{C}}}\text{–NH–} + \text{H}_2\text{N–}\underset{\text{R}'}{\overset{\text{H}}{\text{C}}}\text{–}\overset{\text{O}}{\overset{\|}{\text{C}}}\text{–}$$

$$\xrightarrow{\text{H}_2\text{O}} \text{E–OH} + \text{HOOC–}\underset{\text{H}}{\overset{\text{R}}{\text{C}}}\text{–NH–} \quad \text{(degradation)}$$

$$\xrightarrow{\text{H}_2\text{N–}\underset{\text{H}}{\overset{\text{R}''}{\text{C}}}\text{–}\overset{\text{O}}{\overset{\|}{\text{C}}}\text{–}} \text{E–OH} + \text{–NH–}\underset{\text{H}}{\overset{\text{R}}{\text{C}}}\text{–}\overset{\text{O}}{\overset{\|}{\text{C}}}\text{–}\underset{\text{H}}{\text{N}}\text{–}\underset{\text{R}''}{\overset{\text{H}}{\text{C}}}\text{–}\overset{\text{O}}{\overset{\|}{\text{C}}}\text{–} \quad \text{(synthesis)} \qquad (1)$$

Whether the degradative or the synthetic reaction is more important will depend among other things on the substrate (nucleophile) concentration and the pH, as discussed earlier. The higher the nucleophile concentration, the more readily it can compete with the water (ca. 55.5*M*) in the reaction. More than just concentration and nucleophilicity are involved, however, since D-amino acid esters do not become incorporated (Table III). This would imply that the nucleophile may be bound during the reaction as well. Addition of organic solvents (such as acetone) will decrease the water concentration as well as "drive" the reaction by decreasing the solubility of the synthesized product. pH affects the reaction because the acyl-enzyme intermediate generally reaches higher concentrations at lower pHs (for α-chymotrypsin *see* Ref. *26*), yet the amino form (not the protonated amino group) is the correct active form of the nucleophile.

If plastein formation results in increase in molecular size, a lower-molecular-weight product must occur at the same time; this product can be either an oligopeptide or a free amino acid. In addition to the above workers (*44*), Yamashita et al. (*45*) have observed that free amino acids appear gradually during the plastein reaction, supporting the occurrence of transpeptidation. Though this mechanism clearly explains an increase in molecular weight as a result of the plastein reaction, it need not lead to a substantial increase in overall size. Hofsten and Lalasilis (*41*) generally did not find an overall increase in size using a whey protein hydrolysate as measured by gel chromatography. However, when a hydrolysate fraction estimated to have a molecular weight of 1000–20,000 was subjected to the plastein reaction, a product was obtained with an upper limit of molecular weight range ~ 30,000 (*41*). In this case a significant amount of a product smaller than the original substrate was formed simultaneously. Tsai et al. (*40*) investigated the molecular weight distribution of a plastein reaction product from a soybean protein hydrolysate by means of 7.5% polyacrylamide-gel electrophoresis in a phenol–acetic acid–water system. They found that the product contained both high-molecular-weight and low-molecular-weight substances—the former, though very small in quantity, appeared to have a molecular weight of ~ 30,000. Further work is needed, preferably with improved solvent systems, because most plasteins are partially insoluble in conventional aqueous solvents used for electrophoresis.

Not only transpeptidation, but also a certain amount of peptide–peptide condensation, is possibly involved in the plastein reaction. With some of the proteases used for plastein formation, especially α-chymotrypsin (*26, 52, 53, 54*), the acyl-enzyme intermediate can be formed at pHs $\simeq$ 5 from the reversal of the degradative reaction (Equation 1: E–OH + HOOC–CHR–NH $\leftrightharpoons$ E–O–OCCHR–NH– + H_2O). Once the acyl-enzyme intermediate is formed, the acyl group can be transferred to a nucleophile resulting in peptide bond synthesis.

Virtanen's group, on the basis of results from terminal amino group analyses, has concluded that the condensation reaction occurs during plastein formation involving a zein hydrolysate and pepsin (*38, 46, 47*). Wieland et al. (*48*) have isolated two plastein-forming oligopeptides from Witte peptone and demonstrated that these condense when incubated with pepsin under conditions used for the plastein reaction. Subsequently, Determann (*49*), in a model experiment using the synthetic pentapeptide H·Tyr-Leu-Gly-Glu-Phe·OH, obtained the following results on incubation of the pentapeptide with pepsin at pH 4:

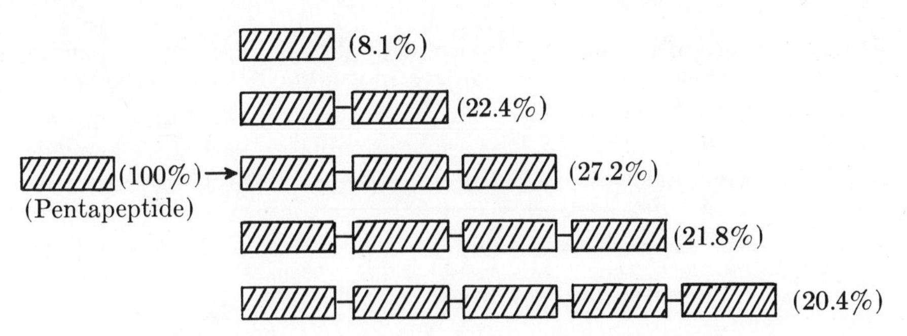

Thus, the di-, tri-, tetra-, and pentamer of the pentapeptide were formed in the indicated relative yields. Yamashita et al. (*50*), using as substrate a soybean globulin hydrolysate in which all the carboxyl groups were labelled with ^{18}O, observed that $H_2{}^{18}O$ is liberated at a rapid rate within the first hour during the plastein formation by α-chymotrypsin at pH 5.3 (Figure 2). At the same time there is a corresponding decrease in the ninhydrin reactive groups. *N*-Acetylation of the soybean globulin hydrolysate or reduction of the –COOH to –CH_2OH significantly prevented plastein formation (*50*). Tanimoto et al. (*51*) attempted the preparation of plastein with chemically modified α-chymotrypsin. When His-57 of this enzyme was methylated, the plastein-producing ability decreased as the degree of methylation increased. Similarly, the diisopropylphosphorylation of Ser-195 led to loss of plastein formation.

Although from an energetic point of view condensation would be a less favorable contributor to the plastein formation than transpeptida-

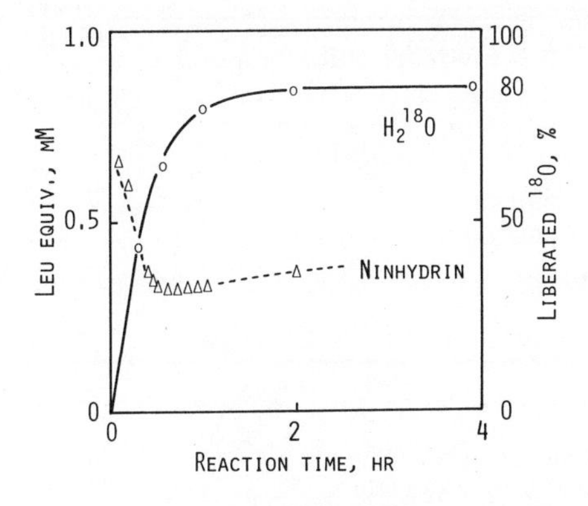

Agricultural and Biological Chemistry

Figure 2. Liberation of water and decrease in ninhydrin response during the resynthesis reaction by α-chymotrypsin at pH 5.3 (45)

tion, it can occur, as indicated by the data above. Recently a Japanese group (*55*) attempted, with fair success, to synthesize biologically active peptides, e.g., angiotensin, secretin, eledoisin, etc., through fragment condensation by various proteases. For example, protected Val^5–angiotensin II, shown below, was efficiently synthesized from its three peptide

```
     NH2 NO2     Bzl      Bzl
      |   |       |        |
Boc-Asp-Arg-Val-Tyr-Val-His-Pro-Phe-OEt
     1   2   3   4   5   6   7   8
```

fragments by the papain-catalyzed condensation between Arg-2 and Val-3 and between Tyr-4 and Val-5. Interestingly enough, the reaction conditions adopted in this case (*55*) were almost identical to those required for plastein formation.

Further investigation is needed on methods of removing the condensation products from the reaction in order to drive the condensation to completion. This might be achieved through changes in solvent composition to minimize solubility of the product through specific binding of the product, etc.

Properties of Plasteins. Plasteins are generally characterized by their low solubility in water. If during the resynthesis reaction a part of the product becomes insoluble, this serves as a driving force. Aso et al. (*56, 57*) investigated some physicochemical properties of a water-insoluble fraction of a plastein produced from a soybean globulin hydrolysate. This fraction interacted with 1-anilino-8-naphthalene sulfonate (ANS) to give a new and larger ANS emission spectrum at 450 nm (Table II).

Table II. Hydrophobic Properties of a Plastein Compared with Its Substrate

	Substrate[a]	*Plastein*
Fluorescence (nm) caused by ANS[b]	—	450
Fluorescence (nm) caused by tryptophan	356	348
NMR spectrum (ppm) of the methylene protons of SDS[c]	1.33	1.27
Amount of bound *n*-heptane (mg/g substrate or plastein)	12.7	51.7
Heat assembly	No	Clear

[a] Peptic hydrolysate of soybean globulin.
[b] 1-Anilino-8-naphthalene sulfonate.
[c] Sodium dodecylsulfate.

The fraction also reacted with sodium dodecylsulfate (SDS) to modify the NMR signal caused by the SDS methylene protons; the absorbance spectrum became broad and shifted upfield by 0.06 ppm. The water-insoluble plastein fraction showed a much higher affinity for a hydrocarbon (n-heptane) than the soybean globulin hydrolysate. The tryptophan-originating fluorescence of the plastein fraction was blue-shifted (by 8 nm) compared with that of the hydrolysate. This probably indicates that the tryptophan residues in the plastein molecules are exposed to a more hydrophobic environment. A temperature effect was distinctly observed for the plastein; when the plastein was suspended in cold water and heated gradually, a larger amount of visible particles appeared at a higher temperature. Plastein formation is sometimes accompanied by gelling. Hofsten and Lalasidis (*41*) considered the gel formation observed in their system to be related to some noncovalent rearrangement of peptides in the reaction system although they have not ruled out rearrangement of peptide sequences in the system. Eriksen and Fagerson (*58*) speculated that the gel formation is an entropy-driven process, the increase in entropy of water acting as the driving force after an initial concentration of suitable peptides has been formed by either condensation or transpeptidation or both. At present it is quite unclear whether this phenomenon is closely related to the hydrophobic interaction among peptides formed by the plastein reaction. The plastein yield, as far as this is determined on the basis of trichloroacetic acid insolubility, is actually enhanced when a highly hydrophilic substrate (i.e., casein hydrolysate) and a highly hydrophobic one (i.e., zein hydrolysate) are used in combination with each other (*59, 60*).

Plasteins are characterized also by their insolubility in aqueous ethanol or acetone (*48, 49, 61*). It is thus feasible to separate plasteins from lower-molecular-weight products by precipitation with ethanol or acetone (*61*). Either dialysis or membrane filtration is useful for separation as well.

Applications of the Plastein Reaction

Model Experiment. Any L-amino acid ester (D-amino acid esters do not work; Table III), when added to a reaction medium containing a protein hydrolysate, is generally incorporated during the resynthesis reaction under the appropriate reaction conditions. In this way it is feasible to prepare a plastein whose amino acid composition has been altered. Therefore, the plastein reaction, accompanied by new amino acid incorporation, would be expected to be much more valuable than when plastein formation is carried out in the absence of new amino acids.

In order to obtain basic information with regard to amino acid incorporation, Aso et al. (*62*) investigated a model system in which ethyl hippurate (i.e., N-benzoylglycine ethyl ester) was used as substrate instead of a protein hydrolysate. Ethyl hippurate (50mM) was first incubated with papain (1.7×10^{-2}mM) at pH 6.0 and 37°C. After 15

Table III. Extent of Incorporation by Papain of Various Kinds of Amino Acid Ethyl Esters into a Protein Hydrolysate after 2 hr[a]

Amino Acid Ethyl Ester	*Velocity (μmol/mg papain/min)*	*Amino Acid Ethyl Ester*	*Velocity (μmol/mg papain/min)*
Gly-OEt	0.007	L-Met-OEt	0.115
L-Ala-OEt	0.016	D-Met-OEt	0.000
D-Ala-OEt	0.000	L-Tyr-OEt	0.120
L-α-Abu-OEt	0.058	L-Phe-OEt	0.127
L-Val-OEt	0.005	L-Trp-OEt	0.132
L-Nva-OEt	0.122	L-Glu-α-OEt	0.025
L-Leu-OEt	0.119	L-Glu-α,γ-diOEt	0.109
L-Ile-OEt	0.005	L-Lys-OEt	0.043
L-Nle-OEt	0.125	N^{ϵ}-Ac-L-Lys-*OEt*	0.096

[a] Unusual abbreviations: Abu = α-aminobutyric acid, Nva = norvaline, Nle = norleucine, Et = ethyl, and Ac = acetyl. The protein hydrolysate was produced from ovalbumin by pepsin (*64*).

min an amino acid ester (5m*M*) was added, and incubation continued under the same conditions. Analysis of the reaction and products indicated that the following reaction occurred:

$$\text{Bz-Gly-Papain} + \text{AA-OR} \rightarrow \text{Bz-Gly-AA-OR} + \text{Ethanol} + \text{Papain}$$

where AA-OR refers to the amino acid ester. The amino acid ester acted as a nucleophile or aminolysis reagent to react with the acyl-enzyme intermediate, Bz-Gly-Papain, synthesizing a peptide bond to form Bz-Gly-AA-OR. Various L-amino acid ethyl esters produced different velocities depending on the amino acid side-chain structure; a more hydrophobic side chain was more effective except for the case of the β-branched chain as in isoleucine. This observation is similar to that of Alecio et al. (*63*). Aso et al. (*62*) also found that even the β-branched amino acid, isoleucine, could be effectively acylated when esterified with a hydrophobic alcohol such as *n*-hexanol. Similarly, *n*-hexyl esters of glycine and alanine were much more reactive than their ethyl esters. Data for several amino acid ethyl esters are shown in Figure 3.

The data above indicate that more is involved in the reaction than just the amino acid ester serving as a nucleophile in competition with water. A specificity of the enzyme toward the amino acid ester also exists. Perhaps a better understanding of the reaction can be obtained by considering some of the properties of papain. Considerable data are available on the structure of papain, in particular its active site. His-159 is thought to be involved in catalysis as a general acid-base in which Cys-25 becomes acylated in an intermediate step (*64*). Schechter and Berger (*27, 28*) postulated from data of studies on the specificity of papain for synthetic substrates that binding of substrate occurs by interaction with one or more of the seven subsites designated S_1, S_2, S_3, and S_4 toward the N-terminal from the scissile point of a substrate peptide and S_1', S_2', and S_3' toward the C-terminal from the scissle point. A substrate with amino acid side chains, R_1, R_2, R_3, R_4, R_1', R_2', and R_3' is postulated to fit the corresponding subsites in the manner shown in Figure 4 and subsequently undergo hydrolysis. A possible explanation of the data in Figure 3 would require that the side chain of the amino acid ester fit into the S_1' subsite (and possibly the alcohol chain into S_2') since the amino acid ester specifically serves as a nucleophile to attack the carbonyl carbon of the thioester part of the acyl-papain (Bz-Gly-Papain in the above experiment). Without this specificity of combining at subsites S_1'

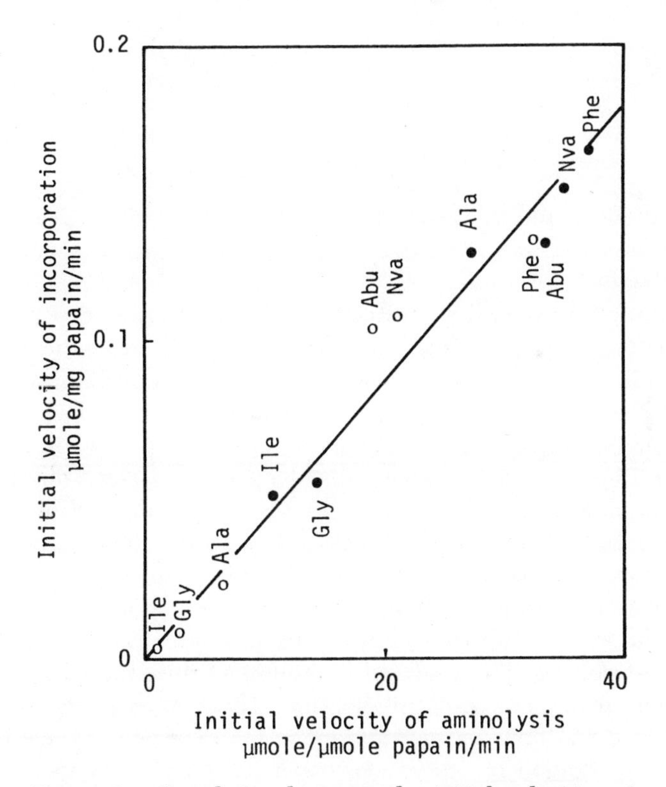

Figure 3. Correlation between the initial velocities of the papain-catalyzed aminolysis of ethyl hippurate by amino acid esters (abscissa) and the initial velocities (extent after 2 hr) of their incorporation during the plastein formation from an ovalbumin hydrolysate by papain (ordinate). Open circles: ethyl esters. Filled circles: n-*hexyl esters. Unusual abbreviations: Abu = α-aminobutyric acid, Nva = norvaline.*

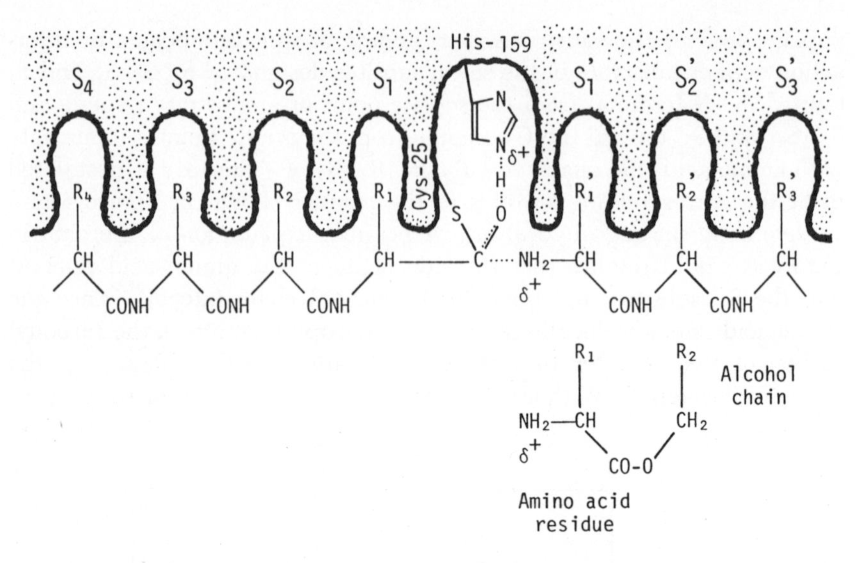

Figure 4. Schematic showing the fit of a peptide substrate and an amino acid ester to the subsites of papain

and S_2' there should be no effect of the alcohol constituent, and D-amino acid esters should also function in this reaction which is not the case (Table III).

Protein Hydrolysates. Instead of ethyl hippurate, a peptic hydrolysate of ovalbumin was used as substrate for the resynthesis reaction (*64*). This substrate (300 mg) was dissolved in water, adjusted to pH 6.0 with NaOH and to 0.9 ml with additional water. An amino acid ester was added to produce a 22.2m*M* solution and the mixture preincubated at 37°C for 15 min. Papain (3 mg), dissolved in 0.1*M* L-cysteine (0.1 ml), was combined with the above-mentioned preincubation mixture and incubation carried out at 37°C. After 2 hr, 0.1*N* NaOH (10 ml) was added to stop the enzymatic reaction and the resulting solution allowed to stand for 3 hr to hydrolyze completely the remaining amino acid ester as well as the ester group from the peptide product. The free amino acid produced from the base-catalyzed hydrolysis of the amino acid ester was determined with an amino acid analyzer. The amount of the amino acid incorporated was obtained by subtracting the determined value from the initial concentration of amino acid ester. The data obtained with the same L-amino acid esters as used in the model experiment (above) are plotted along the ordinate of Figure 3. An excellent correlation is found between the data from the model experiment and those from this experiment using a protein hydrolysate. In Table III data are shown for the extent of covalent incorporation after 2 hr of various amino acid ethyl esters into the protein hydrolysate. There is a close relationship between the extent of incorporation after 2 hr and the hydrophobicity of the amino acid side chain, except for the cases of the β-branched amino acids, valine, and isoleucine. Provided the amino acid ester acts as a nucleophile in the protein hydrolysate as shown for the model system (Figure 3), the amino acid would be incorporated at the C-terminal ends of plastein molecules.

Yamashita et al. (*65*) incorporated L-methionine into a soybean protein hydrolysate by means of the plastein reaction with papain. A 10:1 mixture of a peptic hydrolysate of soybean protein isolate and L-methionine ethyl ester was incubated in the presence of papain, the conditions being similar to those mentioned above. The methionine content of the plastein was 7.22 wt %, nearly seven times the original methionine content of the soybean protein isolate. To determine the location of the incorporated methionine residues, the plastein was treated with carboxypeptidase A. Methionine was liberated much faster than any other amino acid. A second portion of the plastein was methylated and then treated with lithium borohydride to reduce the COOH to CH_2OH. Hydrolysis of the chemically treated plastein with 6*N* HCl gave aminols in satisfactory yields. Subsequently, the aminols were converted to their DNP-derivatives, which were separated by thin layer chromatography. These experiments, together with some others, showed that 84.9% (molar basis) of the C-terminals of the plastein molecules were occupied with methionine, whereas only 14.4% of the N-terminals contained methionine.

Nutritional Improvement of Soybean Protein. An important use of the plastein reaction would be to incorporate limiting essential amino acids thereby improving protein nutritional quality. A study was first conducted to evaluate the in vitro and in vivo digestibilities of a plastein prepared from soybean protein. The plastein was readily digested, and its nutritive value, tested in rats, was comparable with that of soybean protein itself (*66*). In order to increase the sulfur-containing amino acid level of soybean protein (*67, 68*), a soybean protein isolate was hydrolyzed with pepsin to about 80% splitting of the bonds susceptible to pepsin. The resulting hydrolysate was neutralized and freeze-dried. Dried hydrolysate (10 g) was added to L-methionine ethyl ester (1 g) and the mixture dissolved in 10m*M* L-cysteine (25 ml) containing papain (100 mg). (The purpose of the L-cysteine is to activate papain.) After incubation at 37°C for 24 hr, the entire reaction mixture was dialyzed against running water to obtain a plastein as a non-dialysate in a yield of about 70% on a dry-matter basis. The methionine content of the plastein was 7.22%; no free methionine was present. Prior to a feeding test with rats, the plastein was diluted with soybean protein to a methionine level of 2.74%. This sample gave a PER value of 3.38 ± 0.08, compared with a PER value of 2.40 ± 0.05 for casein as a control. By

Table IV. Amino Acid Compositions of

Amino Acid	*Soybean* Protein	*Soybean* Plastein M[a]	*Soybean* Plastein G[b]
Lysine	5.28	4.73	3.05
Histidine	2.04	2.20	1.22
Arginine	5.94	5.61	3.82
Aspartic acid[d]	8.70	7.76	11.76
Threonine	2.63	2.11	3.08
Serine	3.53	2.75	4.46
Glutamic acid[e]	15.00	10.20	41.93
Proline	4.32	2.18	2.88
Glycine	4.38	2.55	3.64
Alanine	3.98	2.65	2.51
Valine	3.36	4.29	4.34
Isoleucine	3.00	5.72	2.37
Leucine	5.17	7.26	3.65
Aromatic amino acids	7.03	9.46	3.38
tyrosine	2.83	3.52	0.88
phenylalanine	4.20	5.94	2.50
Tryptophan	1.34	1.30	0.70
S-containing amino acids	2.94	9.96	2.80
methionine	1.18	7.98	1.20
half-cystine	1.76	1.98	1.60

[a] Methionine-incorporated plastein.
[b] Glutamic acid-incorporated plastein.
[c] Lysine-, methionine-, and tryptophan-incorporated plastein.

taste panel evaluation the methionine-incorporated plastein was bland in taste and almost completely free from any sulfur flavor (*68*).

Arai et al. (*69*) have prepared a similar plastein from soybean protein on an enlarged scale and purified it by precipitation with 70% ethanol. Table IV shows the amino acid composition of this purified plastein in comparison with that of the soybean protein used in preparation of the protein hydrolysate.

Improvement of the Lysine Level of Gluten. Although L-lysine ethyl ester was incorporated less effectively than L-methionine ethyl ester into a protein hydrolysate (Table III), it is possible to prepare a gluten plastein enriched with a large amount of lysine (*70*). Commercially available gluten was fully hydrated in 0.01*N* NaOH, and a fungal alkaline protease was used to degrade the gluten to 85% of maximum hydrolysis by this protease. Various amounts (1.0, 2.0, 3.0, 4.0, and 5.0 g) of L-lysine ethyl ester were added to the gluten hydrolysate (10 g), and each of the mixtures was incubated with papain under the following conditions: reaction system, 20% acetone (pH 6.0) containing 10m*M* L-cysteine;

Plasteins and Their Starting Proteins (wt %)

Spirulina		R. capsulatus		T. repens L	
Protein	*Plastein LMT*[c]	*Protein*	*Plastein LMT*[c]	*Protein*	*Plastein LMT*[c]
4.59	7.75	5.37	7.39	6.06	8.23
1.77	1.91	2.35	2.44	1.94	2.01
6.50	6.70	6.27	6.05	3.66	3.84
8.60	11.87	8.57	10.21	11.10	10.45
4.56	5.42	5.07	4.36	5.60	5.93
4.20	4.43	3.16	4.63	4.17	4.00
12.60	14.68	10.03	9.77	16.00	15.54
3.90	3.62	4.26	3.45	3.75	3.57
4.75	4.76	4.53	5.98	5.01	4.79
6.80	5.80	8.74	8.32	6.23	6.05
4.69	6.00	6.59	6.56	7.45	7.00
6.03	6.32	4.96	5.30	4.98	4.55
8.02	8.98	8.45	8.47	8.86	8.24
8.92	8.96	8.09	8.16	9.61	9.00
3.95	3.98	3.21	3.53	4.11	3.88
4.97	4.98	4.88	4.63	5.50	5.12
1.40	2.72	2.05	2.56	1.51	2.73
1.77	8.75	3.73	9.06	1.82	8.13
1.37	8.22	2.97	8.29	0.85	7.14
0.40	0.53	0.76	0.77	0.97	0.99

[d] Aspartic acid plus asparagine.
[e] Glutamic acid plus glutamine.

substrate concentration, 35% (w/v); enzyme-substrate ratio, 1/100 (w/w); temperature, 37°C; and incubation time, 48 hr. After incubation each reaction mixture was ultrafiltered (Amicon UM-05) to obtain a plastein having a molecular weight > 500. The lysine content in the plastein depended on the amount of L-lysine ethyl ester used; there was a sigmoidal curve relationship between lysine ethyl ester used and amount of incorporation. The curve reached a plateau when 5.0 g of the ester was mixed with 10 g of the gluten hydrolysate. The observed covalent lysine content in the plastein was ca. 16% in this case.

Individual Improvement of Lysine, Threonine, and Tryptophan Levels of Zein. Aso et al. (*71*) incorporated individually the three essential amino acids, lysine, threonine, and tryptophan, in which zein is deficient. Commercially available zein was hydrolyzed with pepsin to solubilize 90% of the protein. A 10:1 (w/w) mixture of the soluble hydrolysate and one of the ethyl esters of L-lysine, L-threonine, and L-tryptophan was incubated with papain in 10m*M* L-cysteine under the following conditions: substrate concentration, 50% (w/v); enzyme–

substrate ratio, 3:100 (w/w); pH 6.0; temperature, 37°C; and incubation time, 48 hr. The lysine-, threonine-, and tryptophan-incorporated plasteins were obtained as water-insoluble fractions in yields of 30.0, 26.0, and 39.9%, respectively. The lysine, threonine, and tryptophan contents in the respective plasteins were 2.14, 9.23, and 9.71%, whereas those in the zein used as starting material were 0.20, 2.40, and 0.38%, respectively.

Simultaneous Improvement of Lysine, Methionine, and Tryptophan Levels of Proteins from Photosynthetic Origin. Peptic hydrolysates of proteins extracted from *Spirulina maxima* and *Rhodopseudomonas capsulatus,* after purification on a Sephadex G-15 column to remove pigments (*72*), were used as substrates for the plastein reaction. Peptic hydrolysate was also prepared from a leaf protein extracted from *Trifolium repens L,* a type of white clover, and also used as a substrate. Plasteins in which controlled amounts of lysine, methionine, and tryptophan had been simultaneously incorporated were produced (*72*). Table V gives the amounts of reagents used and weight of plastein recovered. In each case the plastein reaction was performed with papain under conditions similar to those given in the production of a plastein from a soybean protein hydrolysate (*68*). After incubation each reaction mixture was ultrafiltered to obtain a plastein fraction having a molecular weight > 500. The yields of the respective plasteins are shown in Table V and their amino acid compositions in Table IV. The extents of incorporation (molar basis) were: 22.3–22.5% for L-lysine ethyl ester, 73.1–89.9% for L-methionine ethyl ester, and 86.0–96.9% for L-tryptophan ethyl ester. The extent of incorporation is in general agreement with that expected from the results given in Table III.

The purpose of the above study of incorporating lysine, methionine, and tryptophan was to improve the essential amino acid patterns of the algal, bacterial, and leaf proteins. Figure 5 compares the relative essential amino acid compositions of the three proteins and those of the plasteins; in each case the FAO/WHO-suggested relative composition for the

Table V. Material Balances in Plastein Production from Photosynthetic Protein Resources (g)

	Spirulina maxima	Rhodo-pseudomonas capsulatus	Trifolium repens L
Starting material, dried	10	10	10
Decolored protein hydrolysate	4.74	3.90	2.00
L-Lysine ethyl ester · 2HCl	1.18	0.65	0.33
L-Methionine ethyl ester · HCl	0.54	0.35	0.25
L-Tryptophan ethyl ester · HCl	0.10	0.04	0.04
Plastein	4.96	4.08	2.03

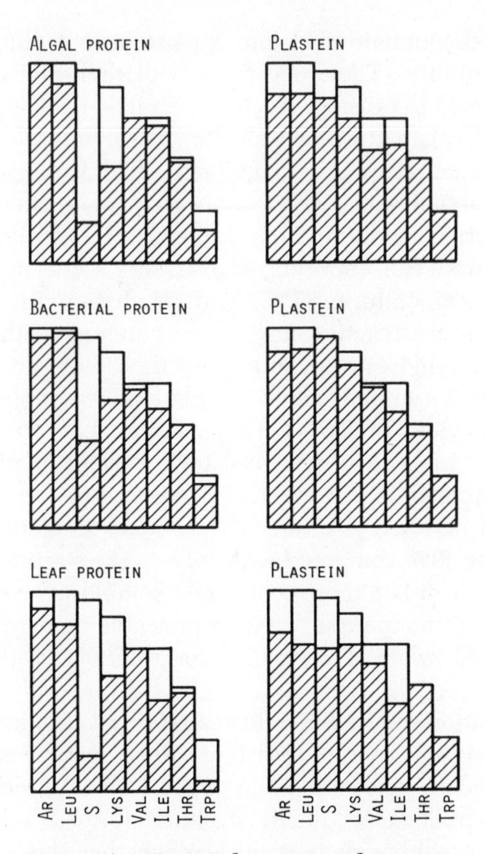

Figure 5. Essential amino acid patterns of proteins extracted from Spirulina maxima, Rhodopseudomonas capsulatus, *and* Trifolium repens L *and those of plasteins produced from these proteins, with the FAO/WHO suggested pattern in each background* (white)

human adult (*10*) is shown in the background. Whereas the essential amino acid composition of the starting proteins are more or less different from those recommended by FAO/WHO, especially with respect to the relative levels of methionine, lysine, and tryptophan, there is no great difference in this respect between the essential amino acid composition of each plastein and the FAO/WHO composition.

Improvement of Solubility of Soybean Protein. Denatured soybean protein is relatively insoluble in aqueous media, which is a disadvantage in processing this protein. This problem might be solved by incorporating a large amount of certain hydrophilic amino acid into soybean protein through its enzymatic degradation and resynthesis. Yamashita et al. (*73,*

74) incorporated glutamic acid into a plastein and obtained a product with greater solubility. They used the α, γ-diethyl ester of glutamic acid which was expected to be more reactive than its α-monoethyl ester (Table III). A 2:1 (w/w) mixture of a peptic hydrolysate of soybean protein isolate and α,γ-diethyl ester of L-glutamic acid was used as substrate. The plastein reaction was carried out with papain under the following conditions: reaction medium, 20% acetone (pH 5.5) containing 10m*M* L-cysteine; substrate concentration, 52.5% (w/v); enzyme-substrate ratio, 2:100 (w/w); temperature, 37°C; and incubation time, 24 hr. After incubation the entire reaction mixture was treated with aqueous alkali to hydrolyze the ethyl ester moieties and then dialyzed in a cellophane membrane against running water. A plastein was obtained as a non-dialysate in the yield of 9.57 g (dry-matter basis) from 10 g of the material starting soybean protein isolate; the yield of plastein is 6.38 g based on a combined 10 g weight of a mixture of the hydrolysate and L-glutamic acid α,γ-diethyl ester. The content of glutamic acid in the plastein was over 40% compared with 15% in the starting protein (Table IV). This plastein was almost completely soluble in water in the range of pH 1–9 at room temperature. No appreciable amount of an insoluble fraction appeared when a neutral water solution of the plastein was heated at 100°C for 1 hr, whereas a similar treatment of a native soybean globulin preparation resulted in a decrease of its solubility by about 70%. This glutamic acid-incorporated plastein had an average molecular weight of about 6200 and an isoelectric point of pH 1.5–4.0. A circular dichroism study indicated that some α-helix structure formed as a result of the plastein reaction. Although it is unclear whether the α-helix formation resulted from incorporation of a large amount of L-glutamic acid, Noguchi et al. (*75*) have identified the oligomeric L-glutamic acids, Glu-Glu, Glu-Glu-Glu, Glu-Glu-Glu-Glu, and Glu-Glu-Glu-Glu-Glu, in a similar plastein hydrolyzed exhaustively with Pronase.

Preparing a Low-Phenylalanine Plastein as a Dietetic Food Material. Feeding a low-phenylalanine diet is the only way of controlling *phenylketonuria,* a well known infant disease (*76*). The present method is to use an amino acid mixture which has most of the phenylalanine removed. Although there may be no nutritional or physiological difference between such an amino acid mixture and a corresponding protein, there is a distinct difference in eating quality. Yamashita et al. (*77*) prepared a bland peptide mixture having a low phenylalanine content by the following method. A fish protein concentrate (FPC) and a soybean protein isolate (SPI) were used as starting materials. Each was partly hydrolyzed with a very small amount of pepsin under conventional conditions and subsequently with Pronase under an unconventional condition at pH 6.5 to liberate phenylalanine, tyrosine, and tryptophan in

Table VI. Amino Acid Compositions of Low-Phenylalanine High-Tyrosine Plasteins from Soybean and Fish Proteins (wt %)

Amino Acid	*Plastein from Soybean Protein*	*Plastein from Fish Protein*
Lysine	3.83	10.11
Histidine	1.41	1.76
Arginine	4.21	4.22
Aspartic acid[a]	18.00	13.67
Threonine	4.39	4.20
Serine	4.67	3.58
Glutamic acid[b]	33.56	27.17
Proline	2.11	4.25
Glycine	3.89	3.94
Alanine	2.56	4.82
Valine	3.24	3.81
Isoleucine	3.83	2.81
Leucine	2.43	3.69
Aromatic amino acids	8.19	7.87
tyrosine	7.96	7.82
phenylalanine	0.23	0.05
Tryptophan	2.80	2.98
S-containing amino acids	2.76	3.31
methionine	0.94	1.90
half-cystine	1.82	1.41

[a] Aspartic acid plus asparagine.
[b] Glutamic acid plus glutamine.

greater amounts than other amino acids. After these three amino acids were removed by means of their peculiar adsorption to Sephadex G-15, the peptide fractions were combined and freeze-dried. Necessary amounts of L-tyrosine ethyl ester and L-tryptophan ethyl ester were added to the freeze-dried fraction, and the mixture was incubated with papain to produce a plastein. By ultrafiltration the plastein was obtained as a fraction having a molecular weight > 500; the yields were 69.3 g from 100 g FPC and 60.9 g from 100 g SPI. The amino acid compositions of the two plasteins are shown in Table VI. A taste panel described both plasteins as almost completely bland in taste. It will be necessary to evaluate the therapeutic effect of this type of dietetic food material.

Removal of Unwanted Compounds

In biological systems many proteins interact specifically with various chemical compounds (*78*). There is also a less-specific type of interaction by which even denatured proteins can bind miscellaneous nonprotein compounds. Such a phenomenon is often observed in the area of food processing. Most food proteins contain nonprotein impurities such as odorants, taste substances, coloring materials, lipids and related compounds, etc. These impurities sometimes affect the acceptability of food proteins and even their safety. Some of these impurities form strong complexes with protein molecules so that they are not completely removed by chemical and physical treatments under conventional conditions. Protein degradation is expected to be effective in loosening the interaction and consequently in liberating the protein-bound impurities. Proteases are particularly useful for this purpose since these usually permit treatments of food proteins under mild conditions (*3, 4, 5*).

The resulting protein hydrolysates, though free of nonprotein impurities, are accompanied by another problem in most cases—a problem caused by a bitterness resulting from protease action on proteins. As discussed below, the resynthesis reaction leading to plastein formation

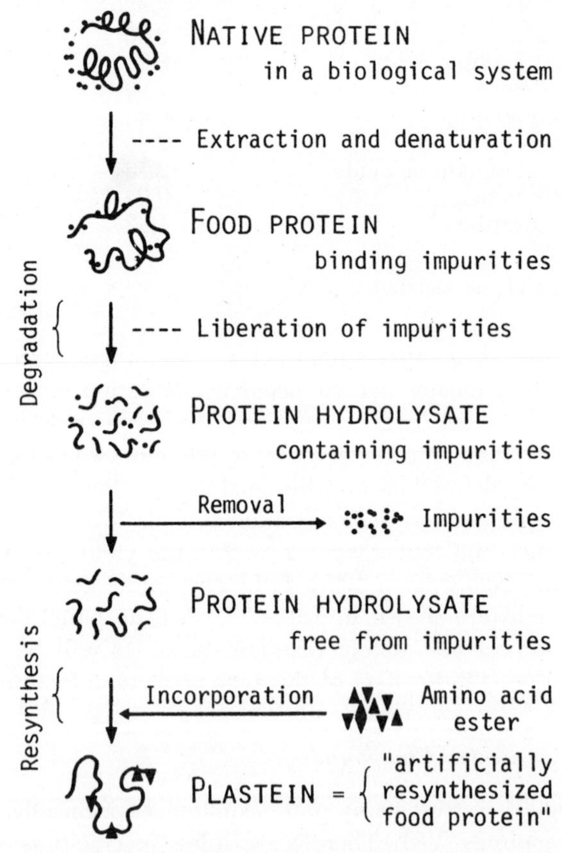

Figure 6. A combined process of enzymatic protein degradation and resynthesis for producing a plastein with an improved acceptability and an improved amino acid composition

has been found very effective for the debittering of protein hydrolysates (*3, 4*). As already discussed, it is feasible to incorporate essential amino acids during the resynthesis and produce a plastein whose amino acid composition has been altered. Therefore, a process is proposed in combination with the enzymatic protein degradation and the enzymatic resynthesis for food protein improvement, as shown in Figure 6. The details of removing impurities are discussed by means of examples of treating soybean protein and related materials.

Deodorization. Volatile flavor components of soybean have been investigated in detail (*79, 80, 81, 82*). Arai et al. (*83*) have studied the interaction of denatured soybean protein with 1-hexanol and 1-hexanal which are the typical beany flavor compounds of raw and processed soybeans. These protein-bound compounds are liberated by treating the denatured soybean protein with pepsin (*83*). Noguchi et al. (*84*) observed that not only 1-hexanol and 1-hexanal but also other flavor compounds are effectively liberated and removed from a soybean protein isolate during treatment with an acid protease (Molsin). A subsequent study has ascribed this effect to the activity of aspergillopeptidase A, an endopeptidase, which has been identified as a main constituent of Molsin (*85*). Fujimaki et al. (*86, 87*) examined several protease preparations for their usefulness in deodorization and reported that a pepsin treatment followed by ether extraction is most effective for deodorizing some protein preparations of soybean and fish.

Defatting. When tofu, a protein curd made from whole soybean, was treated with Molsin, in addition to volatile compounds, nonvolatile fatty materials were liberated including triglygerides, fatty acids, phosphatidyl choline, phosphatidyl ethanolamine, phosphatidyl inositol, sitosteryl-D-glucoside, genistein, saponins, etc. (*84*).

Generally, compared with storage proteins such as soybean globulin, single-cell proteins (SCP) are rather difficult to refine. Fujimaki et al. (*61*) found that an SCP preparation from *n*-paraffin-assimilating yeast (*Candida*) contained significant amounts of lipids and related substances which were not extracted with ether. However, when this SCP preparation (100 g) was treated with pepsin (1 g) at 37°C for 24 hr, an ether-extractable fraction was obtained in the yield of 0.27 g. Analyses showed that this fraction contained various fatty acids including odd-carbon-number acids (15:0, 17:0, and 17:1), a series of *n*-paraffin homologues (C_{11}–C_{24}), several polycyclic hydrocarbons (such as anthracene, phenanthrene, and pyrene), and a kind of ubiquinone.

Decoloring. Much attention is now given to proteins from photosynthetic origin (*88*), especially those from blue-green algae (*89*), nonsulfur purple bacteria (*90*), and green leaves (*91*), as possible food protein sources. These proteins contain large amounts of photosynthetic pigments. Arai et al. (*72*) investigated the removal of the pigments from *Spirulina maxima,* a blue-green alga, and from *Rhodopseudomonas capsulatus,* a non-sulfur purple bacterium. Each was treated with sufficient ethanol to obtain a residue largely free from color. The residue was ground in a mechanical mill and then treated with aqueous alkali to extract protein. The extract was dialyzed against running water to obtain a non-dialysate as a fraction containing the protein. This was placed on a Sephadex G-15 column and eluted with 10% ethanol. The protein fraction eluted from the column still contained pigments. When the non-dialysate was treated with pepsin and subsequently subjected to a similar Sephadex treatment, a colored zone was distinctly separated from a main peptide zone. Completely decolorized protein hydrolysates were obtained from the alga and the bacterium.

Bitter Peptides and Debittering. We (*92, 93, 94, 95*) have investigated the bitterness occurring during the peptic degradation of soybean protein and have identified the following bitter peptides: Gly-Leu, Leu-Phe, Phe-Leu, Leu-Lys, Arg-Leu, Arg-Leu-Leu, Tyr-Phe-Leu, Ser-Lys-Gly-Leu, pyrrolidone carboxyl-Gly-Ser-Ala-Ile-Phe-Val-Leu, and a tetracosapeptide bearing -Gln-Tyr-Phe-Leu as the C-terminal structure. Hata's group (*96, 97*) has identified in a tryptic hydrolysate of casein the following three bitter peptides: Gly-Pro-Phe-Pro-Val-Ile, Phe-Phe-Val-Ala-Pro-Phe-Pro-Glu-Val-Phe-Gly-Lys, and Phe-Ala-Leu-Pro-Glu-Tyr-Leu-Lys. A bitter peptide, Arg-Gly-Pro-Pro-Phe-Ile-Val, has been found in casein treated with a microbial alkaline protease (*98*). Some tri- and tetrapeptides produced by chemical synthesis (*99*) had the following order of bitterness: Phe-Gly-Gly = Gly-Phe-Gly < Gly-Gly-Phe; and Leu-Gly-Gly-Gly = Gly-Leu-Gly-Gly = Gly-Gly-Leu-Gly < Gly-Gly-Gly-Leu. Generally, the bitterness tends to increase when a peptide is enriched with hydrophobic amino acid residues, especially when one of them is located at the C-terminal.

Arai et al. (*85*) decreased the bitterness of a soybean protein hydrolysate by treatment with either carboxypeptidase A or with *Aspergillus* acid carboxypeptidase (*100*). This is a method for debittering based on further degradation of bitter peptides with exopeptidases. Debittering can also be obtained by the resynthesis reaction with endopeptidases. When a peptic hydrolysate of soybean protein was concentrated and incubated with α-chymotrypsin at a slightly acidic pH, the bitterness gradually decreased as plastein formation proceeded (*101, 102*). Yamashita et al. (*45*) reported that during the plastein reaction the two typical bitter peptides, Gly-Leu and Leu-Phe, are decreased in the incubation mixture without being broken down to the free amino acids. A sensory test has confirmed that the plastein obtained by precipitation from aqueous ethanol is almost completely bland in taste (*87*).

Conclusion

An enzymatic protein degradation and resynthesis process is effective in removing unwanted impurities from food proteins and reconstituting a protein-like substance called plastein which may serve as an acceptable food material. In particular, it is feasible to incorporate controlled amounts of essential amino acids during the resynthesis reaction and thereby prepare plasteins having improved nutritional quality. What is now of special importance for practical application of the proposed process is how to recover a maximum amount of the enzyme from the process because cost of the plastein production is dependent primarily on the cost of the enzyme. One approach might be the application of immobilized enzyme technology to the process. It is imperative that the chemical and food engineer examine this process to determine its commercial and economic feasibility. It is also important that the safety of plastein as a human food be established.

Acknowledgments

We are deeply indebted to our collaborators Keiichi Aso and Shin-ya Tanimoto for their hard work and for the many ideas which they contributed to the studies made in this laboratory and also for their permission to include materials unpublished at the time of writing.

Literature Cited

1. Kurihara, K., Beidler, L. M., *Science* (1968) **161,** 1241.
2. Morris, J. A., Cagan, R. H., *Biochim. Biophys. Acta* (1972) **261,** 114.
3. Fujimaki, M., Kato, H., Arai, S., Yamashita, M., *J. Appl. Bacteriol.* (1971) **34,** 119.
4. Fujimaki, M., Arai, S., Yamashita, M., "Proceedings of the International Symposium on Conversion and Manufacture of Foodstuffs by Microorganisms," Saikon Publishing Co., Japan, 1971, p. 19.
5. Fujimaki, M., Kato, H., Arai, S., Yamashita, M., U.S. Patent 3,585,047, 1971.
6. Sawjalow, W. W., *Z. Physiol. Chem.* (1907) **54,** 119.
7. Tauber, H., *J. Amer. Chem. Soc.* (1951) **73,** 1288.
8. Borsook, H., *Advan. Protein Chem.* (1953) **8,** 127.
9. Haurowitz, F., Horowitz, J., *J. Amer. Chem. Soc.* (1955) **77,** 3138.
10. Joint FAO/WHO *Ad Hoc* Expert Committee, "Energy and Protein Requirements," World Health Organization, Geneva, 1973.
11. Arai, S., "Proceedings of the Symposium on New Techniques for Use of Enzymes in Food Technology," Japanese Assoc. Food Technol., Tokyo, Japan, 1974, p. 31.
12. Lis, M. T., Crampton, R. F., Matthews, D. M., *Br. J. Nutr.* (1972) **27,** 159.
13. Burston, D., Addison, J. M., Matthews, D. M., *Clin. Sci.* (1972) **43,** 823.
14. Altschul, A. M., *Nature* (1974) **248,** 643.
15. Blow, D. M., *in* "The Enzymes," Vol. 3, 3rd ed., P. D. Boyer, Ed., p. 205, Academic, New York, 1971.

16. Keil, B., *in* "The Enzymes," Vol. 3, 3rd ed., P. D. Boyer, Ed., p. 263, Academic, New York.
17. Killheffer, J. V., Jr., Bender, M. L., *Crit. Rev.* (1973) **1**, 149.
18. Dixon, M., Webb, E. C., "Enzymes," 2nd ed., p. 226, Longman Green & Co., London, 1967.
19. Sanger, F., Tuppy, H., *Biochem. J.* (1951) **49**, 481.
20. Silman, H. J., Sela, M., *in* "Poly-α-Amino Acids," G. D. Fasman, Ed., p. 608, Marcel Dekker, New York, 1967.
21. Neumann, H., Sharon, N., Katchalski, E., *Nature* (1962) **195**, 1002.
22. Fruton, J. S., *Adv. Enzymol.* (1970) **33**, 401.
23. Fruton, J. S., *in* "The Enzymes," Vol. 3, 3rd ed., P. D. Boyer, Ed., p. 119, Academic, NewYork–London, 1971.
24. Morihara, K., Oka, T., *FEBS Lett.* (1973) **33**, 54.
25. Morihara, K., Oka, T., Tsuzuki, H., *Arch. Biochem. Biophys.* (1969) **135**, 311.
26. Bender, M. L., Kézdy, F. J., *Annu. Rev. Biochem.* (1965) **34**, 49.
27. Schechter, I., Berger, A., *Biochem. Biophys. Res. Commun.* (1967) **27**, 157.
28. Schechter, I., Berger, A., *Biochem. Biophys. Res. Commun.* (1968) **32**, 898.
29. Robinson, D., Jencks, W. P., *J. Amer. Chem. Soc.* (1965) **87**, 2462.
30. Flory, P. J., "Statistical Mechanics of Chain Molecules," p. 30, John Wiley, New York, 1969.
31. Archer, M. C., Ragnarsson, J. O., Tannenbaum, S. R., Wang, D. I. C., *Biotechnol. Bioeng.* (1973) **15**, 181.
32. Mihalyi, E., Harrington, W. F., *Biochim. Biophys. Acta* (1959) **36, 447.**
33. Northrop, J. H., *J. Gen. Physiol.* (1920) **2**, 595.
34. Northrop, J. H., *J. Gen. Physiol.* (1922) **4**, 487.
35. Greenberg, D. M., Winnick, T., *J. Biol. Chem.* (1940) **135**, 781.
36. Inoue, K., Voynick, I. M., Delpierre, G. R., Fruton, J. S., *Biochemistry* (1966) **5**, 2473.
37. Yamashita, M., Tsai, S.-J., Arai, S., Kato, H., Fujimaki, M., *Agric. Biol. Chem. (Tokyo)* (1971) **35**, 86.
38. Virtanen, A. I., *Makromol. Chem.* (1951) **6**, 94.
39. Determann, H., Bonhard, K., Koehler, R., Wieland, T., *Helv. Chim. Acta* (1963) **46**, 2498.
40. Tsai, S.-J., Yamashita, M., Arai, S., Fujimaki, M., *Agric. Biol. Chem. (Tokyo)* (1974) **38**, 641.
41. Hofsten, B. v., Lalasidis, G., *J. Agric. Food Chem.* (1976) **24**, 460.
42. Tsai, S.-J., Yamashita, M., Arai, S., Fujimaki, M., *Agric. Biol. Chem. (Tokyo)* (1972) **36**, 1045.
43. Tanimoto, S., Arai, S., Yamashita, M., Fujimaki, M., *Abstracts of Papers*, p. 152, Annual Meeting of the Agricultural Chemical Society of Japan, Sapporo, Japan, 1975.
44. Horowitz, J., Haurowitz, F., *Biochim. Biophys. Acta* (1959) **33**, 231.
45. Yamashita, M., Arai, S., Matsuyama, J., Kato, H., Fujimaki, M., *Agric. Biol. Chem. (Tokyo)* (1970) **34**, 1492.
46. Virtanen, A. I., Kerkkonen, H. K., Laaksonen, T., Hakala, M., *Acta Chem. Scand.* (1949) **3**, 520.
47. Virtanen, A. I., Laaksonen, T., Kantola, M., *Acta Chem. Scand.* (1951) **5**, 316.
48. Wieland, T., Determann, H., Albrecht, E., *Ann. Chem.* (1960) **633**, 185.
49. Determann, H., *Ann. Chem.* (1965) **690**, 182.
50. Yamashita, M., Arai, S., Tanimoto, S., Fujimaki, M., *Biochim. Biophys. Acta* (1974) **358**, 105.
51. Tanimoto, S., Yamashita, M., Arai, S., Fujimaki, M., *Agric. Biol. Chem. (Tokyo)* (1972) **36**, 1595.

52. Johnson, C. H., Knowles, J. R., *Biochem. J.* (1966) **101**, 56.
53. Foster, R. J., Shine, H. J., Niemann, C., *J. Amer. Chem. Soc.* (1955) **77**, 2378.
54. Bender, M. L., Kemp, K. C., *J. Amer. Chem. Soc.* (1957) **79**, 116.
55. Isowa, Y., Ohmori, M., Sato, M., Mori, K., *Abstracts of Papers*, p. 483, 35th Meeting of the Chemical Society of Japan, Sapporo, Japan, 1955.
56. Aso, K., Yamashita, M., Arai, S., Fujimaki, M., *Agric. Biol. Chem. (Tokyo)* (1973) **37**, 2505.
57. Aso, K., Yamashita, M., Arai, S., Fujimaki, M., *J. Biochem. (Tokyo)* (1974) **76**, 341.
58. Eriksen, S., Fagerson, I. S., *J. Food Sci.* (1976) **41**, 490.
59. Yamashita, M., Arai, S., Tsai, J.-S., Fujimaki, M., *Agric. Biol. Chem. (Tokyo)* (1970) **34**, 1593.
60. Arai, S., Yamashita, M., Aso, K., Fujimaki, M., *J. Food Sci.* (1975) **40**, 342.
61. Fujimaki, M., Utaka, K., Yamashita, M., Arai, S., *Agric. Biol. Chem. (Tokyo)* (1973) **37**, 2303.
62. Aso, K., Yamashita, M., Arai, S., Fujimaki, M., *Abstracts of Papers*, p. 153, Annual Meeting of the Agricultural Chemical Society of Japan, Sapporo, Japan, 1975.
63. Alecio, M. R., Dann, M. L., Lowe, G., *Biochem J.* (1974) **141**, 495.
64. Aso, K., Yamashita, M., Arai, S., Fujimaki, M., *Abstracts of Papers*, p. 435, 47th Meeting of the Biochemical Society of Japan, Okayama, Japan, 1974.
65. Yamashita, M., Arai, S., Aso, K., Fujimaki, M., *Agric. Biol. Chem. (Tokyo)* (1972) **36**, 1353.
66. Yamashita, M., Arai, S., Gonda, M., Kato, H., Fujimaki, M., *Agric. Biol. Chem. (Tokyo)* (1970) **34**, 1333.
67. Arai, S., Yamashita, M., Fujimaki, M., *Cereal Foods World* (1975) **20**, 107.
68. Yamashita, M., Arai, S., Tsai, S.-J., Fujimaki, M., *J. Agric. Food Chem.* (1971) **19**, 1151.
69. Arai, S., Aso, K., Yamashita, M., Fujimaki, M., *Cereal Chem.* (1974) **51**, 143.
70. Fujimaki, M., Mori, K., Yamashita, M., Arai, S., *Abstracts of Papers*, p. 388, Annual Meeting of the Agricultural Chemical Society of Japan, Kyoto, Japan, 1976.
71. Aso, K., Yamashita, M., Arai, S., Fujimaki, M., *Agric. Biol. Chem. (Tokyo)* (1974) **38**, 679.
72. Arai, S., Yamashita, M., Fujimaki, M., *J. Nutr. Sci. Vitaminol. (Jpn)* (1976) **22**, 447.
73. Yamashita, M., Arai, S., Kokubo, S., Aso, K., Fujimaki, M., *Agric. Biol. Chem. (Tokyo)* (1974) **38**, 1269.
74. Yamashita, M., Arai, S., Kokubo, S., Aso, K., Fujimaki, M., *J. Agric. Food Chem.* (1975) **23**, 27.
75. Noguchi, M., Yamashita, M., Arai, S., Fujimaki, M., *Abstracts of Papers*, p. 399, Annual Meeting of the Agricultural Chemical Society of Japan, Sapporo, Japan, 1975.
76. Wang, P. W. K., Hsia, D. Y. Y., *in* "Modern Nutrition in Health and Disease," R. S. Goodhart, M. E. Shils , *Eds.*, p. 1012, Lea & Febiger, Philadelphia, 1973.
77. Yamashita, M., Arai, S., Fujimaki, M., *J. Food Sci.* (1976) **41**, 1029.
78. Lauffer, M. A., *in* "Proteins and Their Reactions," H. W. Schultz, A. F. Anglemier, *Eds.*, p. 87, AVI, Westport, Conn., 1964.
79. Fujimaki, M., Arai, S., Kirigaya, N., Sakurai, Y., *Agric. Biol. Chem. (Tokyo)* (1965) **29**, 855.

80. Arai, S., Koyanagi, O., Fujimaki, M., *Agric. Biol. Chem. (Tokyo)* (1967) **31**, 868.
81. Mattick, L. R., Hand, D. B., *J. Agric. Food Chem.* (1969) **17**, 15.
82. Smith, A. K., Circle, S. J., *in* "Soybean: Chemistry and Technology," Vol. 1, A. K. Smith, S. J. Circle, *Eds.*, p. 341, AVI, Westport, Conn., 1972.
83. Arai, S., Noguchi, M., Yamashita, M., Kato, H., Fujimaki, M., *Agric. Biol. Chem. (Tokyo)* (1970) **34**, 1569.
84. Noguchi, M., Arai, S., Kato, H., Fujimaki, M., *J. Food Sci.* (1970) **35**, 211.
85. Arai, S., Noguchi, M., Kurosawa, S., Kato, H., Fujimaki, M., *J. Food Sci.* (1970) **35**, 392.
86. Fujimaki, M., Kato, H., Arai, S., Tamaki, E., *Food Technol.* (1968) **22**, 889.
87. Fujimaki, M., Yamashita, M., Arai, S., Kato, H., *Agric. Biol. Chem. (Tokyo)* (1970) **34**, 1325.
88. Gordon, J. F., *in* "Proteins as Human Food," R. A. Lawrie, *Ed.*, p. 328, AVI, Westport, Conn., 1970.
89. Clément, G., *in* "Single-Cell Protein," R. I. Mateles, S. R. Tannenbaum, *Eds.*, p. 306, M.I.T. Press, Cambridge, Mass., 1968.
90. Kobayashi, M., Tchan, Y. T., *Water Res.* (1973) **7**, 1219.
91. Gore, S. B., Mungikar, A. M., Joshi, R. N., *J. Sci. Food Agric.* (1974) **25**, 1149.
92. Fujimaki, M., Yamashita, M., Okazawa, Y., Arai, S., *Agric. Biol. Chem. (Tokyo)* (1968) **32**, 794.
93. Fujimaki, M., Yamashita, M., Okazawa, Y., Arai, S., *J. Food Sci.* (1970) **35**, 215.
94. Yamashita, M., Arai, S., Fujimaki, M., *Agric. Biol. Chem. (Tokyo)* (1969) **33**, 321.
95. Arai, S., Yamashita, M., Kato, H., Fujimaki, M., *Agric. Biol. Chem. (Tokyo)* (1970) **34**, 729.
96. Matoba, T., Nagayasu, C., Hayashi, R., Hata, T., *Agric. Biol. Chem. (Tokyo)* (1969) **33**, 1662.
97. Matoba, T., Hayashi, R., Hata, T., *Agric. Biol. Chem. (Tokyo)* (1970) **34**, 1235.
98. Minamiura, N., Matsumura, Y., Fukumoto, J. Yamamoto, T., *Agric. Biol. Chem. (Tokyo)* (1972) **36**, 588.
99. Sato, S., Oka, T., Okai, H., Sadamori, H., *Abstracts of Papers,* p. 252, Annual Meeting of the Agricultural Chemical Society of Japan, Fukuoka, Japan, 1970.
100. Ichishima, E., *Biochim. Biophys. Acta* (1972) **258**, 274.
101. Yamashita, M., Arai, S., Matsuyama, J., Gonda, M., Kato, H., Fujimaki, M., *Agric. Biol. Chem. (Tokyo)* (1970) **34**, 1484.
102. Fujimaki, M., Yamashita, M., Arai, S., Kato, H., *Agric. Biol. Chem. (Tokyo)* (1970) **34**, 483.

RECEIVED January 26, 1976.

7

Functionality Changes in Proteins Following Action of Enzymes

T. RICHARDSON

Department of Food Science, University of Wisconsin—Madison, Madison, Wis. 53706

Various exogenous and endogenous enzymes alter the functional properties of food proteins. Proteolytic enzymes are the principal hydrolases that act on proteins to alter solubilities, to decrease gel formation and viscosity in solution, to modify foaming and emulsification properties, to yield desirable rheological and physical properties, and in the case of cheesemaking to induce gel formation. Other hydrolases such as phosphoprotein phosphatases, lactase, and myosin ATPase exert a direct or indirect effect on the physical properties of proteins. A limited number of oxidoreductases such as glucose oxidase, lipoxygenase, ascorbic acid oxidase, and dehydroascorbic acid reductase have been shown to alter the functional properties of proteins. Generally, a product of the oxidoreductase reaction is the modifying agent.

Once the nutritional value and safety of a proteinaceous food product have been established, the functional properties of the proteins in foods become of paramount importance. The functional properties of a protein are based on those physicochemical attributes that lend desirable physical characteristics to a food. Some of the functional properties of proteins are listed in Table I (*1*). Certain proteins, either indigenous or added, can impart a desirable character to foods and are partially responsible for certain textural and organoleptic properties of those foods. Thus, in the development of better and new food products, protein functionality is of prime importance.

A variety of enzymes acting on a protein directly or indirectly can affect its functional behavior. The effect of the enzyme can be beneficial or detrimental to the food product. In very general terms, the great

Table I. Some Functional Properties of Proteins in Food Systems (*1*)

1. Emulsification	Emulsifying capacity and emulsion stability —sausages, mayonnaise, soups, breads
2. Water absorption and binding	Hydrogen bonding of water to polar groups on protein to prevent syneresis or loss of water —breads, soups, cakes
3. Gelation, curd formation, coagulation	Gel structures act as a matrix in holding moisture, lipids, polysaccharides and other ingredients—cheese, yogurt, imitation meats
4. Viscosity	The viscosity of a food system may be increased by the introduction of proteins or a protein-containing foodstuff—soups, gravies, porridge
5. Thickening, dough forming, elasticity	Proteins have a definite role in determining the strength of doughs and other structures especially via sulfhydryl and disulfide groups
6. Adhesion, cohesion	Useful in binding particles in structured forms —sausages, luncheon meats, baked goods, imitation meats
7. Whippability, foamability, aeration	Proteins in solution reduce the surface tension of the water and consequently aid in formation of food foams. Measure foam persistence, foaming power, and foam stiffness—whipped toppings, frozen desserts, chiffon mixes, angel food cakes, meringues
8. Fat absorption	Binding of fat is important in sausages and doughnuts
9. Solubility as a function of pH	Allows selective precipitation of proteins to form structured products—imitation meats, cheese, yogurt. Also increased solubility of proteins or hydrolysates to supplement acidic beverages

Evaluation of Novel Protein Products

number of enzymes that affect the functionality of proteins belong to the hydrolases. As one might expect, the proteases, acting directly on the proteins, play a key role in determining the functionalities of a wide variety of proteins. On the other hand a limited number of hydrolases may have an indirect effect on protein functionality, such as the role adenosine triphosphatase plays in the physical structure of meat proteins. The oxidoreductases are the second group of enzymes that can affect the functionality of proteins. However, for the most part, with the oxidoreductases protein functionality is often altered indirectly by a product of the enzyme reaction. For example, lipid hydroperoxides produced by lipoxygenase can oxidize protein sulfhydryl groups to disulfides thus altering the structure of the protein.

Although there are no clear demonstrations of other enzyme groups (e.g., lyases, isomerases, ligases) affecting the functionality of proteins in foods, it is conceivable that they do. For example, protein disulfide isomerases could rearrange disulfide bonds in proteins to change their physical properties.

This paper reviews the effects of enzymes on the functional properties of proteins, protein concentrates, and protein isolates. In addition, selected food systems are discussed to relate the action of endogenous and exogenous enzymes to protein functionality in the food.

Effect of Enzymes on the Functionality of Proteins, Protein Isolates, and Protein Concentrates

There are few studies specifically designed to determine the effects of enzymes on the functionality of highly purified proteins. The need for basic research in this area is enormous. However, it is readily evident that functional properties such as solubility and viscosity of a purified protein can be altered drastically by proteolytic enzymes.

Some examples of the effects of enzymes on the functional properties of purified proteins may prove instructive. A striking example of how the functionality of a protein can be altered by enzymic treatment is the action of chymosin (rennin), pepsins, and other proteases on the κ-casein fraction of cow's milk (*2*). Cheese is derived by treating the milk with the protease thereby causing the casein in milk to coagulate to form a curd. Casein is actually a heterogenous mixture of caseins, primarily α_s-, β-, and κ-caseins. κ-Casein itself is heterogenous in terms of its carbohydrate content, some with and some without carbohydrate. Pure α_s- and β-casein are readily precipitated by Ca^{2+}. However, by virtue of its physicochemical properties, κ-casein stabilizes α_s-, β-, and minor casein components in aggregates termed casein micelles which are approximately 40–200 nm in diameter (*3*). The chymosin and various other proteolytic enzymes act primarily on the κ-casein to destroy its micelle stabilizing properties. Chymosin has only a minor effect on the other casein components during milk coagulation. In the presence of calcium ions the destabilized micelles coagulate to form cheese curd.

Purified κ-casein is attacked by chymosin at pH 6.7 in the absence of calcium ion to form insoluble *para*-κ-casein and a soluble macropeptide with the specific cleavage of a phenylalanyl–methionine peptide bond. This serves as an example of how the cleavage of a single peptide bond by an enzyme can drastically alter the functionality of a protein system. This system is discussed in more detail in the subsequent section on functionality of proteins in the milk system.

An additional example of how enzymes can affect the physical properties of purified proteins relates to the enzymic dephosphorylation of β-casein (*4*). Bovine β-casein contains five phosphate groups per monomer as phosphoseryl residues. Purified β-casein from bovine milk was dephosphorylated by a phosphoprotein phosphatase. Both the native β-casein and 65% dephosphorylated β-casein self-associated to form polymers at 35°C. However, the dephosphorylated β-casein had a larger sedimentation coefficient (S_{35}:22.5) than that of the native β-casein (S_{35}:18.2). Also the sedimentation pattern of the dephosphorylated β-casein was more polydisperse than the hypersharp pattern of the native β-casein. These properties were accentuated with 95% dephosphorylated β-casein. The decrease in the negative charge caused by the loss of phosphate apparently favors the self-association of β-casein resulting from hydrophobic bonding.

Also the calcium ion-dependent precipitation of β-casein at 35°C was almost completely lost after 65% dephosphorylation. The extent of binding of calcium ion was 5.3 mol per native β-casein monomer compared with 0.4 mol per monomer of 95% dephosphorylated β-casein. Thus the binding of calcium ions by the phosphate groups of β-casein is essential for its precipitation. Thus the enzymic removal of phosphate groups from phosphoproteins can markedly affect the physical properties of these proteins. The enzymic dephosphorylation of the caseins in cheese may be an important rate limiting reaction in the textural changes of cheese during aging (*5*).

Enzymic treatment of less-well-defined protein systems results in an alteration of the functional properties of the proteins. One of the primary functional properties is solubility since a protein generally must be in solution before it can exert many of its other functionalities. Unfortunately, many protein concentrates are either naturally insoluble or rendered insoluble at pH values in the neutral range by processing treatments which denature the proteins. Many protein concentrates tend to be insoluble at acidic pH values around 5 (*6*) and require alkaline conditions for solubilization. Supplementation of an acidic soft drink or citrus juice with protein, for example, would require solubilization of the protein at acidic pH values. Consequently, proteolytic enzymes have been employed to solubilize proteins from various sources and thus alter their solubilities through proteolysis and modify their other functional properties.

Roozen and Pilnik (*7*) studied the preparation of acid-soluble enzymic hydrolysates of a soy protein concentrate suitable for enrichment of orange juice. Seventeen commercial proteolytic enzyme preparations were used on denatured and native 4% w/v aqueous solutions of soy protein concentrate. The degree of hydrolysis (estimated from trichloro-

acetic acid solubility) and the flavor score (evaluated organoleptically on a seven-point scale) were tabulated. Denaturation improved digestion of the protein, but had little effect on the flavor of the hydrolysate. Flavor scores varied widely; three microbial acid proteases gave bland-flavored hydrolysates. In some cases, treatment with two proteases (either simultaneously or successively) improved the flavor. Studies of orange juice fortified with the hydrolysates showed that samples containing 5% peptides had a very unpleasant taste, but samples containing the bland hydrolysates at a concentration of 2% were tolerable.

Modification of fish proteins by proteolytic enzymes to increase their solubilities illustrates a variety of techniques and approaches. Basically, three general enzymic methods have been used to prepare fish proteins or hydrolysates with altered solubilities and other functionalities. These methods include: (a) the enzymic solubilization of fish protein concentrate prepared by hot solvent extraction of fish, (b) the enzymic modification of myofibrillar proteins extracted from fish with 0.6*M* NaCl, and (c) the proteolysis of whole fish to prepare biological fish protein concentrate (FPC).

FPC prepared by hot solvent extraction of whole fish is, for the most part, denatured, essentially insoluble at pH values below 12.0–12.4 (*6, 8*), and largely devoid of functional properties. FPC can be solubilized to a great degree with enzymes (*8*). Enzymic hydrolysis employing trypsin or two bacterial proteinases for 24 hr at 40°C and optimum pH with increasing concentrations of enzyme resulted in a rapid increase of protein solubilization at an enzyme:substrate ratio of 1:20 w/w. At this point the extent of protein solubilization was 86–90%. Subsequent increases in solubilization required an inordinate amount of enzyme perhaps indicative of product inhibition. FPC from red hake (*Urophycis chuss*) could be solubilized in batch processes employing eight different commercial proteases with varying specificities (*9*). Among the relative activities of the various proteases for FPC solubilization, porcine pepsin and pronase (a fungal protease) were particularly effective with enzyme: FPC ratios of 0.5% w/w giving 95% solubilization in 24 hr of hydrolysis at 37°C at optimum pH.

Macromolecular substrates such as proteins offer unique opportunities in processing modes with enzymes. Ultrafiltration membrane reactors (*10*) can be used to retain the protein substrate and the proteolytic enzyme in the reactor, while the hydrolytic products escape through the membrane to be collected. Using an ultrafiltration reactor, Cheftel (*11*) was able to solubilize 95% of FPC in 24 hr using pronase digestion.

This system of preparing protein hydrolysates may be of general utility; however, Roozen and Pilnik (*12*) treated a soybean isolate in an ultrafiltration reactor and experienced problems of a thickened retentate in the reactor after 6 hr of operation and a buildup of rejected peptides on the membrane which slowed the flux of hydrolysate. With the soybean proteins ultrafiltration digestion yielded largely middle molecular weight peptides soluble at the isoelectric point, with an absence of bitter taste and of bound odorants.

In addition to studies on enzymic solubilization of solvent-extracted FPC, the relative effectiveness of more than 20 commercially available proteolytic enzymes in the production of hydrolysates from a specially prepared haddock protein substrate has been determined (*13*). The concentration of enzyme, at its optimum temperature and pH, required to effect a 60% digestion in a 24-hr period was the inverse measure of enzyme activity. Pronase exhibited the greatest activity per unit weight. In general the microbial proteases ranked low in relative activity. Porcine pepsin, papain, and pancreatin combined good activity with moderate cost.

The myofibrillar proteins of fish are those proteins soluble in 0.6*M* NaCl (*14*). These proteins have been modified with Rhozyme P-11 at an enzyme-to-protein ratio of 1:75 at 30°C, pH 6.6 for 1 hr. The modified myofibrillar protein was quantitatively recovered as a hexametaphosphate complex at pH 3. Residual lipids were removed by solvent extraction of the complex. After a solution of the phosphate–protein complex was neutralized to pH 7, it was spray-dried or freeze-dried to yield a freely soluble product which was 93.5% protein, 0.15% lipid, and 1.4% phosphate.

In contrast to hot solvent extraction, biological methods for producing FPC have been reviewed by Hale (*15*). The process involves the enzymic digestion of a whole fish slurry at controlled pH and temperature (*16*). Bones and scales are removed by screening. Insoluble solids and oil are removed by centrifugation, and the clarified hydrolysate is spray-dried to yield a water-soluble product consisting of peptides and some free amino acids. The enzyme systems utilized in preparing biological FPC may be endogenous autolytic enzymes and/or commercial exogenous enzymes. On whole red hake (*Urophycis chuss*) at pH 8.5, an alkaline bacterial protease in conjunction with endogenous enzymes active at that pH gave the highest yield of dry solubles. The economics of this process were discussed in some detail.

Protein concentrates other than FPC have also been treated with proteolytic enzymes to modify their solubilities. Arzu et al. (*17*) hydrolyzed cottonseed protein from defatted cottonseed flour with nine proteases from animal, plant, or microbial sources. In these studies the enzymes were incubated with the substrate at their optimum pH values for 5 hr at 45°C. At a level of 0.25 g enzyme per 10 g protein, two bacterial proteinases and bromelain were most active. An interesting aspect of this study was the inhibition of enzymic activity by gossypol, a phenolic compound indigenous to certain cottonseeds. At a level of 0.022% w/v in the reaction mixture, gossypol inhibited a bacterial proteinase approximately 30%. One should be aware of other types of proteolytic inhibitors that may occur in plant and other protein concentrates (*18*).

Hermansson et al. (*19*) determined the effects of pepsin and papain treatments on the solubility of rapeseed protein concentrate (RPC). The original RPC had a solubility of 11–12% at pH values between 4.0 and 7.0. However, hydrolysis with pepsin for 24 hr increased the solubility to 82–90% over this same pH range. On the other hand papain digestion increased the solubility to only 38–43%.

Puski (*20*) treated a soy protein isolate with various levels of a neutral protease from *Aspergillus oryzae* and observed an increased soluble nitrogen at the isoelectric point and in 0.03*M* $CaCl_2$ where the untreated control was insoluble.

Partial proteolysis of soybean proteins with endopeptidases has been used to remove flavor compounds and related fatty materials from soybean curd and defatted soybean flour (*21*). Certain soybean protein concentrates possess an undesirable beany and oxidized flavor. Treatment of soybean curd and defatted soybean flour with endopeptidases such as aspergillopeptidase A released off-flavor compounds such as 1-hexanal and 1-hexanol which could be removed from the hydrolysate by solvent extraction. The enzymically digested products had less odor, taste, and color than the starting material and were more stable to oxidative deterioration.

In enzymic digestions, the structures of the released peptides will, of course, depend upon the specificity of the particular protease. Often the peptides exhibit a very undesirable bitter flavor. For example, Fujimaki et al. (*22*) have characterized seven bitter peptides in peptic hydrolysates of soybean proteins. Almost all the bitter peptides had leucine at the N or C termini, and the bitterness of the peptides could be reduced by treatment with exopeptidases such as carboxypeptidase A.

Another approach to debittering enzymic digests is the formation of plasteins from the peptide fragments by reversal of the proteolytic process (*23, 24*). Fujimaki et al. (*23*) envisioned a process whereby off-flavors can be removed from enzymic digests of soybean proteins by solvent extraction. The resultant hydrolysate containing bitter peptides can be debittered by treatment with a second proteolytic enzyme to reverse the hydrolytic process yielding high molecular weight plasteins. Specifically, a strongly bitter peptic hydrolysate can be neutralized to pH 7, concentrated to 20% w/v and incubated with α-chymotrypsin at 37°C for 24 hr. The α-chymotrypsin resynthesizes high molecular weight

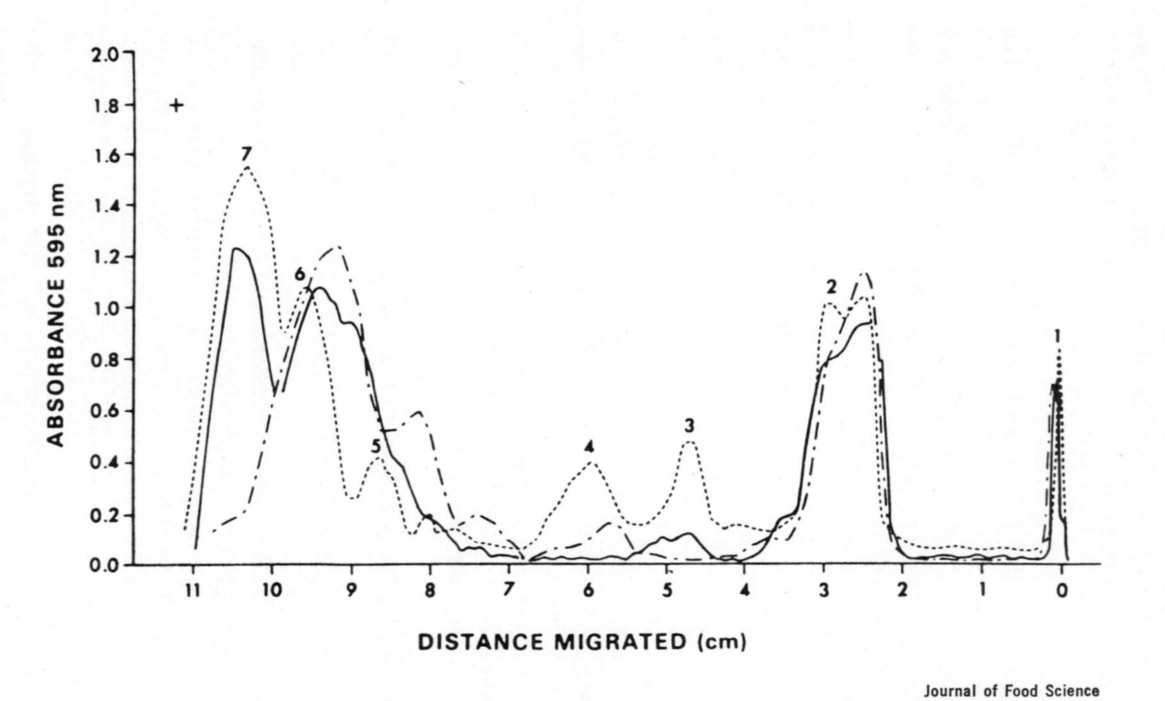

Figure 1. Electrophoretic patterns of enzyme-treated egg albumen after 18 hr of incubation at 34°C. Symbols: · · · · ·, stored control; ——, papain treatment; · — · —, bromelain treatment. For the stored control peaks are identified as follows: (1) immobile protein (still in the slot); (2) conalbumin; (3) globulin G_2; (4) globulin G_3; (5) ovalbumin A_1; (6) ovalbumin A_2; (7) ovalbumin A_3. (25)

plastein polymers from the polypeptides. The polymeric material is virtually devoid of bitter flavor.

There have been a limited number of studies on the effects of enzymic modification of protein concentrates on functional properties other than solubility. Studies on functional properties, as modified by enzymic treatments, emphasize foam formation and emulsifying characteristics of the hydrolysates. Treatment of chicken egg albumen alters the functional properties of the egg proteins in terms of foam volume and stability and the behavior of the proteins in angel food cakes (*25*). Various proteolytic enzymes were used to degrade the egg albumen partially. However, proteolytic enzyme inhibitors indigenous to the egg proteins repressed hydrolysis of the egg proteins compared with casein.

Table II. Performance of Enzyme-treated Egg Albumen on Foam Volume and Stability (*25*)

	Foam Volume[a]		*Foam Stability (ml)*[b]	
Treatment	*ml*[c]	*Ratio T/C*[d]	*1 hr*	*2 hr*
Fresh control	590	0.94	$<$ 10–10	13–22
Stored control	630		20–49	35–60
Papain	669	1.06	68–72	80–82
Ficin	694	1.10	70–73	82–84
Trypsin	715	1.13	51–64	66–75
Bromelain	719	1.14	64–75	74–84
Protease (*A. oryzae*)	729	1.16	60–72	76–83

[a] Each value is a mean of two replications; each replicate consists of three observations.
[b] Each range is derived from two replications; each replicate consists of three observations.
[c] Means connected by the same vertical line are not significantly different, $P < 0.01$.
[d] T/C represents treatment/stored control.

Journal of Food Science

Since proteolytic enzyme inhibitors are rather widespread in nature (*18*), the phenomenon of repressed proteolysis in the modification of protein functionality could be of general significance.

As evidenced by gel electrophoretic patterns (Figure 1) the proteolytic enzymes modified some, but not all, of the components of the egg albumen system. The globulin fractions appeared to be especially susceptible to the action of proteolytic enzymes. Changes in the ovalbumin region were also evident. The results of enzyme treatments on foam volume and stability of whipped egg albumen are presented in Table II. The trypsin, bromelain, and protease treatments produced a significantly greater volume of foam than did the stored control. However, all enzyme treatments resulted in inferior foam stabilities. Both the rate and amount of foam collapse were greater in the enzyme-treated albumen samples. Because of the importance of the globulins in the foaming of egg whites (*26*), the proteolysis of the globulins in this study may have resulted in certain inferior properties of the foams.

Performance of the enzyme-treated albumen in angel food cakes is summarized in Table III. Cake volumes and ratios of treatment to stored controls for all enzyme treatments were comparable to or better than the stored control. However, the cakes made from the enzyme-treated albumen had a coarse, gummy texture. Also, off-odors and flavors could be detected in the cakes prepared from albumen treated with bromelain and protease.

Treatment of a soy protein isolate with a neutral protease from *A. oryzae* had a marked effect on its foaming and emulsification (*20*).

Table III. Performance of Enzyme-treated Egg Albumen in Angel Food Cake Volume (25)

	Cake Volume[a]	
Treatment	*ml*[b]	*Ratio T/C*[c]
Stored control	466	
Papain	467	1.00
Fresh control	476	1.02
Bromelain	482	1.03
Trypsin	500	1.07
Protease (*A. oryzae*)	504	1.08
Ficin	509	1.09

[a] Each value is a mean of three replications; each replicate consists of three observations.
[b] Means connected by the same vertical line are not significantly different, $P < 0.01$.
[c] T/C represents treatment/stored control.

Journal of Food Science

Partial hydrolysis resulted in increased emulsion capacity and decreased emulsion stability. Proteolysis of the soy protein isolate yielded whipped hydrolysates with increased foam volume over the untreated controls, but the foam stabilities of whipped hydrolysates were nil.

The whipping properties and emulsification capacity of whey proteins treated with pepsin, pronase, or Prolase (EB-21) (a bacterial protease) have been studied recently (*27*). At their optimum pH values, the Prolase was least proteolytic and the pronase most proteolytic based on release of α-amino nitrogen and gel filtration patterns. Pepsin preferentially attacked the β-lactoglobulin component of the whey proteins yielding large polypeptide fragments. The emulsifying capacity of the whey proteins, defined as the amount of oil (grams) emulsified per mg protein, decreased with increasing proteolysis. At no time were the emulsifying capacities of the hydrolysates as good as the zero time control for the whey proteins. This indicates that proteolysis was detrimental to emulsifying capacity. In all enzymic treatments the specific volumes of foams (ml/g protein) prepared from hydrolysates increased over the untreated controls after 30 min of proteolysis. However, continued proteolysis by pepsin and pronase resulted in specific volumes of foams less than the control. Whey proteins treated with Prolase, however, retained a greater specific volume of foam compared with the controls over the 4-hr treatment period.

Although the volume of foam was increased by limited proteolysis, the stabilities of the foams, defined as the time required for one half of the weight of the foam to drain from the foam as free liquid, were greatly decreased by this same limited proteolysis. Presumably the increased initial polypeptide concentration in the hydrolysates favors more air incorporation. However, the polypeptides apparently do not have the surface activity required to give a stable foam. The decrease in foam stability becomes evident in the first 30 min of the enzyme reaction. Further hydrolysis results in peptides which lack the ability to stabilize the air cells of the foam. Thus a limited hydrolysis may be advantageous for utilizing whey proteins in foams since the specific volumes of the foams were increased 25% by such treatment. The decrease in foam stability which results from limited hydrolysis can be compensated for by adding stabilizers such as carboxymethyl cellulose (*19, 27*).

The emulsion stability and foaming properties of rapeseed protein concentrate (RPC) after digestion by pepsin or papain have been determined (*19*). In general, the foam volume of the whipped enzymic hydrolysates was 40–70% higher than that of the original RPC. Furthermore the foam stabilities or foam persistence of the enzymic hydrolysates were greater than the original RPC. For example, a pepsin hydrolysate digested 24 hr, but containing carboxymethyl cellulose (CMC) used as a peptide precipitant, yielded a foam more than twice as stable as the original RPC.

Emulsion stabilities of the peptic digest containing CMC were 6–11 times more stable than the original RPC. By comparison the emulsion stabilities of the peptic digest without CMC were only 2½–3 times that of the original RPC. Heating the emulsion at pH 5.5 had little effect on the emulsion stability, an important factor for sausage emulsions. Furthermore the same peptic–CMC hydrolysate possessed about 25% more swelling capacity than the original RPC. Obviously, the presence of a stabilizer such as CMC has a marked effect on swelling and the emulsion and foam stabilities of rapeseed protein hydrolysates. Either the CMC precipitated an especially surface active group of polypeptides or the presence of CMC resulted in a greater combined effect with the polypeptides in the hydrolysate.

The effect of enzymic modification on the foaming and emulsification properties of fish proteins has been studied in several laboratories (*8, 14, 28*). Hermansson et al. (*8*) observed that solvent-extracted FPC modified with an alkaline bacterial protease yielded a whipped foam volume approximately 70% greater than untreated FPC. However, the stability of the foam from the enzyme-modified FPC was about 50% less than that of the untreated FPC. The foam volume of the enzyme-modified FPC was essentially equal to that of egg white, but the foam stability of the FPC hydrolysate was substantially less.

Spinelli et al. (*14*) determined the emulsion stability and emulsion capacity of polypeptides recovered from hydrolysates of fish myofibrillar protein using hexametaphosphate. In general the emulsion capacity and emulsion stability increased through 30 min of proteolysis but then declined. Even the unhydrolyzed protein–phosphate complex yielded better emulsion stability and capacity values compared with sodium caseinate. The effect of the residual hexametaphosphate in the hydrolysate conceivably could have a beneficial effect on emulsifying characteristics of the modified fish myofibrillar proteins.

Groninger and Miller (*28*) succinylated myofibrillar fish protein and then treated the succinylated proteins with bromelain to obtain an acylated mixture of polypeptides. Both the extents of succinylation and enzymic hydrolysis affected the volume of foam of the whipped protein. Optimum foam volume and stability occurred when 54% of the ϵ-amino groups of lysine were acylated and the succinylated protein was mildly hydrolyzed at pH 7, 25°C for 10 min using a bromelain:protein ratio of 1:100 (w/w).

Table IV shows the effect of protein concentration on foam volume and stability compared with egg white and soy isolate. Foam volumes for the hydrolyzed, succinlyated fish protein were highest at 3 g protein/100 ml, but decreased at 4 g/100 ml. Compared with egg white and soy isolate, the hydrolyzed succinylated fish protein yielded a more stable, but less voluminous foam.

The modified fish protein formed foams over the pH range of 3.0–9.0. In the isoelectric range at an approximate pH of 4.5 the foaming properties of the modified proteins were not greatly affected, although the volume and stability of the foam were decreased slightly. Additives affected the volume and stability of foams prepared from the modified protein. Salt up to 2% w/v increased the foam volume, but decreased foam stability. Sucrose at 25% w/v did not affect foam volume, but substantially increased the stability. Generally, oils added before or after whipping had an adverse effect on the foam system. However, the

Table IV. Effect of Protein Concentration on Foam Stability and Comparison of the Stabilities of the Foam of Egg White and Soy with That of Hydrolyzed, Succinylated Protein[a] (*28*)

Concentration (g/100 ml)	*Initial Foam Volume (ml)*	*Stability* — *Volume of Fluid Leakage in Milliliters*			
		Initial	*15 min*	*30 min*	*60 min*
Hydrolyzed succinylated protein					
1	800	2	5	12	30
2	650	0	0	0	2
3	900	0	0	0	0
4	750	0	0	0	0
Egg white					
4	1100	10	29	45	58
Soy isolate	3000	4	28	56	74

[a] Foams were prepared in 100 ml of water.

Journal of Food Science

addition of a surfactant allowed the system to tolerate ca. 5 g of vegetable oil with a slight deterioration in foam volume. Samples of modified dry fish myofibrillar proteins could be stored for four months in air, unprotected from light at ambient temperature without deterioration in aeration or organoleptic properties. The hydrolyzed succinylated protein was incorporated into a dessert topping, a soufflé, and both a chilled and frozen dessert to yield foods that were acceptable to a taste panel. Thus, the modified fish protein was subjected to a variety of test conditions which indicated that the hydrolyzed succinylated proteins were compatible with a wide range of conditions.

Recently Puski (*20*) determined the effects of proteolysis on the water absorption, viscosity, and gelation properties of a soy protein isolate. The water absorption of enzymically hydrolyzed soy protein isolate increased in direct relation to the enzymic treatment. Apparently as the number of polar amino and carboxyl groups increased, the uptake of water increased proportionately. Limited proteolysis of the soy protein isolate resulted in a decided decrease in the viscosity and gelling properties.

Autolytic activities in tissues can have a significant effect on the yields of proteins in the preparation of concentrates from animal and plant tissues. Koury et al. (*29*) observed that fish flesh and viscera contain highly active proteolytic enzymes. In preparing solvent-extracted FPC from hake, these workers observed that if the fish was held for 20 min at 50°C at a pH of 5.5, the yield of FPC decreased ca. 9%. Apparently the endogenous proteolytic enzymes in fish hydrolyze the proteins into subunits that do not coagulate by the 2-propanol extraction method. Autolysis of fish is an important consideration in the preparation of FPC. In a process for the preparation of leaf protein concentrates (LPC) from alfalfa, Huang et al. (*30*) found that treatment of the leaf juices with polar solvents reduced hydrolytic reactions and increased yields of LPC. The action of natural proteolytic enzymes at room temperature for 24 hr reduced the amounts of proteins precipitated from the juice by 17.8–29.4%.

For several reasons, it is difficult to generalize about the foregoing studies. Obviously, the extent of hydrolysis of the various protein concentrates by enzymes depends on variables such as the purity and specific activities of the enzymes used, the time, temperature, pH, reaction volumes, the amount and specificities of enzymes used, and the physical and chemical nature of the substrate. In some studies the activities of the enzymes were standardized against a common substrate, but in others constant ratios of protein to enzymes with varying activities were used. In addition, many of the other variables were not controlled in the same way, so valid comparisons cannot be made among the various studies.

Furthermore, the functionality tests used to evaluate the properties of the modified proteins are empirical and not standardized. Each test seems to have as many variations as there are investigators. A real need in this area of food research is the development of a series of standardized methods for studying the functionality of proteins.

In spite of the above limitations and the relatively small number of studies involved, it is evident that the functionality of proteins as protein concentrates can be modified substantially by enzymes. Solubilities of proteins can be altered (generally increased) by hydrolytic reactions; viscosities and gelation of protein solutions can be decreased dramatically by limited proteolysis; the volume of foams of whipped hydrolysates is generally greater than that of the parent protein, but the foam stability is usually less. Emulsification properties of hydrolysates are affected in different ways depending on the protein. From the very limited number of studies on partial hydrolysis of food proteins it appears that water binding is increased as a result of proteolysis.

Effects of Endogenous and Exogenous Enzymes on the Functionality of Proteins in Selected Food Systems

In the complex environment existing in many food systems the effects of various enzymes on the physical properties of foods are often complicated by a variety of interacting factors. This is especially true where the numerous endogenous enzymes act upon tissue components to alter their physical properties. In addition to the endogenous enzymes, the processor may add exogenous enzymes for a specific purpose. Of course this type of enzyme reaction can be more closely controlled than the array of reactions catalyzed by endogenous enzymes. The following discussion will illustrate the enzymic alterations in the functional properties of proteins in food systems such as meat, bread dough, and milk.

Meat Proteins. The conversion of muscle to meat is characterized by a complex series of biochemical reactions largely involving endogenous enzymes. The ultimate objective of the meat processor is to produce an aged, tender piece of meat. However, before this can happen there occurs a progression of biochemical reactions that converts the muscle tissue to meat. In many cases, the enzymic reactions necessary for the formation of meat are poorly understood. For a detailed discussion of the biochemical reactions in meat chemistry the reader is referred to texts by Lawrie (*31*) and Price and Schweigert (*32*). The purpose of the present discussion is to pinpoint a few of the enzymic reactions which alter the functionality of muscle proteins.

Tenderness is one of the most capricious variables and is also generally acknowledged as one of the most important properties of meat. The enzymic reactions that contribute to meat tenderness are apparently numerous and incompletely understood. Thus at least four separate mechanisms may be at work during the enzymic changes of postmortem aging (*33*). These include: (a) specific biochemical alterations in the myofibrillar proteins, (b) catheptic activity of lysosomal enzymes, (c) degradation of collagen cross-links, and (d) general microbial action. The first three mechanisms involve endogenous enzymes and will be outlined in the subsequent discussion.

Basically, there are three major groups of proteins in muscle tissue: (a) the sarcoplasmic proteins of the muscle cell cytoplasm, (b) the myofibrillar proteins, soluble at high ionic strengths, that make up the myofibril or contractile part of the muscle, and (c) the stromal proteins comprised largely of the connective tissue proteins, collagen, and elastin. The myofibrillar proteins and the stromal proteins are fibrous and elongated; they form viscous solutions with large shear resistance. These properties coupled with other lines of indirect evidence indicate that the physical properties of the myofibrillar and stromal proteins are directly related to the texture and tenderness of meat (*34*).

The myofibril is comprised of relatively few proteins compared with the 200 or more different proteins that exist in a muscle cell. Table V lists the known myofibril proteins of vertebrate skeletal muscle and the approximate amount of each protein in the myofibril. Note that myosin and actin comprise up to 75% of the myofibrillar proteins and together are necessary and sufficient for in vitro contraction. Contraction of

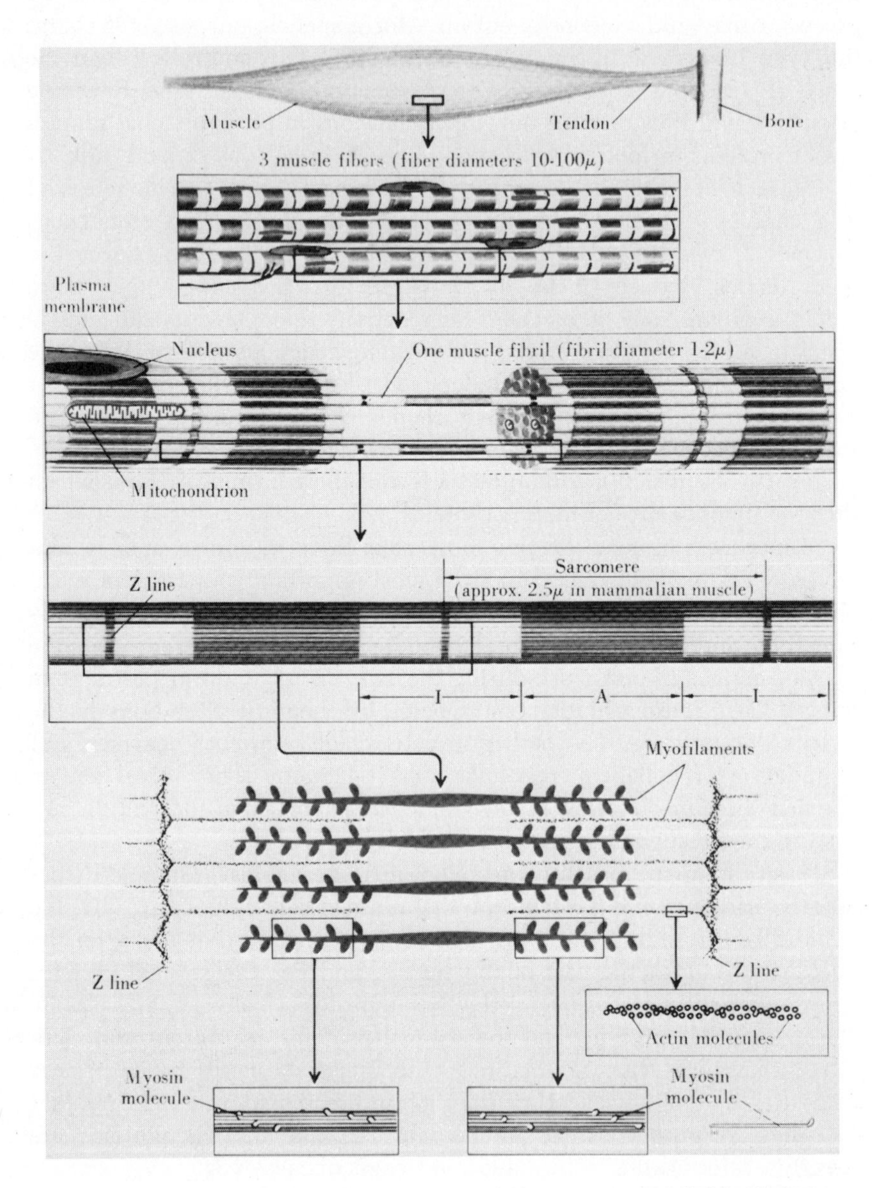

Journal of Food Technology

Figure 2. Schematic showing the structure of mature mammalian skeletal muscle at different levels of organization ranging from an entire muscle (top of figure) to the molecular architecture of the myofilaments (35)

Table V. The Known Myofibrillar Proteins of Vertebrate Skeletal Muscle (34)

Protein		*Wt % of Myofibril*
Myosin	Reconstituted actomyosin.	50–55
Actin	Both necessary and sufficient for in vitro contraction.	15–20
Tropomyosin	Regulatory proteins	5–8
Troponin		5–8
α-Actinin		2–3
β-Actinin		0.5–1
Component C		2–3
M-Line proteins		3–5

Proceedings of the Meat Industry Research Conference

purified actin and myosin is initiated by adenosine triphosphate (ATP) and continues until the ATP is exhausted by hydrolysis. Here is an example of how the enzymic (ATPase) activity of myosin can affect the physical properties of muscle proteins. In the living organism ATP is always present in the muscles; consequently, the regulatory proteins (Table V) are necessary to turn contractions "on" and "off" in the presence of ATP.

The myofibrillar proteins listed in Table V are assembled in a specific array of interdigitating thick and thin filaments as shown in Figure 2. This figure illustrates the structure of skeletal muscle at different levels of organization from an intact muscle at the top of the figure to molecular details at the bottom of the figure. The alternating dark and light bands of the myofibril (fourth diagram from top of Figure 2) as seen under the phase contrast microscope consist of the thick and thin filaments, respectively. Myosin and component C are located exclusively in the thick filament, whereas actin, tropomyosin, troponin, and β-actinin are located in the thin filaments. The thin filaments are attached to the transverse Z-disk which is composed largely of α-actinin.

Muscle contraction is accomplished by the sliding together or telescoping of the interdigitating filaments shown in Figure 2. This sliding mechanism is made possible by the cross-bridges (heads of the myosin molecules) extending out from the myosin molecules (*see* Figure 2). During contraction the cross-bridges cyclically extend outward to attach to the thin filament and then swivel or rotate to translocate the thin filament toward the center of the sarcomere. The cross-bridges then dissociate from the thin filament and prepare to repeat the cycle. Thus, for a single twitch each cross-bridge may perform many cycles of attaching to actin in the thin filament, swiveling, and then dissociating from the actin. In the living system, at any one time only 15–20% of the total cross-bridges in a single sarcomere are attached at any instant, whereas the remaining cross-bridges are in various other stages of the cross-bridge cycle. This asynchronous action of the cross-bridges results in a smooth contractile process. ATP has two roles in muscle contraction. First, the hydrolysis of ATP supplies the energy necessary for muscle contraction. Second, ATP at physiological concentration prevents or dissociates the interaction between myosin and the actin of the thin filament.

In living, resting muscle every myosin cross-bridge is "energized" and contains one molecule of ADP resulting from the hydrolysis of ATP. Thus the energy from ATP hydrolysis is stored in some unknown way in the "energized" cross-bridge. In living, resting muscle the intracellular Ca^{2+} levels are very low ($< 10^{-7}M$) which prevents the actin in the thin filament from interacting with the myosin-ADP "energized" cross-link. Thus in living, resting muscle and in muscle shortly after death, when the ATP has not been depleted, the actin–myosin interaction is not evident, and the interdigitating filaments are free to slide past one another offering minimal resistance; such a muscle is therefore tender.

However, a nerve impulse to the muscle triggers a release of Ca^{2+} from the sarcoplasmic tubular system, where it is ordinarily bound, which increases the intracellular Ca^{2+} concentration to 10^{-5}–$10^{-6}M$. This level of Ca^{2+} allows the actin in the thin filament to accept the energized-ADP-myosin cross-bridge to initiate contraction. As each cross-bridge completes the swivel part of its cycle, it loses the bound ADP and immediately accepts a molecule of ATP that is always supplied to living muscle. The ATP immediately causes a dissociation of the actin–myosin complex, and the myosin catalyzes the hydrolysis of ATP to yield the myosin—ADP energized state again, ready to repeat the cycle.

In physiological relaxation the cycle is interrupted when the Ca^{2+} is rebound and the thin filament is unable to interact with the ADP–energized-myosin. However, in rigor mortis ATP is depleted, and the cross-bridges can no longer dissociate; this results in extended binding of the actin filament to the myosin. Since, in this case, the interacting thin and thick filaments are not free to glide past one another and they remain closely associated, the muscle is inextensible and tough because greater energy is required to sheer the thin–thick filament aggregate and the increased number of cross-links between the thin and thick filaments.

Available evidence indicates that myofibrillar proteins undergo at least two changes that should affect tenderness of meat. First, from the foregoing discussion, formation of actin–myosin cross-linkages can lead to toughness of meat. During the early postmortem period the ATP concentration declines to nearly zero. At the same time the intracellular membranes lose their ability to hold Ca^{2+}, and the intracellular concentration of Ca^{2+} increases to $10^{-5}M$ or greater. Consequently, cross-bridges cycle until the ATP concentration becomes too low to dissociate the cross-links between the thin and thick filaments. The attached cross-bridges begin to accumulate, and when the ATP concentration falls to less than 0.2m*M*, nearly all the cross-bridges cease cycling and remain attached to actin. This increased cross-linking of actin and myosin results in rigid muscles with increased toughness. The foregoing discussion exemplifies a situation in which the ATPase activity of myosin dramatically affects the interaction between actin and myosin which results in a drastic change in the physical properties of the muscle proteins.

Thus, during the first 6–24 hr postmortem, the enzymic depletion of ATP results in rigor mortis with attendant toughness of the muscle. However, as a result of the aging process there is at least a partial reversal of this "actomyosin toughening." The tenderization of meat as a result of aging is a complex and poorly understood phenomenon. A great deal of research effort has been expended in determining the factors which may be important in increasing the tenderness of meat. Some of the factors that may be important in meat tenderization upon aging include: (a) modification of the sulfhydryl groups in the actin–myosin complex, (b) alteration in the actin–myosin complex by the postmortem decline in muscle pH, and (c) gradual degradation of the Z-disk during postmortem storage resulting from a calcium-activated enzyme indigenous to muscle tissue (*34, 36*). Complete disintegration of the Z-disk by the enzyme would result in sarcomere fragments 2.0–2.5 μm long and disintegration of the muscle tissue. However, complete removal of the Z-disk rarely occurs, even after prolonged storage, and electron microscope examination is necessary to detect Z-disk degradation in meat handled through ordinary market channels. However, even partial disintegration of the Z-disk causes weakening of the myofibril, which results in a reduction of the tensile strength of the myofibril.

This endogenous proteolytic enzyme or calcium-activated sarcoplasmic factor (CASF), as it is sometimes called, has been purified and characterized (*34, 37, 38*). The action of purified CASF from rabbit muscle on rabbit muscle myofibrils (*39*) and on purified bovine myofibrillar proteins (*40*) indicates that the CASF attacks primarily the α-actinin component of the Z-disk. Apparently α-actinin is the only myofibrillar protein known to be present in the Z-disk, and it probably acts like a cross-link joining the ends of F-actin fibers in adjacent sarcomeres (*40, 41*). However, the CASF action on α-actinin is rather limited, requiring long incubation times to observe proteolytic activity (*40*). Furthermore, CASF seems to remove Z-disks from myofibrils by means of a very specific proteolytic activity that releases α-actinin without extensively degrading it (*39*). It has recently been shown that the partially purified CASF proteinase functions as a meat tenderizer (*42*). A solution of CASF muscle proteinase was used to reconstitute freeze-dried beef. After storage of the reconstituted meat for five days, the Z-disks of the myofibrils had disintegrated, and the meat when cooked became more tender than a control sample without added proteinase. Thus the CASF proteinase acts as a meat tenderizer. The work with the CASF proteinase illustrates graphically how an endogenous enzyme can alter the functional properties of muscle proteins in meat.

A valuable concept in studying the toughness of meat was proposed by Whitaker, Locker, Marsh, and others (*43, 44, 45, 46*). They proposed that toughness could be resolved into at least two components. These were considered to be "background" toughness related to connective tissue and other stromal proteins and "actomyosin" toughness, which we have just discussed, related to the myofibrillar proteins. This simple concept has in many cases allowed the resolution of seemingly conflicting results. Several studies have implicated the connective tissue content of meat to be highly correlated with toughness of meat. On the other hand, about as many studies have indicated no such relationship (*34*). Consideration of the background toughness concept as only one component of toughness provides a rational means of resolving the apparently conflicting studies. For example, a meat that had a high background toughness would contain relatively high amounts of connective tissue, and there would be a good correlation between toughness and connective tissue. On the other hand, in those muscle tissues where connective tissue content is low or where the actomyosin component is large, correlation of toughness with connective tissue content will be low.

Connective tissue is found throughout the body, but especially in muscle tissue. Connective tissue is a complex matrix of ground substance, proteins, and protein fibers, which include collagen, elastin, reticulin, and mucopolysaccharide protein complexes (*47*). Since collagen is a major component of connective tissue, a great deal of attention has been paid to its relationship to toughness. For example, some muscles contain more collagen than others, and the amount of collagen in various muscles is inversely related to tenderness (*48, 49, 50*). In addition, the percent soluble collagen decreases with an increase in the age of the animal, and this decrease is associated with decreased tenderness of the meat (*50, 51, 52, 53, 54*). Apparently, collagen in older animals becomes more cross-linked, resulting in increased toughness (*55, 56, 57*). Thus it is thought by some researchers in the field (*33, 47*) that enzymic degradation of collagen may play a role in the tenderization of meat that occurs during the postmortem period. In support of this hypothesis, it has been shown (*33*) that during the aging period larger, less charged molecules of intramuscular connective tissue disappear, whereas smaller, higher charged molecules appear. This is illustrated in Figure 3 which shows the influence of aging on the elution patterns of intramuscular connective

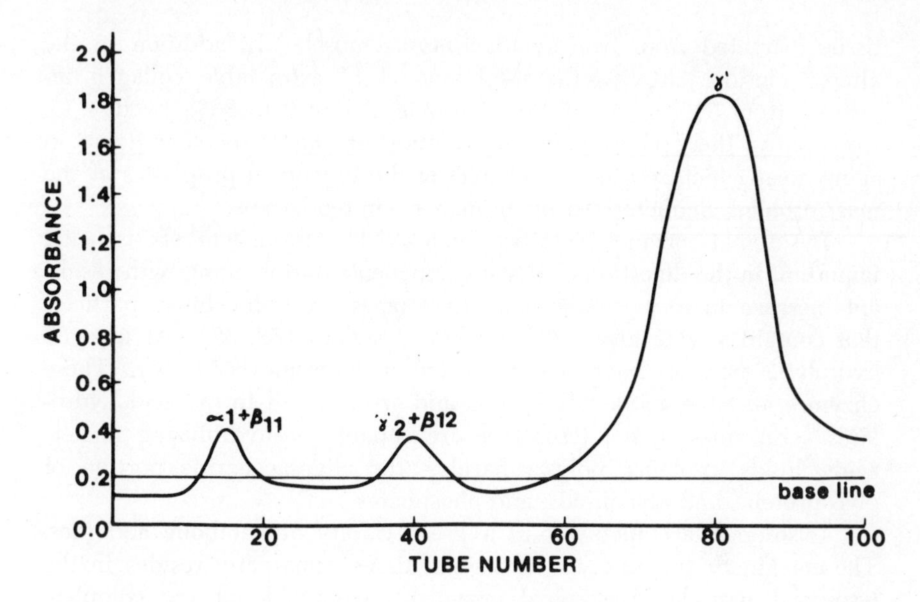

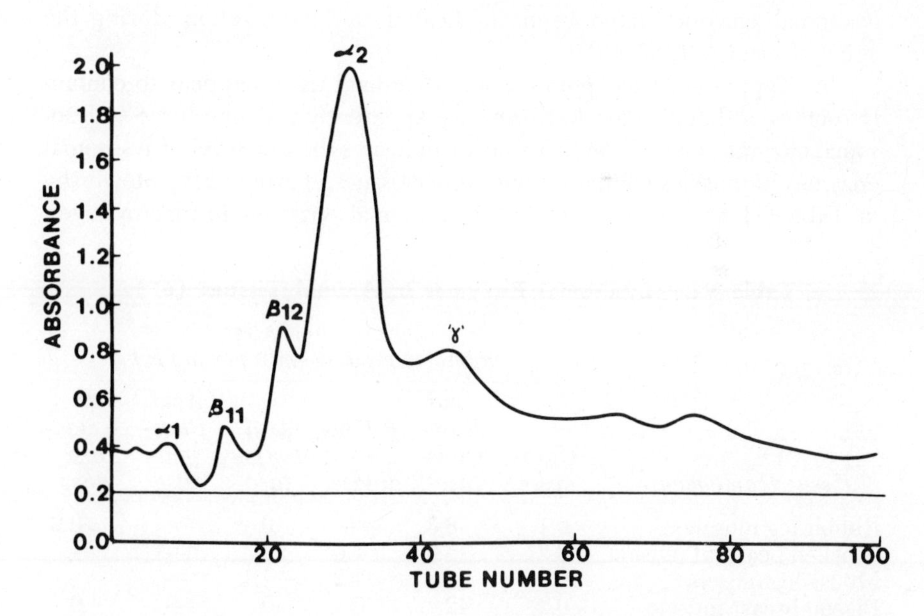

Proceedings of the Meat Industry Research Conference

Figure 3. Typical elution patterns (from carboxymethyl cellulose columns) of intramuscular connective tissue extracted from beef semitendinosus *muscle. A. (above) unaged; B. (below) 13 days postmortem* (33).

tissue extracted from beef *semitendinosus* muscle. In addition to the altered elution patterns, the total amount of extractable collagen increased from 59% of the total in muscle at day 0 to 88% by day 13. Apparently, there is enzymic degradation of the connective tissue in aging meat which results in a change of the functional properties of the meat proteins characterized by an increase in tenderness.

Lysosomal enzymes have been invoked by various workers as being important in the digestion of tissue components during aging with resultant increase in meat tenderness. Lysosomes are sub-cellular particles that contain a wide array of hydrolytic enzymes (*58, 59*). At least 37 hydrolytic enzymes have been detected in lysosomes (*58, 59*). These enzymes all have acidic pH optima and are referred to as "acid hydrolases." Enzymes of the lysosomes are capable of hydrolyzing nucleic acids, lipids, proteins, polysaccharides, the oligosaccharide portion of glycoproteins and glycolipids, and phosphates.

Lysosomes are involved in a great variety of catabolic activities. The machinery for autolysis of animal tissue apparently resides in the lysosomal particle. Lysosomal enzymes are capable of the complete hydrolysis of proteins to amino acids (*59*). The hydrolytic reactions of lysosomal enzymes have been implicated in tenderization during the aging of meat (*31, 47, 60–66*).

As Tappel (*67*) has pointed out, all animal tissues appear to contain lysosomes, although muscle tissues are apparently a poor source of lysosomal enzymes (*60, 67, 68*). Table VI indicates the low level of lysosomal enzymes in muscles compared with other tissues. Particularly noteworthy in Table VI are the high levels of lysosomal enzymes in macrophages.

Table VI. Lysosomal Enzymes in Animal Tissues (*67*)

	Specific Activities (mμmoles product/min per mg N)					
Tissue Homogenate	*Cathepsin*	*Acid Ribonuclease*	*β-Glucurondidase*	*Aryl Sulfatase*	*Acid Phosphatase*	*β-Galactosidase*
Rabbit leg muscle	0.1	8.3	—	0	—	0.2
Chicken pectoral muscle	0.21	5.0	1.6	0	—	—
Mouse leg muscle	0.8	43	4.2	1.1	—	—
Pigeon breast muscle	0.2	42	0	0	—	0
Rabbit liver	1.1	33	—	4.9	—	31
Pigeon liver	0.9	37	126	9.8	12	5.2
Sheep liver	3.6	113	201	82	25	95
Sheep spleen	0.6	170	640	19	10	7.8
Sheep lung macrophages	12	248	497	181	59	2730

In fact, the lysosomal enzymes in muscle tissue may be derived solely from invading macrophages (phagocytic cells). There is controversy concerning whether muscle actually contains lysosomes since a number of morphological studies failed to demonstrate these particles in muscle tissue.

However, consideration of the sedimentation characteristics and equilibrium densities of particles containing acid hydrolases has led Stagni and de Barnard (*69*), Milanesi and Bird (*70*), and Canonico and Bird (*71*) to conclude that normal skeletal muscle from rats and fish contains two distinct sources of lysosomes. One group is apparently derived from macrophages and other connective tissue cells, whereas a second group represents lysosome-like particles from muscle cells. The group of lysosomes from muscle cells appears to be the larger of the two groups and is characterized by relatively larger amounts of cathepsin D and acid phosphatase.

Canonico and Bird (*71*) feel that lysosomes in muscle tissue exist in a morphological configuration that is different from other tissues; this may explain the failure of some researchers using microscopic techniques to observe lysosomes.

The importance of muscle lysosomal enzymes to the food scientist stems from their apparent involvement in the aging of meats. For example, lysosomal cathepsins are possibly involved in the proteolytic degradation of muscle proteins (*31*). Lysosomal enzymes exhibit latency; that is, they are retained in the lysosomal particle and released only when the particle membrane is damaged. In this regard the lysosomal enzymes are liberated and activated when the particle membranes are weakened by the postmortem drop in pH. Also, lysosomes are very subject to cryoinjury (*72*), and freezing and thawing of tissues such as muscle releases lysosomal enzymes resulting in autolysis.

There are four major proteolytic catheptic enzymes apparently associated with the lysosomal particles in muscle tissues (*73, 74*). These are:

(a) Cathespin A—an exopeptidase which resembles carboxypeptidase A in specificity.

(b) Cathepsin B—resembles papain in specificity.

(c) Cathepsin C—requires chloride ion; has unique specificity; catalyzes successive cleavage of susceptible dipeptides from the amino terminal end of various peptides.

(d) Cathepsin D—an endopeptidase with a specificity similar to pepsin.

Cathepsin D seems to be the major protease present in muscle (*67*).

Lysosomal enzymes have been studied in the muscles of chickens (*68, 74, 75, 76, 77, 78*), guinea pigs (*79*), rats (*69, 80*), beef (*47, 60, 61,*

63, 69, 81, 82, 83), lamb (*84*), fish (*70, 84–89*), crab (*88*), shrimp (*90*), and ham (*91*).

In general, in beef muscle free catheptic activity reaches its maximum between four and six days postmortem (*81, 92*). At the same time major textural properties of meat show large changes between four and 13 days postmortem (*33*). Lysosomal enzymes have been implicated in the tenderization of meats by a number of workers.

Valin (*61*) has observed the release of lysosomal enzymes in beef muscle during the aging of meat. The free hydrolytic activities were measured in the sarcoplasmic proteins of press juice obtained by ultracentrifugation of *sternoephalicus* muscle fragments in a slightly hypertonic sucrose solution buffered at pH 7.4. Free specific activities of the enzymes increased during aging of meat, but eight days postmortem the total free activities were still less than 15% of the total activities.

Ono (*60, 82*) followed the release of β-galactosidase from lysosomes of beef muscles as a function of age. From these studies it was apparent that the release of lysosomal enzymes during aging was essentially complete after four days. Furthermore, when muscle fiber bundles were soaked in crude cathepsin, the enzyme(s) produced textural and structural alterations in fresh tissue resembling those found in naturally aged meat under the same conditions of pH and temperature (*93*).

Kim (*94*) has demonstrated the degradation of myofibrillar proteins purified from porcine *semimembranosus* muscle by porcine leukocyte lysosomal proteinases. Troponin treated with leukocyte lysosomal proteinases mixed with tropomyosin did not result in the normal increase in viscosity of this system. The emulsifying capacity of actomyosin treated with lysosomal proteinases at 37°C and pH 7 for 12 hr was higher than that of actomyosin incubated without enzymes. On the other hand, actomyosin treated with papain had low emulsifying capacity because of extensive degradation. Control actomyosin incubated at 37°C, pH 7 for 12 hr formed a gel after removal of KCl, but actomyosin treated with lysosomal proteinases or papain did not gel.

Although there is no extensive proteolysis of connective tissue proteins in aged meats (*31*), certain cross-links in the collagen molecules are apparently broken. A collagenolytic cathepsin, present in low concentrations in muscular tissue of the rat, has been shown to operate optimally at pH 3.5. At this pH and at 28°C the cathepsin cleaves the telopeptide region of insoluble collagen, liberating α chains (*95*). Its pH optimum would prevent it from acting freely outside cells, but it may operate by some mechanism involving the local establishment of low pH environments.

Dutson and Lawrie (*63*) implicated lysosomal enzymes in the aging of beef. Samples of the *M. longissimus dorsi* were taken from beef carcasses 1 hr, 24 hr, 5 days, 7 days, and 14 days postmortem. Tenderness (shear force), the amount of protein and hydroxyproline in filtered muscle homogenates, the activity of β-glucuronidase in muscle fractions, and the electron histochemical determination of acid phosphatase were measured at these postmortem times.

Shear force was a maximum at 24 hr and decreased thereafter to a value less than that of the 1-hr samples by 14 days. The amount of protein and hydroxyproline in filtered muscle homogenates was a minimum at 24 hr and increased to a maximum at 14 days. The specific activity of free β-glucuronidase (a lysosomal enzyme) increased with postmortem aging, while the specific activity of bound (sedimentable) β-glucuronidase decreased with postmortem aging, suggesting that this enzyme is released during the conditioning period. Electron histochemical determination of acid phosphatase activity showed the activity to be localized in specific areas around the I-band at early postmortem times; the activity was more dispersed throughout the cell at later postmortem times. These workers suggested that the β-glucuronidase hydrolyzed glucose–galactose moieties that are known to exist in collagen thus exerting a tenderizing effect. Recently Dutson (*47*) has demonstrated that the addition of β-glucuronidase to homogenized bovine *longissimus* muscle increased the solubility of collagen in 48-hr postmortem samples.

Melo et al. (*91*) observed that catheptic enzymes in cured hams of different aging periods remained high up to two months but decreased at both four and six months; presumably this was caused by their denaturation from the high salt concentration in the cured hams. These authors suggested that catheptic enzymes are involved in the production of increased free amino acids during the aging of country style hams. Thus endogenous lysosomal enzymes may attack connective tissue and myofibrillar proteins to alter their physical properties and increase tenderness during the early postmortem period.

Based on model studies with bovine spleen catheptic enzymes, West et al. (*96*) have suggested that cathepsins may be involved in the onset of rigor mortis by degrading the sarcoplasmic reticulum resulting in decreased binding capacity for Ca^{2+}. The free Ca^{2+} would then be free to initiate rigor mortis.

Exogenous enzymes have been used since antiquity to tenderize meat. This subject is reviewed thoroughly by Lawrie (*31*). Today, commercial tenderizers contain one or several proteolytic enzymes such as papain, bromelain, ficin, Rhozyme (*aspergillus* spp.), subtilisin, and pronase. Pancreatin collagenase (*97*) and zingibain from ginger rhyzome (*98*) have also been suggested as meat tenderizers. In early work on the effects of tenderizing enzymes on ground beef muscle, Miyada and

Table VII. Percent of Meat Fractions Solubilized by Treatment with Enzymes[a] (*101*)

Enzymes	*Meat Fractions*[b]		
	Water-soluble (%)	*Salt-soluble (%)*	*Insoluble (%)*
Bromelain	16	33	51
Collagenase	13	26	61
Ficin	9	54	37
Papain	25	60	15
Rhozyme P-11	34	45	21
Trypsin	19	38	43

[a] Data refer to the non-heat-coagulable nitrogen in solution after enzyme treatment, as a percent of the total nitrogen of the sample.
[b] To facilitate comparisons between enzymes, data have been adjusted to enzyme levels which would produce the same total amount of solubilized nitrogen from the three fractions.

Journal of Food Science

Tappel (*99*) reported on the hydrolytic activity of various enzymes. They found that the relative hydrolysis of muscle proteins increased in the following order: pepsin, Rhozyme A-4, ficin, papain, bromelain, protease 15, Rhozyme P-11, and trypsin. Wang et al. (*100*) observed that papain was twice as active as ficin towards elastin, a minor component of connective tissue. Ficin and bromelain had equal enzyme activity towards collagen, a major component of connective tissue. Kang and Rice (*101*) studied the effects of various tenderizing enzymes on water-soluble sarcoplasmic proteins, salt-soluble myofibrillar proteins, and the insoluble stromal proteins. Table VII tabulates the results of some of these studies.

Numbers in any one line of Table VII thus reflect the relative degree of solubilization of the several types of protein in meat by an enzyme

Table VIII. Activity of Papaya Proteases on Native

	Water-Soluble Muscle Protein	
	Native	*Denatured*
	(Nitrogen)	
Papain	37	68
Chymopapain	31	47
Papaya peptidase A	28	50
Crude enzyme	26	50

[a] Increase in TCA soluble nitrogen or hydroxyproline per mg of enzyme protein per minute at 40°C.

acting under the conditions specified. In a similar manner, numbers in any one column reflect the relative effectiveness of the several enzymes for each type of protein. However, values on the different lines do not represent relative potencies per gram of enzyme—the amounts of enzymes were adjusted to give solutions similar in overall hydrolytic ability. As indicated in Table VII, the three fractions of meat show a varying degree of susceptibility to enzyme degradation, and also each enzyme has its specific pattern of hydrolytic effect. All of the enzymes tested in this experiment showed substantial activity on the salt-soluble and -insoluble fractions and variable activity on the water-soluble fraction.

Recently, Kang and Warner (*102*) fractionated crude papain into crystalline papain, chymopapain, and papaya peptidase A. Chymopapain constituted the major component of the crude papain, and it proved to be the most thermostable at pH values between 5 and 9. Table VIII shows the activity of the various purified enzymes on native and heat-denatured meat fractions after subtracting blank values. All three enzymes degraded meat fractions, but with somewhat different hydrolytic potencies. Chymopapain showed activity comparable to papain and papaya peptidase A on meat fractions.

The water-insoluble fraction was more readily digested than the water-soluble fraction. Heat denaturation naturally made meat proteins more readily digested by these enzymes, and the effect of denaturation was more significant in the water-soluble fraction than the other fraction. As expected, hydrolysis of the native connective tissue by these enzymes as indicated by increased hydroxyproline was extremely low. Connective tissue after heat denaturation was readily hydrolyzed. Papaya peptidase A showed about a fivefold increase while papain and chymopapain increased their hydrolyses seven- to tenfold.

and Heat-Denatured Meat Fractions at 40°C (*102*)

Water-Insoluble Muscle Protein			
Native	*Denatured*	*Native*	*Denatured*
(Nitrogen)		*(Hydroxyproline)*	
(μg/ml)[a]			
77	80	0.41	4.0
70	82	0.58	4.4
57	69	0.56	2.7
67	81	0.43	4.0

Journal of Food Science

Table IX. Activity of Papaya Proteases on Meat Fractions at

	Water-soluble (μg/ml) Nitrogen		
	40°C	*60°C*	*Programmed*
Papain	37	189	4176
Chymopapain	31	171	3450
Papaya peptidase A	28	136	3027
Crude enzyme	26	111	3219

[a] Activity of 1 mg enzyme at 40° or 60°C per minute was determined by increase in TCA soluble nitrogen and hydroxyproline.

Pre-slaughter injection of the live animal has proved to be the most effective method of introducing proteolytic enzymes into meat so that they penetrate uniformly into the furthest interstices of the tissue (*31*). Oxidized papain (*103*) and proteases whereby the sulfhydryl group in the active site is reversibly blocked (*104*) have been injected into the jugular veins of animals approximately 30 min before slaughter to distribute the enzyme in the tissues. Subsequently, in the reducing environment of the meat, the protected enzymes are activated. However, the enzymes are apparently active only between 50°C and 82°C or while the meat is cooking.

In studying the mechanism of action of enzymes injected antemortem Kang and Warner (*102*) found that ^{14}C- and ^{3}H-labeled, purified proteolytic enzymes were uniformly distributed throughout the muscles when injected into live rats. Apparently, a significant proportion of injected enzyme penetrates the capillary membranes into the extravascular system to exert a tenderizing effect when the meat is cooked. Much of the injected enzyme, however, is lost with the blood. The activity that the tenderizing enzymes exert on the muscle tissue as a function of temperature is illustrated in Table IX. Table IX compares the activity of papaya enzymes at 40°, 60°, and a programmed temperature ranging from 25° to 70°C. The programmed conditions simulate the cooking of meat.

Incubation at 60°C showed higher enzyme activity than at 40°C. With the high temperature and programmed conditions, the increase of breakdown was highest with connective tissue as shown by hydroxyproline increase in the TCA soluble filtrates. Papaya peptidase A, which showed highest specific activity on casein and BAEE, proved to be least effective among three papaya proteases in hydrolyzing meat proteins. These results suggest that the use of meat or meat protein fractions should serve as a better substrate for predicting tenderizing potency of these enzymes than the substrates commonly used for protease assay.

40°C,[a] 60°C,[a] and Programmed Temperature[b] (25°–70°C) (*102*)

Water-insoluble (mcg/ml)					
Nitrogen			*Hydroxyproline*		
40°C	*60°C*	*Programmed*	*40°C*	*60°C*	*Programmed*
77	260	8624	0.41	17.8	344
70	217	7262	0.58	13.7	356
57	190	6047	0.36	13.3	302
67	222	5747	0.43	11.9	369

[b] Total activity of 1 mg enzyme in 38 min period of programmed temperature from 25° to 70°C.

Journal of Food Science

Collagen in connective tissue undergoes unfolding of the triple helix structure in the region of 55°–65°C, rendering itself more susceptible to enzymic degradation. The thermal shrinkage temperature is affected by interaction with electrolytes; therefore the hydroxyproline release in samples incubated at 60°C and programmed 25°–70°C showed a much greater degree of hydrolysis than was obtained at 40°C. The results indicate that papaya proteases degrade meat above and below 60°C, but they degrade connective tissue only after it is denatured.

The extent of shortening that a muscle, free to shorten, undergoes as rigor mortis develops is a direct function of the temperature down to about 14°–19°C (*31*). Below 14°C there is an increasing tendency for this muscle to shorten further. This "cold-shortening" phenomenon is associated with very tough meat upon cooking. Rhodes and Dransfield (*105*) have demonstrated that the toughening of lamb as a result of cold-shortening could be prevented in carcasses given an antemortem injection of papain solution at the normal commercial dose level.

Thus it is quite evident that exogenous proteolytic enzymes alter the functionality of meat proteins, i.e. increase tenderness, by a more or less random attack on the various muscle proteins. However, this alteration occurs primarily during the cooking period because of denaturation of substrate proteins and the effects of increased temperature on enzyme activity.

Exogenous proteolytic enzymes have been shown to alter other functional properties of bovine muscle tissue as well. Du Bois et al. (*106*) studied the effects of proteolysis on emulsification capacity of bovine muscle proteins. Figure 4 shows the effects of papain treatment (1 mg/g protein) on a muscle homogenate prepared from bovine *Longissimus dorsi* five days postmortem. After 5 hr of proteolysis there was a 39% increase in the amount of lipid that could be emulsified. A positive relationship existed in the first 5 hr between the soluble protein

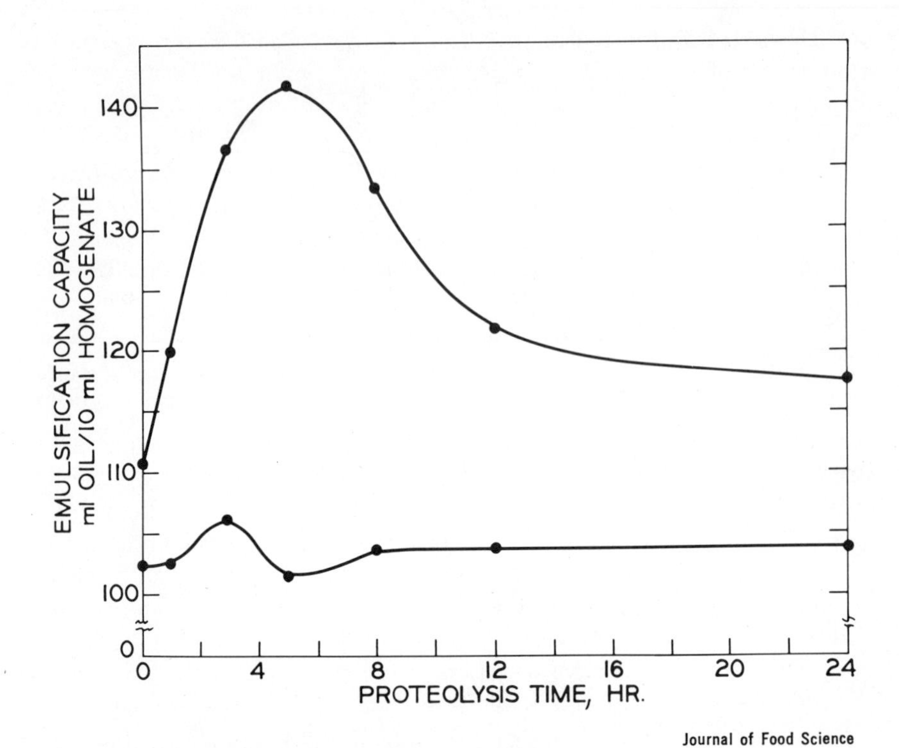

Figure 4. Emulsification capacity of papain-treated and control homogenates. These data are mean values of duplicate determinations on four samples. Bottom curve: control. Top curve: papain treatment. (106)

concentration and emulsion capacity, but after the 5 hr the relationship is negated as the emulsification capacity declines. Perhaps the most attractive explanation for this is that as the mean molecular size of the proteins decreases because of proteolysis an optimum mean size is reached and then passed.

Effects of proteolytic enzymes on the cohesion of chunks of poultry meat have been studied to determine the mechanism of heat-initiated cohesion (*107*). All the proteolytic enzymes tested showed a decrease in binding scores in the following order: collagenase, ficin, elastase, protease, papain, trypsin, and chymotrypsin. The involvement of stromal proteins in binding is suggested.

In general, the treatment of meat with exogenous proteases increases the water-binding activity of the meat proteins (*108*). Increased water binding by meat after treatment with ficin (*109*), trypsin (*110*), microbial proteases (*111, 112*), and fungal proteases (*113*) has been reported. In addition, the brining of beef and pork in brines containing papain, ficin, or bromelain (*114*) resulted in higher water binding.

Hamm (*108*) suggested that this increased water binding of meats as a result of proteolytic enzymic treatment might retard the loss of juices during the heating of meats, particularly canned hams. However, this has not proven to be the case (*115*). Apparently, enzymic action of short duration results in increased water binding by meat proteins; however, the longer and greater enzymic activity resulting from heating means a decrease in water binding capacity.

In general, it is thought that the increased activity of endogenous enzymes resulting from higher temperatures during cooking also plays a significant role in the tenderizing of meats (*66*).

Thus we have seen that endogenous and exogenous enzymes can markedly affect such physical properties of meat proteins as toughness or tenderness, emulsification properties, cohesion, and water binding capacity.

Dough Proteins. Wheat flour dough is a viscoelastic system, the rheology of which is very important in the preparation of bread. The rheology of bread dough can be affected by both exogenous and endogenous enzymes which directly or indirectly affect the physical properties of the dough proteins. The present discussion is limited to: exogenous proteases which exert a direct effect on dough proteins to modify their rheology, exogenous soybean lipoxygenase which seems to act more or less indirectly to modify the physical properties of dough proteins, and endogenous ascorbic acid oxidase and dehydroascorbic acid reductase which exert an effect on the dough proteins in the presence of ascorbic acid.

Endogenous proteases in wheat flour and the use of exogenous proteases in the preparation of bread have been reviewed by Barrett (*116*). Apparently the native proteases of wheat do not play a significant role in baking. They do not modify gluten, nor do they perceptibly affect the use of milk proteins in breadmaking. Several protease preparations are approved for use in modifying dough rheology. The principal protease, of fungal origin, is derived from *Aspergillus orzyae.* Bromelain and papain are also approved for use. Addition of proteinases to doughs improves the handling properties of the doughs and the elasticity and texture of the gluten, and increases loaf volumes substantially.

The proteinases are used in the fermentation stage of bread preparation to allow maximum contact with the flour proteins. Apparently the basic chemistry involved in the rheological modification of the gluten proteins is incompletely understood and will require additional research. However, the gluten proteins are hydrolyzed and the shorter peptide chains can realign in the dough to modify the protein so that shorter mixing times and less energy input are required for maximum extensibility. Since mixing times are critical in present-day bakery operations, the use of exogenous proteases to control the rheological properties of bread dough is a very important practical application.

Thiol and disulfide groups in the gluten proteins of wheat flour are important determinants of the rheology of dough. As one might expect, the reduction of disulfide bridges in the dough by reducing agents tends to weaken the dough structure, whereas the oxidation of thiol groups with the attendant formation of intra- and intermolecular disulfide bridges has a strengthening effect on the dough. The gross result of the effects of oxidizing agents is a tougher, drier, more extensible dough which produces a loaf of bread with a larger volume, brighter crumb, better texture, and improved appearance (*117*). The importance of thiol and disulfide groups in dough rheology has been reviewed by Jones et al. (*118*) and by Bloksma (*119*).

There are two enzyme systems that are thought to affect the thiol disulfide system of wheat flour dough by virtue of an oxidizing effect. Although these enzymes, soy-flour lipoxygenase and endogenous dehydroascorbic acid reductase, apparently improve the rheology of the dough, the exact mechanisms of their action is not fully understood.

Depending upon the source, lipoxygenase is actually a mixture of isozymes with varying substrate specificities (*120*). Some isozymes are specific only for free polyunsaturated fatty acids containing the 1,4-*cis,-cis*-pentadiene double bond system found in naturally occurring polyunsaturated fatty acids such as linoleic or linolenic. In the case of linoleic acid, the principal products of lipoxygenase activity are optically active 9-D-hydroperoxy-10(*trans*),12(cis)-octadecadienoic acid and/or 13-L-hydroperoxy-9(cis),11(*trans*)-octadecadienoic acid. However, some lipoxygenase isozymes will oxidize polyunsaturated fatty acids such as linoleic when they are tied up as esters (e.g., triglycerides). This distinction in substrate specificity is important to the subsequent discussion.

The addition of soy-flour lipoxygenase to dough results in greater "mixing tolerance" and dough extensibility which means that the dough can be mixed longer and machined without damaging the rheological properties required to make good bread (*117*). When dough without added soy-lipoxygenase is mixed in air, oxygen is absorbed and thiol groups are oxidized (*121, 122*). The flour lipids are apparently involved in the oxidation of thiol groups since defatted flour dough did not lose thiol groups as rapidly when mixed in oxygen or air for over 5 min (*122*). Peroxidized flour lipids or peroxidized linoleic acid when added to dough exerted an improving effect on the dough which was correlated with a decrease in thiol groups (*121, 122*). Addition of exogenous soy-lipoxygenase to the dough resulted in even greater levels of lipid peroxidation (*121*). It was proposed that the endogenous lipoxygenase of wheat and/

or the exogenous lipoxygenase of soy-flour improved the dough rheology as a result of the formation of lipid peroxides which in turn oxidized thiol groups of gluten proteins to disulfides (*121, 122*). However, since the endogenous lipoxygenases in wheat apparently are not sufficient of themselves to maximize the improvement in dough rheology, soy-lipoxygenase is used commercially to improve dough rheology (*117*). Presumably, this stems from the broader substrate specificities of the isozymes in soybeans. For example, the wheat lipoxygenase isozymes will peroxidize primarily free fatty acids (linoleate and linolenate) and their monoglycerides, whereas the soybean isozymes peroxidize more complex lipids such as triglycerides (*123, 124, 125, 126, 127*). Thus there is more substrate available for the soybean lipoxygenases which then provides more peroxidized lipids for the oxidation of thiols.

Although flour lipids are essential for the effects of lipoxygenase on dough rheology, some workers doubt the direct involvement of lipid peroxides in the rheological changes of dough (*128, 129*). The recent work of Frazier et al. (*129*) definitely demonstrates the improvement in rheological properties of bread dough by the addition of purified soybean lipoxygenase over and above any endogenous system. Mixture of the doughs under nitrogen or heat denaturation of the soy-flour lipoxygenase resulted in no improvement of the doughs. Frazier and co-workers observed a release of bound lipids during the development of dough in air and in the presence of soy-flour or purified soybean lipoxygenase together with linoleic acid as a substrate. This latter system resulted in desirable rheological characteristics similar to those observed for soy flour. However, these workers found that enzymically prepared peroxides did not affect the physical properties of doughs mixed under nitrogen. They suggested that the release of dough lipids is linked with structural changes in dough proteins and that the improvement in dough rheology is the result of a coupled oxidation of protein thiol groups mediated by "transient lipid oxidation intermediates" as catalyzed by lipoxygenase. Perhaps the singlet oxygen thought to be involved in the mechanism of action of lipoxygenase (*130*) is coupled with the oxidation of thiol groups. In any event, Mapson and Moustafa (*131*) have demonstrated the coupled oxidation of the thiol group of glutathione in the action of lipoxygenase on polyunsaturated fatty acids.

Paradoxically, ascorbic acid, a reducing agent, acts oxidatively to improve dough rheology in the presence of oxygen (*117*). Apparently the reactive species in the oxidation of thiol groups in dough is the dehydroascorbic acid resulting from the oxidation of ascorbic acid. The proposed mechanism for the action of ascorbic acid is as follows (*132*):

```
 O=C———┐
   |   |
HO—C   |                Ascorbic acid              O=C———┐
   ||  O + ½O2  ——————————————————————————►          |   |
HO—C   |                   Oxidase                 O=C   O + H2O
   |   |                                             |   |
  HC———┘                                           O=C   |
   |                                                 |   |
HO—C—H                                              HC———┘
   |                                                 |
  CH2OH                                           HO—C—H
                                                     |
                                                    CH2OH

ascorbic acid                                     dehydroascorbic acid
```

```
 O=C———┐                                           O=C———┐
   |   |                                             |   |
HO=C   |                 (anaerobic)              HO—C   |
   ||  O + 2RSH  ——————————————————————————►         ||  O + R–S–S–R
HO=C   |                     DHA                  HO—C   |
   |   |                  Reductase                  |   |
  HC———┘                                            HC———┘
   |                                                 |
HO—C—H                                            HO—CH
   |                                                 |
  CH2OH                                             CH2OH

dehydroascorbic acid                              ascorbic acid
```

The ascorbic acid is oxidized by ascorbic acid oxidase in the flour dough to dehydroascorbic acid. In turn, the reduction of dehydroascorbic acid is coupled with the oxidation of protein thiol groups by the enzyme dehydroascorbic acid reductase known to be in wheat (*133*).

The lipoxygenase and dehydroascorbic acid reductase enzyme systems in dough illustrate how the rheology of proteins can be modified by either the products of enzymic reactions or the coupled oxidation of substrate and thiol groups of the proteins.

The thiol–disulfide bonding system is important in the physical properties of a number of foods (*134*). Various enzymes catalyze the oxidation of thiols, the rearrangement of disulfide bonds, and the interchange of thiol–disulfide (*135–141*). The significance of these enzymes, some of which are found in baker's yeast (*137*), in the rheology of bread doughs, in other food systems, or in modifying novel proteins such as the keratins remains to be determined.

Milk Proteins. Milk is a complex biological fluid that contains as the principal constituents a wide array of proteins, the milk carbohydrate (lactose), and a variety of polar and nonpolar lipids. The proteins of milk consist primarily of caseins existing as macromolecular aggregates termed "casein micelles" and the soluble whey or serum proteins.

Table X (*142*) lists the composition of the major proteins existing in milk. Even those proteins listed in Table X are heterogeneous in nature. For example, the α_s group of proteins is comprised of α_{s_0}, α_{s_1}, α_{s_2}, α_{s_3}, α_{s_4}, α_{s_5}. Furthermore, all of the proteins listed in Table X exist as polymorphic genetic variants differing in primary structure and separable by polyacrylamide or starch gel electrophoresis. However, the existence of these variants will not have a marked effect on the ensuing discussion.

The caseins exist in milk as polydisperse aggregates ranging in size from ca. 40 to 220 nm (*3*), but the size distribution of micelles depends upon the method of measurement. These casein micelles scatter light and are responsible for the whitish, opaque nature of skim milk. The casein micelles are also associated with a colloidal apatite comprised of calcium–phosphate–citrate (CPC) which has a stabilizing influence on the micelle structure. The colloidal CPC is in equilibrium with soluble CPC in the milk serum phase and is solubilized as the pH is reduced. Thus, as the pH is reduced to the isoelectric point of the caseins (*4.6*), the colloidal CPC solubilizes, and the caseins precipitate (*143*). This phenomenon should be kept in mind during some of the following discussions.

The arrangements of the various casein components in the casein micelles are not known with certainty. However, a number of models have been proposed for the micelle structure, and this matter is currently the subject of much debate (*144*).

The major casein components differ in their behavior toward the calcium ion (*145*). The α_s group of caseins are very susceptible to precipitation by the calcium ion. Beta casein undergoes a temperature-dependent precipitation in the presence of calcium ion, being soluble at 4°C and insoluble at 35°C. κ-Casein, discovered by Waugh and von Hippel (*146*) in 1956, is resistant to precipitation in the presence of calcium ion, and when mixed with appropriate portions of α_s- and β-casein, it forms complexes or micelles with these caseins and stabilizes

Table X. Approximate Composition of Skim Milk Proteins (*142*)

Protein Fraction		*Approx. % of Skim Milk Protein*
α_s-Casein		45–55
β-Casein		25–35
κ-Casein		8–15
γ-Casein		3–7
β-Lactoglobulin	whey proteins	7–12
α-Lactalbumin	whey proteins	2–5

them against precipitation by calcium ion (*146, 147, 148*). The stabilizing action of κ-casein is specifically destroyed by rennin (*146*). Figure 5 (*149*) illustrates the specific release of nonprotein nitrogen from κ-casein, which is the primary point of rennin attack. Thus it is thought that κ-casein plays a key role in stabilizing the structure of native casein micelles. Indeed, it has become evident that κ-casein plays a central role in the functionality of the milk casein system. In the formation of cheese the casein micelle is destabilized by the action of proteolytic

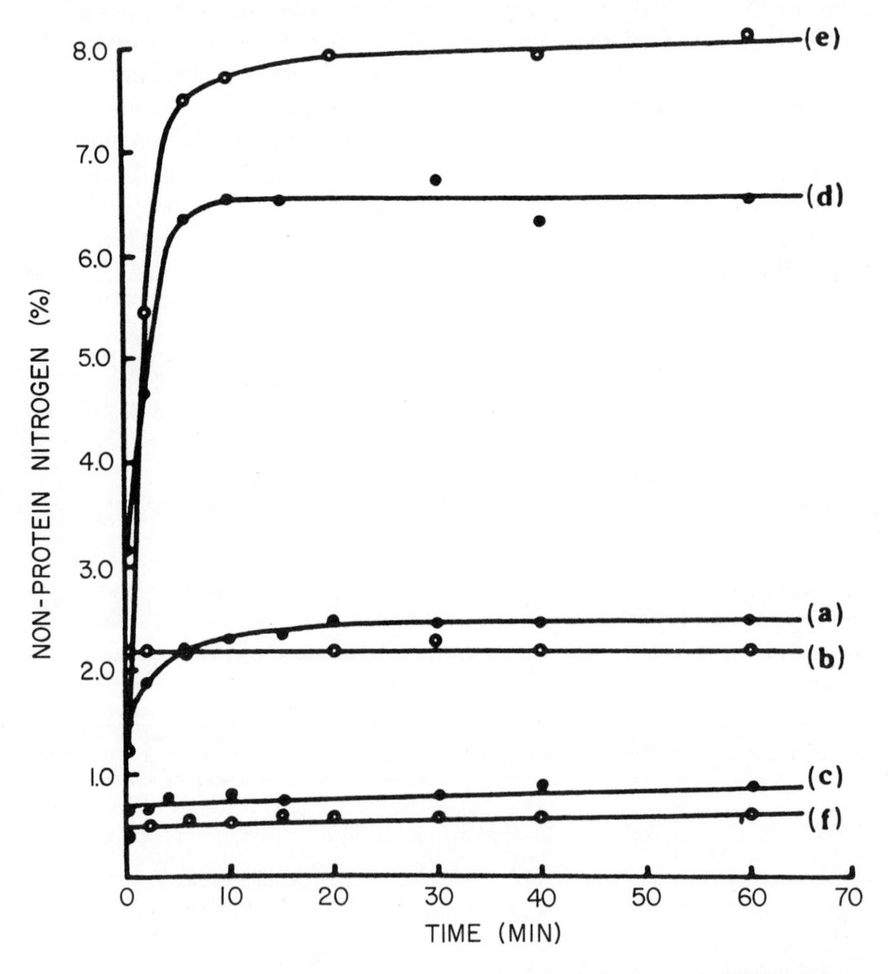

Australian Journal of Biological Sciences

Figure 5. Liberation of nonprotein nitrogen (soluble in 12% TCA) from milk protein fractions by rennin (149). *(a) 2% first cycle casein (largely whole casein); (b) total whey protein; (c) 2% second cycle casein—fraction P (mostly* α_s*- and β-caseins); (d) 1% second cycle casein—fraction S (mostly κ- and β-caseins); (d) 1% κ-casein; and (f) 2% α-casein.*

enzymes, and in the presence of calcium ions the cheese coagulum is formed (*2*). As far back as 1929, Linderstrom–Lang (*150*) proposed that a stabilizing component of the casein system was selectively attacked by rennin (chymosin); this attack led to aggregation of the casein. κ-Casein apparently acts as the stabilizing colloid first proposed by Linderstrom–Lang.

κ-Casein has been isolated and studied in some detail (*2*). It is the only casein component containing carbohydrate. However, the κ-casein is heterogeneous with regard to the carbohydrate content; some molecules contain carbohydrate, and others do not (*151*). In addition, κ-casein exists as polymorphic genetic variants involving amino acid substitutions in the primary sequence (*152*). The primary structure of κ-casein has been determined (*153*).

As shown in Figure 5 the primary proteolytic action of rennin at pH 6.7 and 30°C is on κ-casein, and the other casein components are not materially affected. This pH is far removed from the optimum proteolytic pH (3.8) (*154*) of rennin as is the pH at which milk is generally coagulated in cheesemaking (*2*). This suggests a specific type of action by rennin and other clotting enzymes such as pepsin and certain microbial proteases. Thus enzymes most useful in cheesemaking are those that possess a high clotting activity involving this specific action as opposed to general proteolytic activity. For example, many proteolytic enzymes can clot milk but are too proteolytic at the customary pH values of milk and cheese to be useful because they generate bitter peptides which give an off-flavor to the cheese. Thus the most useful proteolytic enzymes have a high clotting-to-proteolytic enzyme ratio (*2*).

When rennin acts on isolated κ-casein in the absence of calcium ions, an insoluble aggregate forms (*155*), but 23–30% of the total nitrogen remains in solution (*149, 156*). This reaction has been characterized as follows:

$$\kappa\text{-casein} \xrightarrow{\text{rennin}} \underset{\text{(insoluble)}}{p\text{-}\kappa\text{-casein}} + \underset{\text{(soluble)}}{\text{macropeptide}}$$

In the absence of calcium ions but in the presence of other casein components the normally insoluble κ-casein is apparently stabilized by the calcium-sensitive caseins (*157*). Thus calcium ions are required to coagulate whole casein after treatment with rennin. Therefore in the native milk system the micelle-stabilizing power of κ-casein is specifically destroyed by rennin, and in the presence of calcium ions in milk a coagulum is formed (*2*). This offers a dramatic example of how the functionality of an entire protein system can be altered by specific proteolytic action on a component of that system.

κ-Casein molecules are normally held together as aggregates by disulfide bonds (*158, 159, 160*). Reduction of the disulfide bonds and subsequent electrophoresis of the κ-casein isolated from pooled milk illustrated the heterogeneity of κ-casein by the presence of 7–8 discrete bands with all the components contributing to these bands attacked by rennin (*159, 160, 161*). Treatment of S-carboxymethyl (SCM) κ-casein, prepared from reduced κ-casein, with rennin indicated an attack on SCM κ-casein similar to intact κ-casein by the formation of insoluble *p*-SCM-κ-casein with about 25% of the nitrogen remaining in solution (*159*). In addition SCM-κ-casein or S-carboxyamidomethyl (SCAM) κ-casein is heterogeneous, indicating the heterogeneity of the κ-casein component. The heterogeneity of κ-casein arises primarily from the soluble macropeptide which differs with regard to its carbohydrate content and amino acid differences in the peptide moiety (*152, 159*). Thus when the SCM-κ-casein is subjected to electrophoresis, the heterogeneous nature of κ-caseins is demonstrated by the existence of carbohydrate-free as well as carbohydrate-rich κ-caseins. Both the carbohydrate-rich and carbohydrate-free SCM-κ-caseins had essentially the same micelle stabilizing power and were equally attacked by rennin (*159*). Thus the carbohydrate moiety of the macropeptide has little influence on the micelle stabilizing power of κ-casein. This has led to suggestions that the hydrophilic amino acids in the macropeptide portion may aid in stabilizing the micelle (*2, 153, 162*). Consideration of the primary sequence of κ-casein (*162*) has led to the concept that the charged amino acids and carbohydrate portion of the macropeptide are released by proteolytic enzymes exposing a core of hydrophobic amino acids available for hydrophobic interactions to form a coagulum (Figure 6). It has been suggested that κ-casein acts as an amphiphile in stabilizing the micelle (*163*). Much like a detergent, hydrophobic amino acid residues are oriented toward the interior of micellar sub-units, and the hydrophilic residues are at the exterior of micellar sub-units associating with the aqueous phase.

It appears that the heterogeneity of the reaction products results from the heterogeneity of the κ-casein rather than from a random attack on the κ-casein by rennin (*2*). This suggests that perhaps a single peptide bond might be split by rennin. If this is the case, then new C-terminal and N-terminal amino acid residues should be detectable after rennin action. The same C-terminal amino acids were found in κ-casein and in the macropeptide indicating that the macropeptide occupied the C-terminal position of the κ-casein (*164, 165*). Thus a new C-terminal residue should be exposed in *p*-κ-casein and a new N-terminal amino acid in the macropeptide. Carboxypeptidase analysis of *p*-κ-casein indicated a C-terminal phenylalanine or leucine (*165*). Reduction of

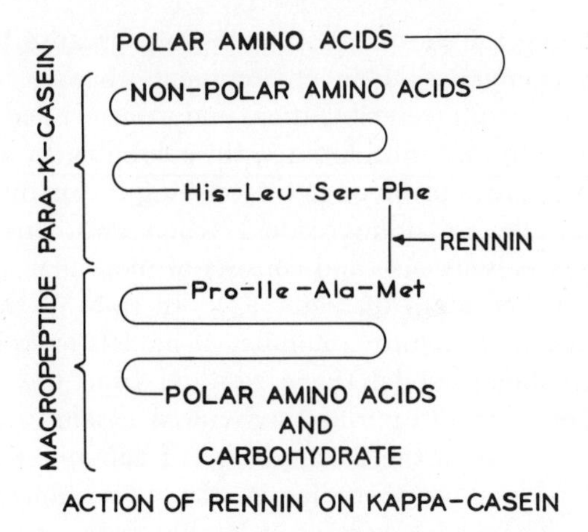

Figure 6. Schematic of a proposed structure of cow κ-casein A with attached carbohydrates (162).

κ-casein by lithium borohydride yielded an insoluble *p*-κ-casein and a macropeptide similar to that from rennin action. The presence of phenylalanol in the precipitate suggested phenylalanine as the C-terminal residue in *p*-κ-casein (*166*). Early work failed to detect an N-terminal residue in the macropeptide (*2*). However, limited acid hydrolysis of the macropeptide allowed the detection of a very labile N-terminal methionine which had previously escaped detection as a result of destruction during hydrolysis (*167*). The existence of an N-terminal methionine in the macropeptide was subsequently confirmed (*168*). The existence of the rennin-susceptible Phe–Met bond was also demonstrated from studies on the primary sequence of κ-casein (*see* Figure 6).

Although the general proteolytic activities of various milk clotting enzymes may vary, their milk clotting activities are apparently predicated on the same specific cleavage of the Phe–Met bond in κ-casein. Apparently rennin, pepsin, chymotrypsin, a microbial protease, proteases from *Endothia parasitica, Mucor pusillus,* and *Mucor miehei* exert the same type of activity on κ-casein (*2, 169*). Enzymes that are currently used commercially for cheesemaking in the United States include rennin, rennin–pepsin mixtures, and microbial proteases from *Endothia parasitica, Mucor pusillus,* and *Mucor miehei.*

The sequence of events following the enzymic action of clotting enzymes on κ-casein to destabilize the micelles is poorly understood. In part, this results from an incomplete understanding of the structure of the casein micelles. Studies of the structure of native micelles as they occur in milk are very few, indeed. Perhaps this stems from the difficulties in manipulating micelles in the native state because their structures are subject to change as their environment changes. For example, micelles isolated by ultracentrifugation and resuspended in a solution designed to simulate milk ultrafiltrate still solubilize to a slight degree (*170*). The complexities involved in establishing a structure or structures for the milk micelle are quite evident when one considers that the micellar system is polydisperse and consists of molecular aggregates that range from 40 to 220 nm in diameter (*3*). In spite of the problems in studying the micellar structure, a number of models have been proposed (*144, 171, 172*). Some models have resulted from the preparation of synthetic micelles from the purified individual caseins and calcium ion (*171, 172, 173, 174*) from the fractionation of subunits of the caseinate system (*172, 175*) and from studies on the native micelles from milk (*176, 177, 178*). The models vary considerably in the structural arrangements of the casein components.

Models employed in any discussion of the mechanism of milk coagulation must satisfy certain basic criteria. For example, caseins will not clot in the absence of Ca^{2+}; therefore provisions must be made for Ca^{2+} in the coagulation mechanism. Also, the structural role of colloidal calcium phosphate in the micelle should be considered. Since κ-casein is the principal substrate for the clotting enzymes, any model must insure its availability to the enzyme. Furthermore, as a result of coagulation whey is expelled from the curd during cheesemaking, a process called "syneresis." Any micellar model should account for the syneresis phenomenon. In addition, the crenated surface of the micelle evident in electron micrographs (*179*) coupled with fractionation data after disruption of the micelles with dissociating agents (*180*) indicate that the micelles consist of discrete subunits.

None of the models satisfy all of the above requirements for micelle structure. However, a discussion of the mechanism of milk coagulation in terms of some of the models may prove illuminating (*2*). A number of models for micelle structure have been proposed; four models requiring various clotting mechanisms are discussed here. Since any mechanism of milk clotting must be predicated on the correct micelle model, it will become evident from the ensuing discussion that much additional research is necessary to fully understand the clotting process.

Waugh and Noble (*181*) have proposed a "coat-core" model whereby the core of the micelle is comprised of calcium α_s-caseinate covered by a coat of calcium α_s- and κ-caseinates. However, the κ-casein in the coat is oriented toward the outer surface of the micelle so that the hydrophilic macropeptide portion of the kappa casein would interact with the aqueous environment of the milk serum to stabilize the integrity of the individual micelles so that the micelles normally do not interact.

At the same time, the hydrophobic portion of the κ-casein interacts with the calcium α_s- and β-caseinates to form coat subunits which interact strongly with the calcium α_s-caseinate in the core. The micelles are pictured as solid, crenated spheres (Figure 7) whereby the available κ-casein for the formation of coat subunits determines the average size of the micelles.

The appealing feature of this model is the occurrence of κ-casein on the surface of the micelle where it would be available for the action of proteolytic enzymes. Presumably, proteases such as rennin would cleave the exposed macropeptide from the κ-casein on the micellar surface yielding patches of *p*-κ-casein on the surface of the micelle. After a certain portion of the coat units is modified, adjacent micelles would cohere, leading to the coagulation of the milk. Approximately 30% of the κ-casein in milk is so-called "soluble" κ-casein (*176*) and is also

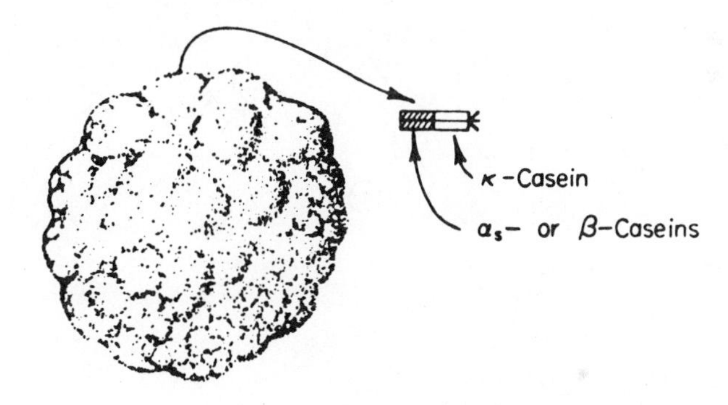

Milk Proteins, Chemistry and Molecular Biology

Figure 7. Solid sphere milk micelle model proposed by Waugh and Noble (174)

probably modified by the proteolytic enzymes to yield *p*-κ-casein which can participate in the clotting process. Calcium ions destabilize the treated micelles, possibly by binding to the modified micelles to reduce electrostatic repulsion (*182*).

Garnier and Ribadeau-Dumas (*178*) have proposed a model for the micelle which depicts an open, sponge-like structure. This model has trimers of κ-casein serving as branch points for polymers of α_s- and β-casein (Figure 8). However, this model has a rather rigid set of structural constraints to yield the open pore structure. This model differs from that of Waugh and Noble in that the structure is sponge-like and the casein components are distributed uniformly throughout the micelle.

There is probably more direct evidence, using native micelles, to support a structure of this type or of the Slattery and Evard (*171*) type than any of the other models. Initial evidence stems from the observation

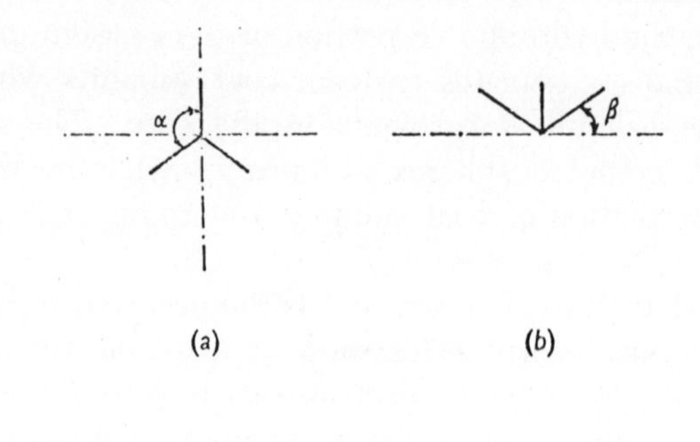

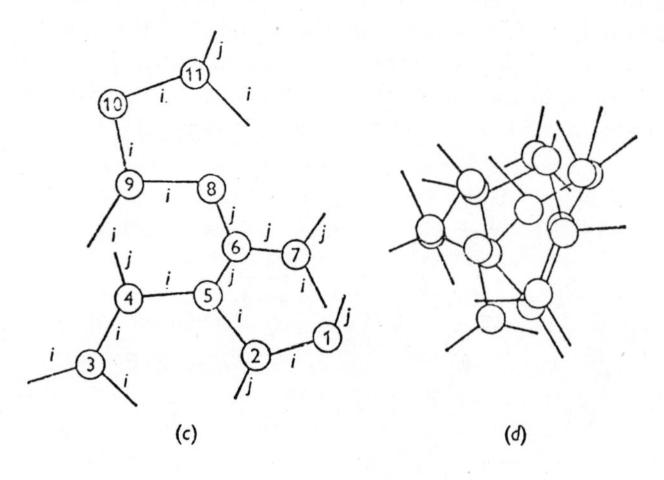

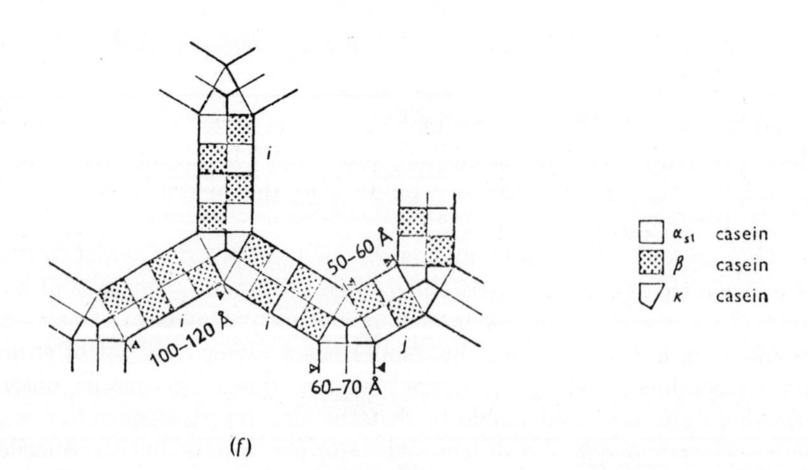

Journal of Dairy Research

Figure 8. Sponge-like micelle model proposed by Garnier and Ribadeau-Dumas (178)

of Garnier and Ribadeau-Dumas (*178*) that the unique C-terminal residues of all the casein components of the casein micelle were available to carboxypeptidase A (MW 34,400). This indicated that the micelle must be porous enough to allow entry of carboxypeptidase A into the interior of the micelle. Since rennin (MW 31,000) is about the same size as carboxypeptidase A, presumably it too can penetrate the micelle. Further evidence for a uniform distribution of the casein components in the micelle has been contributed by Ashoor et al. (*183*) and by Cheryan et al. (*170*). Ashoor et al. (*183*) treated isolated micelles with an insoluble papain superpolymer much larger than the micelles. The insolubilized papain would digest the micelles from the outside inwardly. The ratios of α_s-, β-, and κ-caseins digested in the micelles and in soluble casein were similar, suggesting a uniform distribution of caseins throughout the micelle. Cheryan et al. (*170*) determined the casein components on the surfaces of micelles fixed with glutaraldehyde using insoluble carboxypeptidase A. From analysis of the C-terminal amino acids released from the surface of the fixed micelles it appears that all the major caseins occur on the micelle surface, suggesting a uniform distribution of the caseins throughout the micelle.

In this model, clotting enzymes would readily penetrate to the interior of the micelle attacking κ-casein on the surface as well as in the interior. The patches of p-κ-casein on the surface would lead to hydrophobic interactions between micelles leading to coagulation. The formation of p-κ-casein in the interior of the micelles (Figure 8) presumably would not seriously disrupt the structure of the micelles with regard to alteration of the α_s- and β-casein branches. However, manipulation of the micelles by mechanical handling during cheesemaking may distort the micelles so that the interior patches of p-κ-casein interact hydrophobically to cause the micelle to collapse permanently and to expel the fluid in the interior of the micelle. This would readily explain the phenomenon of whey syneresis following clot formation. A role for calcium ion or colloidal calcium phosphate employing this model in the clotting mechanism was not proposed.

Parry and Carroll (*176*) used ferritin-labeled antibodies to κ-casein to study the distribution of κ-casein on the surface of large micelles. Electron micrographs of antibody-treated micelles indicated the absence of detectable κ-casein on the micelle surface. They proposed that the κ-casein in the large micelles was in the interior of the micelle surrounded by α_s- and β-caseins (Figure 9). In addition, ca. 30% of the κ-casein in milk exists as "soluble" κ-casein in association with small amounts of α_s- and β-caseins (Figure 9). Parry and Carroll proposed that the "soluble" κ-casein was attacked by rennin yielding highly insoluble p-κ-casein. This p-κ-casein presumably interacted strongly with the large

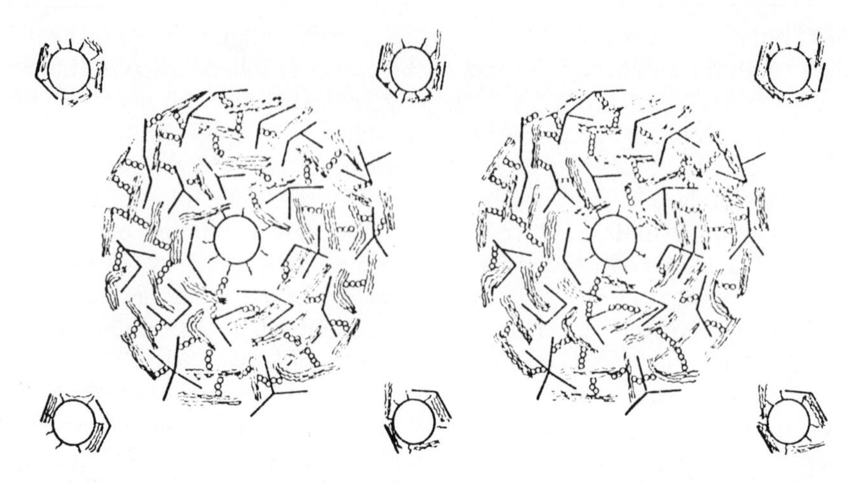

Biochimica et Biophysica Acta

Figure 9. Milk micelle model proposed by Parry and Carroll (176) *showing non-micellar κ-casein.* ○, *κ-casein;* ≡, *α_{s1}-casein;* ——, *β-casein;* ∞∞ , *calcium phosphate.*

micelle surfaces to occlude the large micelles in a three-dimensional network of interlocking strands of *para*-κ-casein. Green (*182*) questioned the accuracy of this mechanism of clotting because pretreatment of milk serum with rennet did not reduce the clotting time when recombined with previously isolated micelles. Furthermore when micelles were washed three times with milk dialyzate, the clotting time did not increase as expected, but decreased. However, Hicks et al. (*184*) treated milk

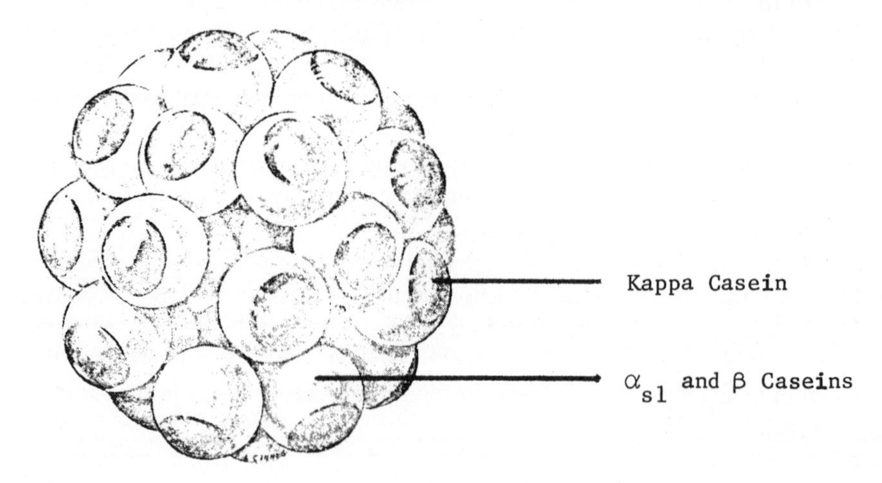

Biochimica et Biophysica Acta

Figure 10. Milk micelle model of Slattery and Evard (171) *containing ca. 40 subunits each comprised of κ-, α_s-, and β-caseins.*

serum containing "soluble" κ-casein with immobilized pepsin. Addition of the treated serum to skim milk or washed micelles resulted in coagulation of the system. These workers suggested the following sequence of events during enzymic coagulation of milk. First, the milk clotting enzyme cleaves a highly acidic macropeptide from κ-casein as suggested by Mackinlay and Wake (*157*). This would leave *p*-κ-casein more positively charged than native κ-casein, as suggested by Green and Crutchfield (*185*), and would decrease electrical repulsive forces between casein micelles (*186*). Secondly, *p*-κ-casein, whether in the serum as suggested by Parry and Carroll (*176*) or on the surface of the micelle, forms bridges interlinking micelles to produce the gelatinous coagulum.

Recently, Slattery and Evard (*171*) proposed a model for the formation and structure of casein micelles from studies devoted to association products of the purified caseins. They proposed that the micelle is composed of polymer subunits, each 20 nm in diameter. In the micellar subunits the nonpolar portion of each monomer is oriented radially inward, whereas the charged acidic peptides of the Ca^{2+}-sensitive caseins and the hydrophilic carbohydrate-containing portion of κ-casein are near the surface. Asymmetric distribution of κ-casein in a micelle subunit results in hydrophilic and hydrophobic areas on the subunit surface. In this situation, aggregation through hydrophobic interaction forms a porous micelle (Figure 10). Micelle growth is limited by the eventual concentration, at the micelle surface, of subunits rich in κ-casein.

From this model it is evident that κ-casein on the surface would be readily available to rennin. The release of the macropeptide by rennin would result in the subsequent reduction in charge and hydrophilicity on the surface. That rennin-treated micelles do not aggregate in the cold is evidence of the fact that hydrophobic interaction is mainly responsible for clotting. The κ-casein areas exposed to rennin action would result in hydrophobic patches separated by one subunit diameter (20 nm), whereby the micelles could cohere to form threads and fibrils characteristic of coagulation.

The model of Slattery and Evard (*171*) explains many of the properties of micelles, including the events associated with clotting by rennin action. For example, it explains the crenated surface of micelles observed on electron micrographs (*179*), the subunit structure of casein micelles (*180*), the porous nature of micelles (*177, 178*) allowing syneresis by continued action of rennin on interior subunits, and the more or less random distribution of the caseins in the micelle (*170, 177, 178, 183*).

Apparently, a decrease in the electrostatic forces holding the micelles apart contribute to clotting (*182, 186*). Green and Crutchfield (*185*) found that renneted micelles have a lower electrophoretic mobility than native ones and suggested that the difference in charge is sufficient for the attractive forces between the micelles to overcome the repulsive forces in renneted micelles, while the reverse is true in native micelles. This is also indicated by the work of Kirchmeier (*187*) who observed that the effective repulsive potential of the micelles was reduced from 47.6 mV for native micelles to 3.5 mV for micelles treated with rennin. Kirchmeier (*187*) proposed that the secondary phase of milk coagulation is essentially an isoelectric precipitation. Since the *p*-κ-casein resulting from rennin action is essentially positive at the pH of milk (*162*) compared with the net negative charges of α_s- and β-caseins, electrostatic interactions would be favored. Hydrophobic interactions between the *p*-κ-caseins on the surface of different micelles would probably also play a role. Green (*182*) proposed that the accelerating effect of Ca^{2+} on the secondary phase of milk clotting can be explained if clotting is primarily an electrostatic process. The Ca^{2+} would bind to the negatively charged groups of the casein, thus reducing the electrostatic repulsion between micelles and favoring their aggregation after rennet treatment.

From the above discussion it is evident that the primary action of proteases on the casein system is fairly well understood. But how this primary action alters the functionality of the casein micelles resulting in micellar coagulation is incompletely understood and will require further research.

USE OF IMMOBILIZED PROTEASES TO CLOT MILK ON A CONTINUOUS BASIS. Immobilized proteases might be valuable in continuous coagulation of milk for cheese manufacture. Since the immobilized enzyme would not remain in the product, it may be possible to substitute a less expensive, less desirable, but more readily available enzyme which normally cannot be used, such as undesirable microbial proteases, instead of commercially available milk-clotting enzymes.

The coagulation of milk by a column or other reactor containing immobilized proteases is made possible by the behavior of milk at low temperatures. The two phases of milk coagulation—enzymic action on κ-casein and the subsequent gelation or clotting—can be separated by manipulation of the temperature. Rate of clotting decreases 15- to 20-fold by lowering the temperature 10°C (*2*), whereas enzymic activity of pepsin-glass has a temperature coefficient of about 1.5 (*188*). Lowering the temperature of the enzyme bed allows sufficient enzymic activity to be retained to prevent coagulation of the skim milk until after emergence from the reactor. Subsequent warming of the milk would cause rapid clotting. The curd could then be processed continuously.

Attempts have been made to immobilize rennin (*189*) and chymotrypsin (*189, 190, 191*) for coagulation of milk, but either the immobilized enzyme activity was very low (*190*) or soluble proteolytic activity continued to leach from the immobilized enzyme preparations (*189*), thus preventing a definitive study.

The work of Ferrier et al. (*192*) and Cheryan et al. (*193*) have definitely established the feasibility of using immobilized proteases, particularly immobilized pepsin, for the continuous clotting of milk. The major problem is the decay in enzymic activity apparently resulting from the coating of the immobilized pepsin with milk proteins creating a steric barrier. However, recently Lee (*194*) prepared a pepsin immobilized on protein-coated glass beads with a usable lifetime of about 200 hr defined as the time for the clotting time of the milk to decay from 1 min to 10 min.

CHANGES IN THE BODY AND TEXTURE OF CHEESE AS A RESULT OF AGING. Alterations in the physical (functional) properties of the caseins occur during the development of the desirable smooth texture of cheese from the rubbery, elastic curd as a result of aging. The development of desirable textural characteristics of aged cheese is the result of very complex biochemical processes which are incompletely understood at this time.

Ernstrom and Wong (*2*) have reviewed the earlier literature on the changes in the cheese proteins as a function of aging. Apparently the two major factors involved in the aging of cheese are the proteolytic enzymes of microorganisms in the starter culture which lyse during aging and residual proteolytic activity of the coagulating enzyme. However, the relative importance in proteolysis during cheese ripening of the coagulating enzymes and proteinases released by lysis of the bacteria has been a matter of controversy. Undoubtedly proteinases from starter microorganisms are involved in the alteration of cheese texture. The present discussion is limited to recent literature on the effects of coagulating enzymes and the milk protease on the proteolysis of the caseins in cheese.

The proteases currently used as coagulants in the cheese industry are known to proteolyze the various casein components. Rennin, porcine pepsin, and proteases from *Endothia parasitica, Mucor miehei,* and *Mucor pusillus* readily attack α_s-, β-, and κ-caseins (*195, 199*).

Only about 7% of the coagulating enzyme activity is retained in the cheese curd (*200*); the remaining activity is expelled in the whey. Recently, Dulley (*200*) observed that cheeses containing normal and double normal amounts of rennet showed very little difference in peptides soluble in dilute trichloroacetic acid after 10 months maturation; this suggests that the amount of rennet did not significantly affect proteolytic breakdown of the cheese proteins. Furthermore, Dulley (*200*) did not observe a significant difference in starch gel electrophoretic patterns of cheese proteins from cheese containing normal and double normal quantities of rennet. Since total proteolysis increased greatly over an aging period of 12 months, Dulley concluded that microbial enzymes seemed to be the major source of proteolysis.

On the other hand, Green and Foster (*201*) prepared starterless, aseptic cheese using glucono-delta-lactone for acid production and rennet or swine pepsin as the coagulants. The aged cheese was analyzed by gel filtration. As shown in Figure 11, during aging the α_s- and β-casein peaks become progressively smaller as ripening proceeds indicating breakdown of these proteins to smaller products. The proteolysis of the caseins was confirmed by starch gel electrophoresis. Between 50 and 60 days of aging the α_s-casein had completely disappeared from the cheeses, but some β-casein was still present. None of the cheeses was contaminated with bacteria to a significant extent at any stage of the experiment. The above data indicate that coagulant enzymes retained in the cheese can have a marked effect on the caseins in cheese. Furthermore, comparison of the rates of proteolysis of starter-containing cheese using swine pepsin (which is largely inactivated during cheesemaking) and rennet as coagulants indicated that rennet at least contributes significantly to protein breakdown in cheese. Gel filtration indicates faster and greater proteolysis in starter-containing cheese with rennet than with swine pepsin. On the other hand, the course of proteolysis observed in cheeses made using starter suggested that the proteinases released from bacterial cells contribute considerably to the protein breakdown. Green and Foster (*201*) claim that their data are consistent with the conclusions of Ohmiya and Sato (*202*), that starter enzymes and rennet are synergistic in their action on casein and its breakdown products, and that starter enzymes could be responsible for at least half of the proteolysis in cheese.

Milk is known to contain a protease indigenous to the casein micelle (*203*). Survival of this protease during the cheesemaking process would allow this proteolytic enzyme to exert its activity. Indeed it has been proposed that the milk protease plays a significant role in the aging of cheese (*204*), and the β-casein is its principal substrate (*204, 205, 206*).

EXOGENOUS ENZYMES THAT INDIRECTLY AFFECT CASEINS. Acid development in the preparation of dairy products such as cottage cheese or yogurt from skim milk results from fermentation by microorganisms or acidulants such as glucono-delta-lactone, lactic acid, or hydrochloric acid. Rand (*207*) and Rand and Hourigan (*208*) have proposed an alternate method of generating acid in skim milk to form an acid gel. Their method employs exogenous enzymes to develop the acid from the lactose in the skim milk. Basically the process involves the use of lactase to hydrolyze the lactose to glucose and galactose. A second enzyme, glucose oxidase, oxidized the glucose to glucono-delta-lactone which spontane-

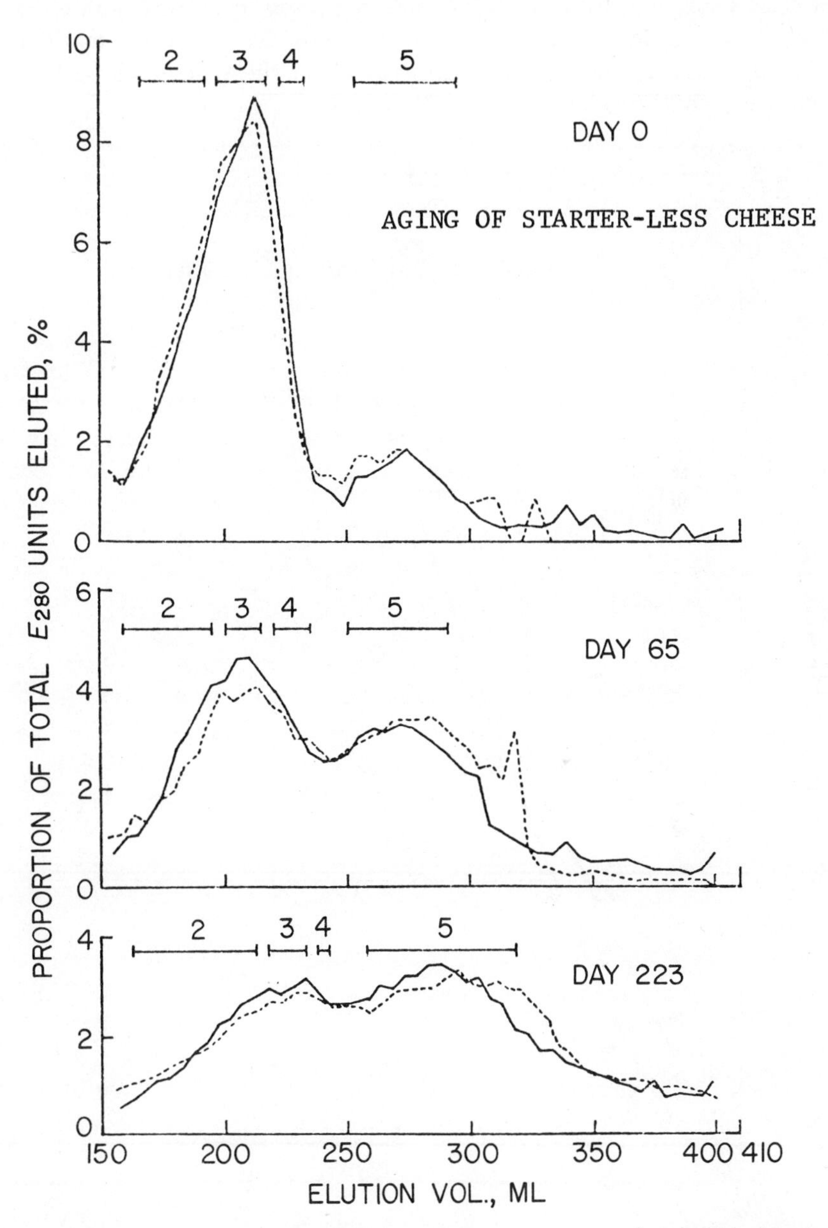

Journal of Dairy Research

Figure 11. Proteolytic breakdown of casein in aging, starter-less cheese as determined by gel filtration (200).

ously hydrolyzes to gluconic acid. As shown in Figure 12 the generation of acid is sufficient to override the buffering capacity of milk and reduce the pH to ca. 4.5 where an acid gel will form under quiescent conditions. Thus the enzyme reaction can indirectly affect the functionality of proteins through its products.

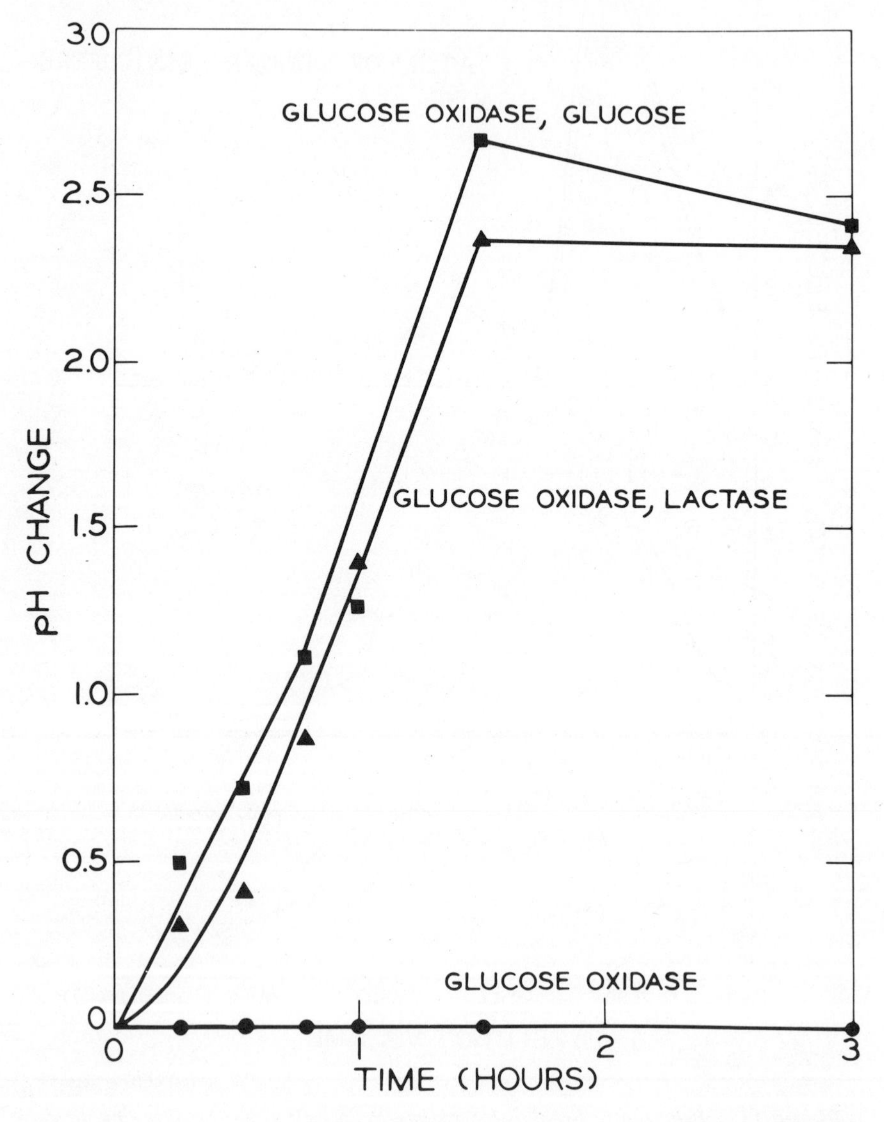

Journal of Food Science

Figure 12. Formation of acid in skim milk by glucose oxidase at 25°C (enzyme conc. 130 units per ml) (206)

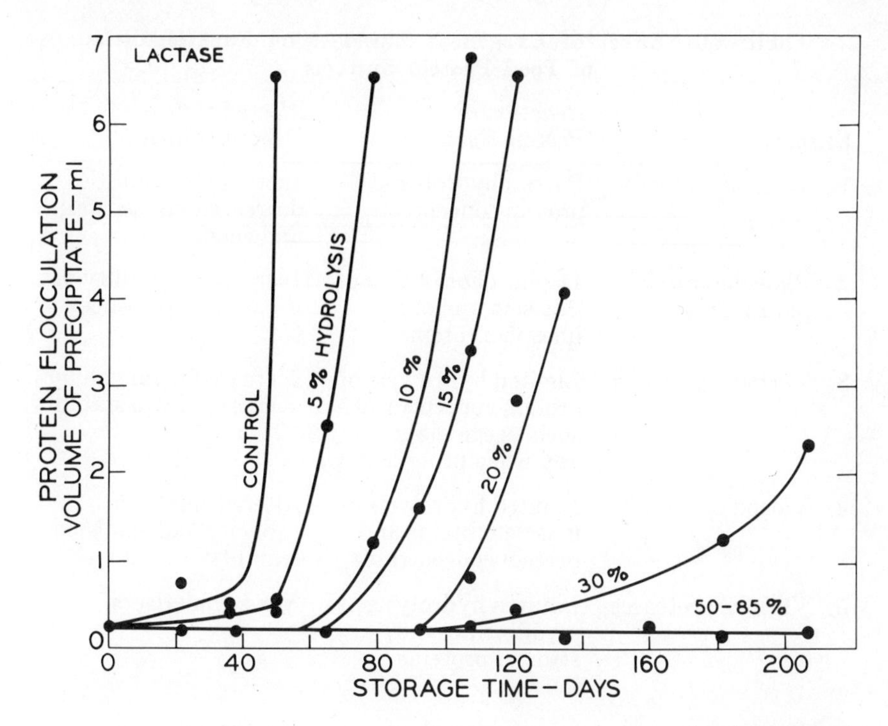

Journal of Dairy Science

Figure 13. The influence of enzymatic lactose hydrolysis on the protein stability of concentrated whole milk (35% solids, stored at −9.5°C) (208)

Lactase can also exert an indirect effect on the casein system in frozen, concentrated milk. In this case, the crystallization of lactose at −18° to −12°C in frozen, concentrated milk has been correlated with destabilization and aggregation of the casein system resulting in undesirable flocculation of the casein (*209*). As shown in Figure 13 enzymic hydrolysis of the lactose by lactase prevents the aggregation of the casein system. This alteration in functionality of the casein system has been postulated to result from the hydrolysis of lactose and prevention of its crystallization which in some way destabilizes the casein system.

Summary and Conclusions

Tables XI and XII give a brief summary of the enzymes discussed (for the most part) in this review that have an effect on the functionality of food proteins. It is readily evident that a number of endogenous and exogenous enzymes alter the physical properties of food proteins and, as a result, markedly affect the functionality of these proteins. However, virtually nothing is known about the properties of proteins that are

Table XI. Effect of Exogenous Enzymes on Functionalities of Food Protein Systems

Enzymes	*Reaction in Protein Systems*	*Altered Functionalities*
1. Various proteases	Partial hydrolysis of protein concentrates	Increased solubilities, decreased viscositites, and gelation
2. Phosphoprotein phosphatase	Dephosphorylation of β-casein and other phosphoproteins	Increased solubility of β-casein in presence of Ca^{2+}
3. Various proteases	Limited hydrolysis of protein concentrates such as egg albumen and whey proteins	Increased foam volume; decreased foam stability
4. Various proteases	Limited hydrolysis of muscle proteins and protein concentrates	Altered emulsion capacity and emulsion stability
5. Various proteases	Limited hydrolysis of myofibrillar and stromal proteins of muscle	Meat tenderizers
6. Rennin, pepsin microbial rennets	Specific attack of Phe–Met in κ-casein to destabilize casein micelle	Gel formation of caseins in cheesemaking; aging of cheese
7. Lactase	Hydrolysis of lactose in frozen concentrated milk	Prevention of casein aggregates
8. Lactase + glucose oxidase	Generation of acid in milk systems	Formation of acid gel in making of cottage cheese or yogurt
9. Fungal proteases	Partial hydrolysis of wheat proteins in bread dough	Reduction of dough viscosity; more pliable dough
10. Soybean lipoxygenase	Generation of hydroperoxides or reactive intermediates which oxidize SH groups in flour proteins to crosslink proteins	Improve rheological properties of dough
11. Various proteases	Partial hydrolysis of beer proteins	Prevention of chill haze in beer resulting from protein–phenolic interactions

Table XII. Effect of Endogenous Enzymes on Functionalities of Food Protein Systems

Enzymes	*Reaction in Protein Systems*	*Altered Functionalities*
1. Myosin ATPase	Hydrolysis of ATP to supply energy for muscle contraction	Tougher contracted muscle
2. CASF sarcoplasmic protease	Attack of α-actinin in myofibril Z-disk to partially degrade Z-disk	Tenderization of meat
3. Lysosomal cathepsins and β-glucuronidase	Limited hydrolysis of myofibrillar and stromal proteins of muscle	Tenderization of meat. Increased water binding
4. Protease in casein micelle	Partial hydrolysis of caseins in cheese	Development of body and texture in cheese
5. Ascorbic acid in presence of ascorbic acid oxidase and dehydroascorbic acid reductase	Resultant dehydro-ascorbic acid is reduced to ascorbic acid by DHA reductase. Reduction is coupled to oxidation of flour protein SH groups	Improvement of rheological properties of dough

important in exerting certain functional roles. The mechanisms of functionality of proteins to a large degree are unknown. There is a great need for basic research on the importance of structural features involved in the various types of protein functionality. In this regard, specific enzymic and chemical modifications of purified proteins would be invaluable to establishing mechanisms of functionality. A basic understanding of functionality would be desirable in the effective utilization of our protein resources.

Furthermore, the practical, empirical tests for evaluating protein functionalities should be standardized. Also, if possible, the empirical functional tests on a protein should be correlated with performance of the protein in given food products.

Literature Cited

1. Briskey, E. J., "Evaluation of Novel Protein Products," A. E. Bender, R. Kihberg, B. Lofquist, and L. Munck, Ed., Pergamon, New York, 1968, p 303.
2. Ernstrom, C. A., Wong, N. P., "Fundamentals in Dairy Chemistry," B. H. Webb, A. H. Johnson, and J. A. Alford, Eds., Avi, Westport, Conn., 1974, p 662.
3. Dewan, R. R., Chudgar, A., Bloomfield, V. A., Morr, C. V., *J. Dairy Sci.* (1974) **57**, 394.
4. Yoshikawa, M., Tamaki, M., Sugimoto, E., Chiba, H., *Agric. Biol. Chem.* (1974) **38**, 2051.
5. Dulley, J. R., Kitchen, B. J., *Aust. J. Dairy Technol.* (1973) **28**, 114.
6. Mattil, K. F., *J. Am. Oil Chem. Soc.* (1971) **48**, 477.
7. Roozen, J. P., Pilnik, W., *Dtsch. Lebensm. Rundsch.* (1974) **70**, 280.
8. Hermannson, A. M., Sivik, B., Skjoldebrand, C., *Lebensm. Wiss. u Technol.* (1971) **4**, 201.
9. Cheftel, C., Ahern, M., Wang, D. I. C., Tannenbaum, S. R., *J. Agric. Food Chem.* (1971) **19**, 155.
10. Weetall, H. H., "Immobilized Enzymes for Industrial Reactors," R. A. Messing, Ed., Academic, New York, 1975, p 118.
11. Cheftel, C., *Ann. Technol. Agric.* (1972) **21**, 423.
12. Roozen, J. P., Pilnik, W., *Proc. Biochem.* (1973) **8** (7), 24.
13. Hale, M. B., *Food Technol.* (1969) **23**, 107.
14. Spinelli, J., Koury, B., Miller, R., *J. Food Sci.* (1972) **37**, 604.
15. Hale, M. B., *Tech. Rep. Nat. Oceanic and Atmospheric Admin. NMFS, SSSRF* (1972) 657.
16. Hale, M. B., *Marine Fisheries Rev.* (1974) **36** (2), 15.
17. Arzu, A., Mayorga, H., Gonzalez, J., Rolz, C., *J. Agric. Food Chem.* (1972) **20**, 805.
18. Whitaker, J. R., Feeney, R. E., "Toxicants Occurring Naturally in Foods," 2nd ed., F. M. Strong, Ed., Natl. Acad. Sci., Washington, D. C., 1973, p 276.
19. Hermansson, A.-M., Olsson, D., Holmberg, B., *Lebens.-Wiss. u Technol.* (1974) **7**, 176.
20. Puski, G., *Cereal Chem.* (1975) **52**, 655.
21. Noguchi, M., Arai, S., Kato, H., Fugimaki, M., *J. Food Sci.* (1970) **35**, 211.
22. Fujimaki, M., Yamashita, M., Okazawa, Y., Arai, S., *J. Food Sci.* (1970) **35**, 215.
23. Fukimaki, M., Kato, H., Arai, A., Yamashita, M., *J. Appl. Bact.* (1971) **34**, 119.
24. Fujimaki, M., Yamashita, M., Arai, S., Kato, H., *Agric. Biol. Chem.* (1970) **34**, 483.
25. Grunden, L. P., Vadehra, D. V., Baker, R. C., *J. Food Sci.* (1974) **39**, 841.
26. MacDonnell, L. R., Feeney, R. E., Hanson, H. L., Campbell, A., Sugihara, T. F., *Food Technol.* (1955) **9**, 49.
27. Kuehler, C. A., Stine, C. M., *J. Food Sci.* (1974) **39**, 379.
28. Groninger, H., Miller, R., *J. Food Sci.* (1975) **40**, 327.
29. Koury, B., Spinelli, J., Weig, D., *Fish. Bull.* (1971) **69**, 241.

30. Huang, K. H., Tato, M. C., Boulet, M., Riel, R. R., Julien, J. R., Brisson, G. J., *Can. Inst. Food Technol. J.* (1971) **4**, 85.
31. Lawrie, R. A., "Meat Science," 2nd ed., Pergamon, New York, 1974.
32. Price, J. F., Schweigert, B. S., Eds., "The Science of Meat and Meat Products," 2nd ed., W. H. Freeman, San Francisco, 1971.
33. Stanley, D. W., *Proc. Meat Ind. Res. Conf.* (1974) 109.
34. Goll, D. E., Stromer, M. H., Olson, D. G., Dayton, W. R., Suzuki, A., Robson, R. M., *Proc. Meat Ind. Res. Conf.* (1974) 75.
35. Novikoff, A. B., Holtzman, E., "Cells and Organelles," Holt, Rinehart and Winston, New York, 1970, p 154.
36. Busch, W. A., Stromer, M. H., Goll, D. E., Suzuki, A., *J. Cell Biol.* (1972) **52**, 367.
37. Dayton, W. R., Goll, D. E., Zeece, M. G., Robson, R. M., Reville, W. J., *Biochemistry* (1976) **15**, 2150.
38. Dayton, W. R., Reville, W. J., Goll, D. E., Stromer, M. H., *Biochemistry* (1976) **15**, 2159.
39. Suzuki, A., Nonami, Y., Goll, D. E., *Agric. Biol. Chem.* (1975) **39**, 1461.
40. Penny, I. F., *J. Sci. Food Agric.* (1974) **25**, 1273.
41. Stromer, M. H., Goll, D. E., *J. Molec. Biol.* (1972) **67**, 489.
42. Penny, I. F., Voyle, C. A., Dransfield, E., *J. Sci. Agric.* (1974) **25**, 703.
43. Whitaker, J. R., *Adv. Food Res.* (1959) **9**, 1.
44. Locker, R. H., *J. Food Sci.* (1960) **25**, 304.
45. Marsh, B. B., *Proc. Meat Ind. Res. Conf.* (1972) 109.
46. Marsh, B. B., Leet, N. G., *J. Food Sci.* (1966) **31**, 450.
47. Dutson, T. R., *Proc. Meat Ind. Res. Conf.* (1974) 99.
48. Cover, S., Ritchey, S. J., Hostetler, R. L., *J. Food Sci.* (1962) **27**, 469.
49. Field, R. A., Pearson, A. M., Schweigert, B. S., *J. Agric. Food Chem.* (1970) **18**, 280.
50. Herring, H. K., Cassens, R. G., Briskey, E. J., *J. Food Sci.* (1967) **32**, 534.
51. Goll, D. E., Bray, R. W., Hoekstra, W. G., *J. Food Sci.* (1963) **28**, 503.
52. Goll, D. E., Hoekstra, W. G., Bray, R. W., *J. Food Sci.* (1964) **29**, 615.
53. Goll, D. E., Bray, R. W., Hoekstra, W. G., *J. Food Sci.* (1964) **29**, 622.
54. Hill, F., *J. Food Sci.* (1966) **31**, 161.
55. Pfeiffer, N. E., Field, R. A., Varnall, T. R., Kruggel, W. C., Kaiser, I. I., *J. Food Sci.* (1972) **37**, 897.
56. Shimokomaki, M., Elsden, D. F., Bailey, A. J., *J. Food Sci.* (1972) **37**, 892.
57. Bailey, A. J., *J. Sci. Food Agric.* (1972) **23**, 995.
58. Brown, H. D., Cahttopadhyay, S. K., "Chemistry of the Cell Interface, Part A," H. D. Brown, Ed., Academic, N. Y., 1971, p 73.
59. Tappel, A. L., "Lysosomes Biol. Pathol," Vol. 2, J. T. Dingle, Ed., North Holland Publ. Co., Amsterdam, 1969, p. 163.
60. Ono, K., *J. Food Sci.* (1970) **35**, 256.
61. Valin, C., *Ann. Biol. Anim. Biochem. Biophys.* (1970) **10**, 313.
62. Rossei, R., Rozier, J., *J. Ann. Nutr. Aliment.* (1968) **22**, B443.
63. Dutson, T. R., Lawrie, R. A., *J. Food Technol.* (1974) **9**, 43.
64. Eino, M. F., Stanley, D. W., *J. Food Sci.* (1973) **38**, 45.
65. Eino, M. F., Stanley, D. W., *J. Food Sci.* (1973) **38**, 51.
66. Laakkonen, E., *Adv. Food Res.* (1973) **20**, 257.
67. Tappel, A. L., "The Physiology and Biochemistry of Muscle as a Food," E. J. Briskey, R. G. Cassens, J. C. Trautman, Eds., U. of Wisc. Press, Madison, 1966, p 237.

68. Wilcox, F. H., *Poult. Sci.* (1971) **50**, 1514.
69. Stagni, N., de Bernard, B., *Biophys. Acta* (1968) **170**, 129.
70. Milanesi, A. A., Bird, J. W. C., *Comp. Biochem. Physiol.* (1972) **41B**, 593.
71. Canonico, P. G., Bird, J. W. C., *J. Cell. Biol.* (1970) **45**, 321.
72. Persidsky, M. D., Ellet, M. H., *Cryobiology* (1971) **8**, 345.
73. Mycek, M. J., *Methods Enzymol.* (1970) **19**, 285.
74. Caldwell, K. A., Grosjean, O. K., *J. Agric. Food Chem.* (1971) **19**, 108.
75. Iodice, A. A., Chin, J., Perker, S., Weinstock, I. M., *Arch. Biochem. Biophys.* (1972) **152**, 166.
76. Harikumar, P., Ninjoor, V., Warrier, B. S., Kumta, U. S., *J. Agric. Food Chem.* (1974) **22**, 530.
77. Harikumar, P., Ninjoor, V., Kumta, U. S., *Fleischwirtschaft* (1974) **54**, 936.
78. Fukushima, K., Gnoh, G. H., Skimano, S., *Agric. Biol. Chem.* (1971) **35**, 1495.
79. Peter, J. B., Kar, N. C., Barnard, R. J., Pearson, C. M., Edgerton, V. R., *Biochem. Med.* (1972) **6**, 257.
80. Goldspink, D. F., Harris, J. B., Park, D. C., Pennington, R. J., *Enzymol. Biol. Clin.* (1970) **10**, 481.
81. Eino, M. F., Stanley, D. W., *J. Food Sci.* (1973) **38**, 45.
82. Ono, K., *J. Food Sci.* (1971) **36**, 838.
83. Lutalo-Bosa, A. J., MacRae, H. F., *J. Food Sci.* (1969) **34**, 401.
84. Warrier, S. B. K., Ninjoor, V., Sawant, P. L., Hirlekar, M. G., Kumta, U. S., *Fleischwirtschaft* (1973) **53**, 980.
85. Bilinski, E., Jonas, R. E. E., Lau, Y. C., *J. Fish. Res. Board Can.* (1971) **28**, 1015.
86. Geist, G. M., Crawford, D. L., *J. Food Sci.* (1974) **39**, 548.
87. Reddi, P. K., Constantinides, S. M., Dymsza, H. A., *J. Food Sci.* (1972) **37**, 643.
88. Wojtowicz, M. G., Odense, P. H., *J. Fish. Res. Board Can.* (1972) **29**, 85.
89. Tokowa, T., Matsumiya, H., *Bull. Jpn. Soc. Sci. Fish.* (1969) **35**, 1099.
90. Eitenmiller, R. R., *J. Food Sci.* (1974) **39**, 6.
91. Melo, T. S., Blumer, T. N., Swaisgood, H. E., Monroe, R. J., *J. Food Sci.* (1974) **39**, 511.
92. Stanley, D. W., Eino, M. F. (1974). Cited in Ref. 33, p 112.
93. Eino, M. F., Stanley, D. W., *J. Food Sci.* (1973) **38**, 51.
94. Kim, M. K., *Diss. Abstr.* (1974) **35B**, 1288.
95. Etherington, D. J., *Biochem. J.* (1972) **127**, 685.
96. West, R. L., Moeller, P. W., Link, B. A., Landmann, W. A., *J. Food Sci.* (1974) **39**, 29.
97. Satterlee, L. D., *J. Food Sci.* (1971) **36**, 130.
98. Thompson, E. H., Wolf, I. D., Allen, C. E., *J. Food Sci.* (1973) **38**, 652.
99. Miyada, D. S., Tappel, A. L., *Food Res.* (1956) **21**, 217.
100. Wang, H., Weir, C. E., Birkner, M., Ginger, B., *Proc. Meat Ind. Res. Conf.* (1967) 69.
101. Kang, C. K., Rice, E. E., *J. Food Sci.* (1970) **35**, 563.
102. Kang, C. K., Warner, W. D., *J. Food Sci.* (1974) **39**, 812.
103. Beuk, I. F., Savich, A. L., Goeser, P. A., Hogan, J. M., U. S. Patent **2,903,362** (1958).
104. Kang, C. K., Warner, W. D., Rice, E. E., U. S. Patent **3,818,106** (1974).

105. Rhodes, D. N., Dransfield, E., *J. Sci. Food Agric.* (1973) **24,** 1583.
106. Du Bois, M. W., Anglemier, A. F., Montgomery, M. W., Davidson, W. D., *J. Food Sci.* (1972) **37,** 27.
107. Schnell, P. G., Vadehra, D. V., Baker, R. C., *Poultry Sci.* (1973) **52,** 1977.
108. Hamm, R., "Kolloidechemie des Fleisches," Paul Parey, Berlin, 1972, pp 213–215.
109. Whitaker, J. R., personal communication, 1951. Cited in Ref. *108,* p 214.
110. Ssolowjew, W. I., "The Ripening of Meat: Theory and Practice of the Process," Vol. I and II, Boston Spa, Yorkshire, England, 1968. Cited in Ref. *108,* p 214.
111. Ssolowjew, W. I., Poklad, A. G., *Dokl. Uch Spets Myas Prom SSSR* (1968) 3. Cited in Ref. *108,* p 215.
112. Mizik, W., *Mjasn Ind. SSSR* (1967) **38,** 34. Cited in Ref. *108,* p 215.
113. Wismer-Pedersen, J., *Symposium on Recent Points of View on the Condition and Meat Quality of Pigh for Slaughter,* Zeist, Netherlands, 1968. Cited in Ref. *108,* p 215.
114. Dordevic, V., Skenderovic, B., Kekik-Simic, N., Sukakov, N., *Technologica mesa* (1966) **7,** 37. Cited in Ref. *108,* p 215.
115. Kormendy, L., Szeredy, I., Mihaly, V., *Nahrung* (1968) **12,** 97.
116. Barrett, F. F., "Enzymes in Food Processing," 2nd ed., G. Reed, Ed., Academic, New York, 1975, p 301.
117. Matz, S. A., "Bakery Technology and Engineering," 2nd ed., AVI, Westport, Conn., 1972.
118. Jones, I. K., Phillips, J. W., Hird, F. J. R., *J. Sci. Food Agric.* (1974) **25,** 1.
119. Bloksma, A. H., *Cereal Chem.* (1975) 52II, 171 v.
120. Richardson, T., "Principles of Food Science Part I: Food Chemistry," O. R. Fennema, Ed., Marcel Dekker, New York, 1976, p 285.
121. Tsen, C. C., Hlynka, I., *Cereal Chem.* (1962) **39,** 209.
122. Tsen, C. C., Hlynka, I., *Cereal Chem.* (1963) **40,** 145.
123. Koch, R. B., Stern, B., Ferrari, C. G., *Arch. Biochem. Biophys.* (1958) **78,** 165.
124. Dillard, M. G., Henick, A. S., Koch, R. B., *J. Biol. Chem.* (1961) **236,** 37.
125. Guss, P. L., Richardson, T., Stahmann, M. A., *J. Am. Oil Chem. Soc.* (1968) **45,** 272.
126. Graveland, A., *J. Am. Oil Chem. Soc.* (1970) **47,** 352.
127. Mann, D. L., Morrison, W. R., *J. Sci. Food Agric.* (1975) **26,** 493.
128. Koch, R. B., *Bakers Dig.* (1956) **30** (2), 48.
129. Frazier, P. J., Leigh-Dugmore, F. A., Daniels, N. W. R., Russell-Eggitt, P. W., Coppock, J. B. M., *J. Sci. Food Agric.* (1973) **24,** 421.
130. Chan, H. W. S., *J. Am. Chem. Soc.* (1971) **93,** 2357.
131. Mapson, L. W., Moustafa, E. M., *Biochem. J.* (1955) **60,** 71.
132. Tsen, C. C., *Cereal Chem.* (1965) **42,** 86.
133. Redman, D. G., *Chem. Ind.* (1974) **10,** 414.
134. Buttkus, H., *J. Food Sci.* (1974) **39,** 484.
135. Janolino, V. G., Swaisgood, H. E., *J. Biol. Chem.* (1975) **250,** 2532.
136. Chandler, M. L., Varandani, P. T., *Biochemistry* (1975) **14,** 2107.
137. Minoda, Y., Kurane, R., Yamada, K., *Agric. Biol. Chem.* (1973) **37,** 2511.
138. De Lorenzo, F., Goldberger, R. F., Steers, E., Jr., Givol, D., Anfinsen, C. B., *J. Biol. Chem.* (1966) **241,** 1562.
139. Kurane, R., Minoda, Y., *Agric. Biol. Chem.* (1975) **39,** 1417.
140. Fuchs, S., De Lorenzo, F., Anfinsen, C. B., *J. Biol. Chem.* (1967) **242,** 398.

141. Chandler, M. L., Varandani, P. T., *Biochim. Biophys. Acta* (1975) **397,** 307.
142. Gordon, W. G., Kalan, E. B., *in* "Fundamentals of Dairy Chemistry," B. H. Webb, A. H. Johnson, J. A. Alford, Eds., AVI, Westport, Conn., 1974, p 90.
143. Parry, R. M., Jr., *in* "Fundamentals of Dairy Chemistry," B. H. Webb, A. H. Johnson, J. A. Alford, Eds., AVI, Westport, Conn., 1974, p 608.
144. Farrell, H. M., Jr., Thompson, M. P., *in* "Fundamentals of Dairy Chemistry," B. H. Webb, A. H. Johnson, and J. A. Alford, Eds., AVI, Westport, Conn., 1974, p 456.
145. Thompson, M. P., Tarassuk, N. P., Jenness, R., Lillevik, H. A., Ashworth, U. S, Rose, D., *J. Dairy Sci.* (1965) **48,** 159.
146. Waugh, D. F., von Hippel, P. H., *J. Am. Chem. Soc.* (1956) **78,** 4576.
147. Zittle, C. A., *J. Dairy Sci.* (1961) **44,** 2101.
148. Zittle, C. A., Walter, M., *J. Dairy Sci.* (1963) **46,** 1189.
149. Wake, R. J., *Aust. J. Biol. Sci.* (1959) **12,** 479.
150. Linderstrom-Lang, K., *C. R. Trav. Lab. Carlsberg* (1929) **17,** 9.
151. MacKinlay, A. G., Wake, R. G., *Biochim. Biophys. Acta* (1965) **104,** 167.
152. Hill, R. J., Naughton, A. M., Wake, R. G., *Biochim. Biophys. Acta* (1970) **200,** 267.
153. Jolles, P., *Mol. Cell. Biochem.* (1975) **7,** 73.
154. Foltmann, B., *Acta Chem. Scand.* (1959) **13,** 1927.
155. Cheesman, G. C., *J. Dairy Res.* (1962) **29,** 163.
156. Beeby, R., Nitschmann, H., *J. Dairy Res.* (1963) **30,** 7.
157. MacKinlay, A. G., Wake, R. G., *in* "Milk Proteins, Chemistry and Molecular Biology," Vol. 2, H. A. McKenzie, Ed., Academic, New York, 1971, p 175.
158. Dennis, E. S., Wake, R. G., *Biochim. Biophys. Acta* (1965) **97,** 159.
159. MacKinlay, A. G., Wake, R. G., *Biochim. Biophys. Acta* (1965) **104,** 167.
160. Woychik, J. H., *Arch. Biochem. Biophys.* (1965) **109,** 542.
161. Schmidt, D. G., *Biochim. Biophys. Acta* (1964) **90,** 411.
162. Swaisgood, H. E., *J. Dairy Sci.* (1975) **58,** 583.
163. Hill, R. J., Wake, R. G., *Nature* (1969) **211,** 635.
164. Jolles, P., Alais, C., Jolles, J., *Biochim. Biophys. Acta* (1961) **51,** 309.
165. Jolles, P., Alais, C., Jolles, J., *Arch. Biochem. Biophys.* (1962) **98,** 56.
166. Jolles, P., Alais, C., Jolles, J., *Biochim. Biophys. Acta* (1963) **69,** 511.
167. Delfour, A., Jolles, J., Alais, C., Jolles, P., *Biochem. Biophys. Res. Comm.* (1965) **19,** 452.
168. De Koning, P. J., Van Rooijen, P. J., Kok, A., *Biochem. Biophys. Res. Comm.* (1966) **24,** 616.
169. Sternberg, M., *Biochim. Biophys. Acta* (1972) **285,** 383.
170. Cheryan, M., Richardson, T., Olson, N, F., *J. Dairy Sci.* (1975) **58,** 651.
171. Slattery, C. W., Evard, R., *Biochim. Biophys. Acta* (1973) **317,** 529.
172. Rose, D., *Dairy Sci. Abstr.* (1969) **31,** 171.
173. Waugh, D. F., Creamer, L. K., Slattery, C. W., Dresdner, G. W., *Biochemistry* (1970) **9,** 786.
174. Waugh, D. F., *in* "Milk Proteins, Chemistry and Molecular Biology," Vol. 2, H. A. McKenzie, Ed., Academic, New York, 1971, p 3.
175. Morr, C. V., *J. Dairy Sci.* (1967) **50,** 1744.
176. Parry, R. M., Jr., Carroll, R. J., *Biochim. Biophys. Acta* (1969) **194,** 138.
177. Ribadeau-Dumas, B., Garnier, J., *J. Dairy Res.* (1970) **37,** 269.

178. Garnier, J., Ribadeau-Dumas, B., *J. Dairy Res.* (1970) **37**, 493.
179. Shimmin, P. D., Hill, R. D., *J. Dairy Res.* (1964) **31**, 121.
180. Carroll, R. J., Farrell, H. M., Jr., Thompson, M. P., *J. Dairy Sci.* (1971) **54**, 752.
181. Waugh, D. F., Noble, R. W., Jr., *J. Am. Chem. Soc.* (1965) **87**, 2246.
182. Green, M. L., *J. Dairy Res.* (1972) **39**, 55.
183. Ashoor, S. H., Sair, R. A., Olson, N. F., Richardson, T., *Biochim. Biophys. Acta* (1971) **229**, 423.
184. Hicks, C. L., Ferrier, L. K., Olson, N. F., Richardson, T., *J. Dairy Sci.* (1975) **58**, 19.
185. Green, M. L., Crutchfield, G., *J. Dairy Res.* (1971) **38**, 151.
186. Green, M. L., *Neth. Milk Dairy J.* (1973) **27**, 278.
187. Kirchmeier, O., *Z. Lebensm. Forsch.* (1972) **149**, 211.
188. Line, W. F., Kwong, A., Weetall, H. H., *Biochim. Biophys. Acta* (1971) **242**, 194.
189. Green, M. L., Crutchfield, G., *Biochem. J.* (1969) **115**, 183.
190. Dolgikh, T. V., Surovtsev, V. I., Kozlov, L. V., Antonov, V. K., Ginodman, L. M., Zvyatintsev, V. I., *Prikl. Biokhim. Mikrobiol.* (1971) **7**, 686.
191. Janusauskaite, V., Kozlov, L. V., Antonov, V. K., *Prikl. Biokhim. Mikrobiol.* (1974) **10**, 410.
192. Ferrier, L. K., Richardson, T., Olson, N. F., Hicks, C. L., *J. Dairy Sci.* (1972) **55**, 726.
193. Cheryan, M., Van Wyk, P. J., Olson, N. F., Richardson, T., *Biotechnol. Bioeng.* (1975) **17**, 585.
194. Lee, H. O., unpublished results.
195. Tam, J. J., Whitaker, J. R., *J. Dairy Sci.* (1972) **55**, 1523.
196. Hovins, G. Deroanne, Cl., Coppens, R., *Lait* (1973) **53**, 610.
197. Rymaszewski, J., Poznanski, S., Reps, A., Ichilczyk, J., *Milchwissenschaft* (1973) **28**, 779.
198. Poznanski, S., Reps, A., Lowalewska, J., Rymaszewski, J., Jedrychowski, L., *Milchwissenschaft* (1974) **29**, 742.
199. Vanderpoorten, M., Weckx, M., *Neth. Milk Dairy J.* (1972) **26**, 47.
200. Dulley, J. R., *Aust. J. Dairy Technol.* (1974) **29**, 65.
201. Green, M. L., Foster, P. M. D., *J. Dairy Res.* (1974) **41**, 269.
202. Ohmiya, K., Sato, Y., *Milchwissenschaft* (1972) **27**, 417.
203. Reimerdes, E. H., Mrovetz, G., Klostermeyer, H., *Milchwissenschaft* (1975) **30**, 271.
204. Noomen, A., *Neth. Milk Dairy J.* (1975) **29**, 153.
205. Creamer, L. K., *J. Dairy Sci.* (1975) **58**, 287.
206. Reimerdes, E. H., Klostermeyer, H., *Milchwissenschaft* (1974) **29**, 517.
207. Rand, A. G., Jr., *J. Food Sci.* (1972) **37**, 698.
208. Rand, A. G., Jr., Hourigan, J. A., *J. Dairy Sci.* (1975) **58**, 1144.
209. Tumerman, L., Fram, H., Cornely, K. W., *J. Dairy Sci.* (1954) **37**, 830.

RECEIVED January 26, 1976. Work supported by the College of Agricultural and Life Sciences, University of Wisconsin, Madison, Wisc.

8

Enzymatic Yeast Cell Wall Degradation

H. J. PHAFF

Department of Food Science and Technology, University of California, Davis, Calif. 95616

The yeast cell envelope presents at least a partial barrier in the effective utilization of the protein of the cell for nutritional or other purposes. Since mechanical cell disruption is inefficient and requires much energy, enzymatic degradation of the alkali-insoluble glucan component of the wall is preferred. The chemical composition of the yeast cell envelope is reviewed. An inner layer of alkali-insoluble β-glucan is responsible for the strength and rigidity of the wall. Its hydrolysis by endogenous yeast enzymes offers advantages, but little is known as yet of the control of these β-glucanases. The remainder deals with cell wall lysis by β-glucanases from snail digestive fluid and from microbial sources. Such lytic enzymes are produced by bacteria, myxobacteria, streptomycetes, and molds. Action patterns of various purified lytic enzymes are reviewed.

Cells of yeast are surrounded by a thick, tough, and rigid wall which is rather difficult to rupture by mechanical means (*1*). Since this symposium is concerned with recovery of protein from the cytoplasm of yeast in substantially unaltered form, this review will deal with the various means of cell-wall removal for the purpose of separating the protein-rich cytoplasm from the whole-cell mass. Since yeast cells are very resistant to changes in osmotic pressure of the surrounding medium, osmotic lysis, which can be used successfully for certain bacteria, is not possible with yeasts. This leaves, as the main approaches, mechanical disruption and enzymatic breakdown of yeast cell walls. The latter approach appears most attractive because of the large energy requirements and relatively low efficiency of mechanical cell breakage. For this reason the emphasis in this review will be on enzymatic degradation of the alkali-insoluble β-glucan component which is responsible for the strength and rigidity of the yeast wall. Such a degradation may be achieved by (a) endogenous β-glucanases produced by the yeast itself, and (b) external enzymes from microbial or other sources. As background it is essential to review present concepts of the structure and composition of yeast cell walls so that the enzymatic requirements can be understood.

Chemistry and Biochemistry of Yeast Cell Walls

A historical review of the development of our knowledge of the yeast cell wall was given by Phaff (*1*). Most information based on chemical studies has been derived, by far, from studies with cell walls from baker's yeast, *Saccharomyces cerevisiae*, and closely related species. The principal components of *Saccharomyces* walls are several types of glucan and a mannan-protein complex which may contain variable proportions of phosphate. A low content of chitin (ca. 1%) may be present depending on the number of times a cell has produced buds. The reason for this is that chitin is present only in the bud scars (ca. 3 μm^2 in area) produced on the surface of a mother cell (*2*), each at a different place on the cell surface.

Some authors have fractionated whole cells to prepare individual wall components. However, in view of the numerous contaminating components from the cytoplasm (especially glycogen), it is preferable in cell wall studies to use clean well-washed cell walls. A commonly used procedure (*see* Scheme 1), adapted from Manners et al. (*3*), shows the steps involved in the separation of the main components. A crucial step is the separation of two components in the alkali-insoluble polysaccharide fraction. Investigators prior to 1968 were not aware that the alkali-insoluble glucan contained a water-soluble β-(1→6)-glucan component (*4, 5*) which apparently can be removed only after numerous successive extractions with dilute acetic acid at 90°C (*3*).

The individual cell wall components of baker's yeast will now be discussed in more detail. For earlier studies the reader is referred to Ref. *1*.

Alkali-insoluble Glucan. Since this is the structural component of the cell wall responsible for the rigidity and tensile strength of the wall, its chemical composition is of special interest. The alkali-insoluble residue, whether obtained by alkali extraction of whole cells or of cell walls, must first be extracted repeatedly with dilute (0.5*M*) acetic acid at 90°C to remove glycogen and a β-(1→6)-linked glucan component (*5, 6*). Although Misaki et al. (*7*) in their preparation of alkali-insoluble glucan followed the procedure of Bell and Northcote (*8*) to remove glycogen by acetic acid extraction followed by autoclaving with 0.02*M*

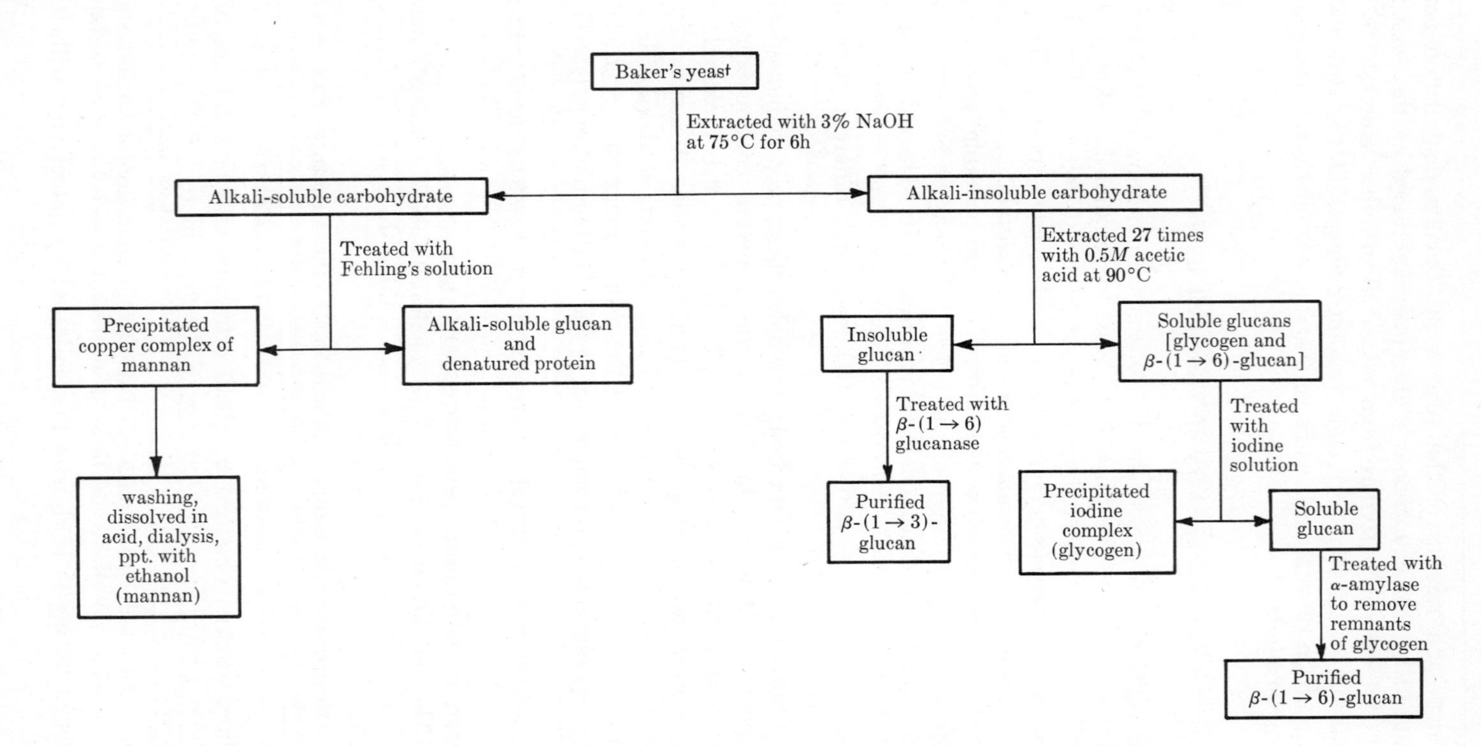

Scheme 1. Fractionation of baker's yeast into individual wall components

Na-acetate at pH 7.0, there is no assurance that all of the β-(1→6)-glucan was removed from the crude starting material. This treatment was followed by dissolving the glucan in dimethylsulfoxide and allowing the glucan to precipitate by dilution with water. This process too does not guarantee that the β-(1→6)-glucan component remains quantitatively in solution after dilution with water. Structural studies on this kind of glucan therefore carry an element of uncertainty since the residual concentration of β-(1→6)-glucan is not known. Variable concentrations of gentiobiose up to gentiotetraose in hydrolysates of this glucan support the view that the glucans had not been sufficiently purified prior to analysis. Manners and co-workers (*3, 6*) subjected the alkali-insoluble glucan to numerous extractions with hot dilute acetic acid (Scheme 1) and to ensure complete removal of the last traces of β-(1→6)-glucan the residue was treated with a fungal endo-β-(1→6)-glucanase to hydrolyze residual contaminating polysaccharide to soluble oligosaccharides. The purified polysaccharide, which represented approximately 85% of the crude alkali-insoluble glucan was shown by Manners and co-workers to be a branched β-(1→3)-glucan containing 3% of β-(1→6)-glucosidic interchain linkages. Its molecular weight was about 240,000 and its degree of polymerization approximately 1500. Its partial structure is shown in Figure 1. However, Manners et al. (*9*) have pointed out that different samples of glucan may differ significantly in degree of branching. Furthermore, there is as yet no conclusive evidence whether the macromolecule has a tree-type structure with multiple branching or a comb-type structure (Figure 1). The latter would have a long chain to which linear side chains with an average of 30 glucose residues are attached at intervals of unknown distances. In this case, since the degree of polymerization of the glucan is about 1500, each molecule would have to have about 50 side chains (*3*). The formation of a highly insoluble polymer with a considerable degree of rigidity is thought to be caused by the linking together of β-(1→3)-linked chains of glucose residues by hydrogen bonding and/or other forces of attraction Indirect support for this view comes from the finding that an algal β-(1→3)-xylan may assume a helical conformation with three chains together forming a triple helix (*10*). On this bases Rees (*11*) has suggested that β-(1→3)-glucan chains of sufficient length could assume similar configurations, since substitution of the equatorial H at C-5 of a D-xylose residue by $-CH_2OH$ should not affect the tendency to form helical structures.

The demonstration by Kopecka et al. (*12*) of a microfibrillar network as an inner layer of the cell wall supports the idea of aggregation of the unit molecules by association of portions of chains from different molecules to form a continuous structure surrounding the cell.

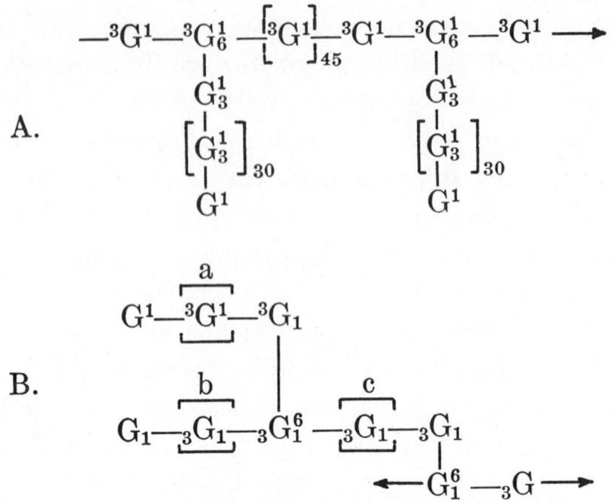

A.

$$—^3G^1—^3G^1_6—\left[^3G^1\right]_{45}—^3G^1—^3G^1_6—^3G^1\rightarrow$$

$$\begin{array}{cc} | & | \\ G^1_3 & G^1_3 \\ | & | \\ \left[G^1_3\right]_{30} & \left[G^1_3\right]_{30} \\ | & | \\ G^1 & G^1 \end{array}$$

B.

$$G^1—\underbrace{\overbrace{^3G^1}^{a}}—^3G_1$$

$$G_1—\overbrace{_3G_1}^{b}—_3G^6_1—\overbrace{_3G_1}^{c}—_3G_1$$

$$\leftarrow G^6_1—_3G\rightarrow$$

Figure 1. Possible structures for alkali-insoluble β-glucan from the cell wall of Saccharomyces cerevisiae *after Manners et al. (3). Part A represents a comb-type structure, while part B represents a tree-type structure. In the former, most or all of the glucose residues comprising the backbone are thought to carry side chains with an average of ca. 30 glucose residues (two side chains shown). In the latter, a + b + c comprise ca. 60 glucose residues. See text for further details.*

The Water-soluble β-(1→6)-Glucan Component. There is less known about the detailed structure of the closely associated β-(1→6)-glucan component, which accounts for about 15% of the alkali-insoluble glucan. Although initial studies on this polysaccharide by Manners and Masson (*13*) did not indicate branching, a subsequent paper (*6*) showed it to have a highly branched structure with a degree of polymerization of about 130–140. This β-(1→6)-glucan contains about 19% of β-(1→3)-glucosidic linkages which may serve both as inter-residue and interchain linkages. About 75% of these were found to be present as interchain linkages at triply linked glucose residues and the remainder as inter-residue bonds. In alkali-insoluble glucan preparations from several other species of yeast, the proportion of the β-(1→6)-glucan component may be significantly greater than in baker's yeast (*9*).

Alkali-soluble Glucan. Several authors have reported the presence of alkali-soluble glucan in yeast cell wall preparations (*14, 15*). This material is extracted together with mannan from whole cells or cell walls with hot alkali (*see* Scheme 1). It is not precipitated with Fehling's solution and can thus be separated from the mannan fraction (*1*). Recent work from Manners' laboratory (*139*) shows that the alkaline extract neutralized with acetic acid forms a gel-like precipitate of alkali-soluble glucan. It can be washed with distilled water and represents approximately 22% of the cell wall. The glucan contained a small proportion of mannose and protein, suggesting that in the cell wall it may have been present as a glucan–mannan–protein complex, as suggested earlier by Kessler and Nickerson (*16*). Chemical studies on the alkali-soluble glucan by Fleet and Manners indicated a primarily β-(1→3)-linked glucan, a D.P. of about ca. 1500 and ca. 8–12% of β-(1→6)-linked glucose residues with ca. 3–4% of branching. The main difference between it and the alkali-insoluble glucan is a number of β-(1→6)-linked glucose residues inserted into the long sequences of β-(1→3)-linked glucose chains, which may be responsible for the increased solubility in alkali. Since some of the β-(1→6)-glucose residues were found to occur in groups of four, the polysaccharide can be cleaved into different soluble subunits by either the action of specific endo-β-(1→3)- or endo-β-(1→6)-glucanases. There is evidence that the alkali-soluble glucan layer is located between an inner layer of alkali-insoluble glucan and an outer layer of a mannan–protein complex. The exposure of the microfibrillar, alkali-insoluble glucan layer by treatment of yeast cell walls with bacterial endo-β-(1→3)- and endo-β-(1→6)-glucanases can now be explained by penetration of these enzymes into the middle layer and hydrolyzing it to soluble units so that the outer mannan layer sloughs off in the process (*12*). Fleet and Manners obtained some evidence that the alkali-soluble glucan is linked to mannan units by glucose residues linked through β-(1→6)-bonds. This type of glucan may thus represent the bonding material which links the outer mannan layer to the innermost alkali-insoluble glucan layer.

Mannan-Protein Complex. The development of our knowledge of the chemistry of yeast mannan has been reviewed (*1, 17, 18*). In this article only the highlights of its structure and recent developments will be discussed. Mannan is usually extracted from whole yeast or from yeast cell walls by heating with dilute NaOH followed by precipitation of the mannan with Fehling's solution (Scheme 1). The precipitate is dissolved in acid, and the mannan-protein complex is recovered by alcohol precipitation after dialysis of the copper salts. A milder extraction procedure involves autoclaving yeast with 0.02*M* sodium citrate at pH 7.0 (*19*). Presumably these extraction procedures break the bonds with which mannan is anchored to the rest of the cell envelope, resulting in a soluble polysaccharide. It has been known for many years that mannan of baker's yeast is composed almost entirely of D-mannose residues which are arranged in the form of an α-(1→6)-linked backbone, about 50 units in length, to which are attached short side chains of α-(1→2)- and α-(1→3)-linked mannose residues. Great advances in structural studies were made by the application of an acid-catalyzed acetolysis reaction, which preferentially cleaves the α-($1 \rightarrow 6$)-linkages of the backbone. This releases the side chains (inclusive of the backbone unit to which they are attached) which can be separated by gel filtration. The lengths of the side chains vary in different yeast species. Similarly there are various side groups attached to the side chains, such as phosphate (in some strains of *Saccharomyces*), mannosyl-phosphate (in *Kloeckera* species) and *N*-acetylglucosamine (in species of *Kluyveromyces*). In some species of yeast, e.g., in the genus *Pichia,* the side chains may contain β-linked mannose residues. The side chains in their various configurations (especially the terminal and side residues) have been shown to be responsible for the different antibodies which are formed in animals which are immunized with a particular strain of yeast. These side chains are therefore the antigenic determinants of the yeast cell (*17*). Recent work suggests that the different side chains in a particular strain of yeast are attached to the backbone in a nonrandom arrangement (*20*). However, the data to date are not yet adequate to support the presence of a single repeating sequence.

The mannan molecule appears to consist of a number of subunits. Mannans isolated by procedures that avoid exposure to alkali have a much higher molecular weight than mannans obtained by alkali extraction (*21*). Native mannan, when exposed to 0.1*N* NaOH at room temperature, breaks up into subunits. This mild treatment could saponify acyl ester bonds but is unlikely to break phosphodiester bonds. It suggests that the subunits are held together by acyl ester bonds or by some structure with the stability of an acyl ester bond (*22*). The size of native mannan molecules and their subunits appears to depend on species differences. Mild alkaline treatment also removes from the mannan–protein complex a number of small mannose oligosaccharides that are attached to serine and threonine of the peptide component. This removal occurs through the mechanism of β-elimination. These short oligosaccharides have the same structure as the oligosaccharide side chains of the mannan molecule (*22, 23*).

The mannan molecule consists of an outer chain [an α-($1 \rightarrow 6$)-linked backbone with oligomannoside side chains, discussed above] and an inner core near the point of attachment to the protein. This linkage fragment of twelve mannose units consists of an α-($1 \rightarrow 6$)-linked backbone with di-, tri-, and tetrasaccharide side chains, which are under different genetic control as the side chains of the outer chain. This linkage fragment is attached by the last mannose unit of the backbone through a β-($1 \rightarrow 4$) bond to *N*-acetylchitobiose, which in turn is linked to asparagine of the peptide chain (*24*). The material constituting the inner core linkage region was later shown to be heterogeneous, consisting of a family of homologous oligosaccharides in which the number of mannose residues varied from 12 to 17 (*25*).

α-(1 $\rightarrow$ 3)-Glucan in Yeast Cell Walls. Polysaccharides which consist almost exclusively of α-(1$\rightarrow$3)-glucosidic linkages have only recently been recognized, but it is now well established that glucans of this type (also referred to as pseudonigeran) play an important structural role in the cell walls of many fungi and of some yeasts (*26, 27, 28*). The α-(1$\rightarrow$3)-glucans so far known are alkali-soluble and have been isolated from fungal mycelium or from isolated wall material by alkali extraction. Neutralization of the alkaline extract yields a white flocculant precipitate of the glucan (*29*). These α-(1$\rightarrow$3)-glucans appear to be unbranched molecules, and the one from *Aspergillus niger* has been reported to have a degree of polymerization of 230–700 (*27, 29*).

An unusual polysaccharide was isolated by Kreger (*30*) from the ascomycetous yeast *Schizosaccharomyces octosporus.* It had an x-ray diffraction spectrum which was different from the β-glucans of yeast. Bacon et al. (*26*) performed chemical analyses of the polysaccharide obtained by neutralizing the alkaline extract of *Schizosaccharomyces pombe* walls, and showed it to be an α-(1$\rightarrow$3)-linked glucan. The occurence of this polysaccharide among ascomycetous yeasts appears to be rare, but in yeast of basidiomycetous origin it is common, although not universal.

Overall Wall Composition. Summarizing the above information, we can say that the cell envelope of baker's yeast represents approximately 15% of the dry weight of the cell and that it is made up of 20–40% mannan, 5–10% protein, 1% chitin, and 30–60% glucan. The medium and growth conditions may influence the mannan/glucan ratio of the cell wall (*31*). The protein content of wall preparations reported in the literature is likely to be influenced by the extent of purification to which the walls were subjected (*1*). The glucan content is more likely to be in the 50–60% range than in the 30–35% range, often reported in the earlier literature. These low values are most likely due to the elimination of about 20–22% alkali-soluble glucan during cell wall fractionation. Figures by authors, who based glucan and mannan contents on the direct determination of glucose and mannose in wall hydrolysates, can be expected to be more reliable than data based on cell wall fractionation (*32*).

Since three of the major wall components are potentially water soluble, the effective removal or weakening of the cell wall must involve enzymatic hydrolysis of the long chains of β-(1$\rightarrow$3)-linked glucose residues in the microfibrillar, alkali-insoluble glucan layer. In some species of yeast an additional structural wall component occurs, which has been identified as a linear α-(1$\rightarrow$3)-glucan.

Enzymatic Hydrolysis of Yeast Cell Walls by Endogenous Enzymes.

That the yeast cell wall is subject to lysis by cell-associated enzymes is evident from phenomena connected with the sexual and vegetative cell cycles. The following events are associated with the cell cycle of yeast. (a) Cell conjugation (fusion) occurs when two haploid cells fuse together to a single diploid cell by lysis of those parts of the walls of the two cells which were in mutual contact with each other before fusion. Since in this case lysis occurs only at the area of contact, the cell must have very strict control over the enzymes involved in the lysis and the welding together of the remainder of the cell walls of the two conjugating cells. (b) In some, but not all, species of ascomycetous yeasts the process of ascosporulation is followed by a rapid and often complete lysis of the wall of the ascus, thus liberating the ascospores. It should be realized that the wall of the ascus is the same structure as the cell wall of the yeast cell before the diploid nucleus underwent reduction division and spore formation. Whether the sudden activity of the lytic enzymes involved in this process represents de novo synthesis, or an activation phenomenon, or the removal of an inhibitor, is not known. (c) During vegetative cell reproduction by budding a certain area of the cell surface undergoes a partial lysis. This probably causes a softening of the rigid fibrillar glucan layer and as a result the cell wall bulges out and a bud begins to form. During this process the cell wall remains intact, since otherwise the cell would burst because of the high internal osmotic pressure. The formation and growth of the bud are thought to involve a limited hydrolysis of the alkali-insoluble glucan layer, a stretching of the wall, and resynthesis and insertion of new glucan fibrils. In the same way cell expansion, i.e. the increase in cell size from a small detached bud to a full-sized cell, must involve carefully controlled, limited glucan hydrolysis coupled with insertion of newly synthesized glucan molecules. (d) Finally, the process of autolysis of yeast may be mentioned. Depending on species or strain of yeast, cells at some stage during their growth cycle may break down due to spontaneous autolysis. This apparently uncontrolled activity of lytic enzymes is favored by unfavorable storage conditions (such as high temperatures) or the presence of certain chemicals (e.g., toluene or chloroform).

Nature of the Endogenous Glucanases in Yeast. As early as 1955, the presence in yeasts of enzyme systems capable of hydrolyzing the β-(1 $\rightarrow$ 3)-glucosidic linkage was implied by the observation that certain yeasts were able to grow on extracts of *Laminaria cloustonii* fronds which contain up to 30% of laminarin on the dry weight basis (*33*). Chesters and Bull (*34*) thereby noted the ability of several yeasts of marine origin to grow in a medium with laminarin as the carbon source. Subsequently they demonstrated laminarinase activity in the culture fluids (*35*).

Later Brock (*36*) reported the partial purification of an intracellular β-glucanase from compressed baker's yeast. The purified enzyme was capable of hydrolyzing both laminarin and pustulan as well as the synthetic, low-molecular-weight substrate *p*-nitrophenyl β-D-glucoside. The ratios of the three activities remained constant throughout the purification procedure, and the three activities moved together during starch block electrophoresis, suggesting that a single protein was responsible for the three activities. Paper chromatographic analysis showing that glucose was the only product of laminarin and postulan hydrolysis demonstrated that the enzyme was an exo-glucanase. A similar exo-β-glucanase was found in intracellular extracts of *Hansenula wingei* (*37*).

Abd-El-Al and Phaff (*38, 39*) later examined various species of yeasts for the presence of β-glucanases. Intracellular extracts of *Saccharomyces cerevisiae, Kluyveromyces fragilis, Saccharomyces elegans,* and *Hansenula anomala* were found to hydrolyze laminarin, pustulan, and *p*-nitrophenyl β-D-glucoside. Glucose was the only product observed during hydrolysis of laminarin and pustulan; this indicates that the enzymes were exo-hydrolases. Exo-β-glucanases were also noted in the culture fluids of *K. fragilis* and *H. anomala.* The intracellular exo-β-glucanases from *S. cerevisiae* and *K. fragilis* and the extracellular exo-β-glucanases from *H. anomala* and *K. fragilis* were extensively purified, and some properties were reported. In all cases the purified enzymes hydrolyzed laminarin, pustulan, and *p*-nitrophenyl β-D-glucoside (PNPG). For a given species, the ratios of these three activities remained constant throughout the purification procedure. Because of this and other evidence, the authors concluded that a single protein exhibited all three activities.

More recently a β-glucanase has been partially purified from the intracellular extracts of *Kluyveromyces lactis* (*40*). Although not studied in detail, this enzyme exhibited properties similar to those previously described for *K. fragilis* (*38*). Although the ratio of activities of exo-glucanases from different yeasts on laminarin and pustulan varies (*38*), an exo-β-glucanase from *Candida utilis* appears to be specific for laminarin only. It had no activity on pustulan or gentiobiose (*41*). However, PNPG was hydrolyzed by this partially purified enzyme.

Abd-El-Al and Phaff (*39*) also found β-glucanase activities in intracellular extracts of *Hanseniaspora valbyensis* and *Hanseniaspora uvarum* and in the culture fluid of *H. valbyensis.* In contrast to the other yeasts, pustulan was not hydrolyzed by these preparations, and laminarin was hydrolyzed yielding oligosaccharides. This suggested that these yeasts produced an endo-β-(1→3)-glucanase, and subsequently such an enzyme was partially purified from the culture fluid of *H. valbyensis* (*39*).

Barras reported the presence of an endo-β-(1→3)-glucanase in crude homogenates of *Schizosaccharomyces pombe* (*42*). Approximately

95–100% of this activity could be sedimented by centrifugation at 1000 × g for five minutes. A clean cell wall sample prepared from the crude homogenate accounted for approximately 54% of the total glucanase activity. Intact cells exhibited roughly 40% of the β-glucan hydrolase detectable in cell homogenates, suggesting that at least 40% of the total enzyme activity had an intimate association with the cell wall in vivo. After treatment of the cells with sodium dodecyl sulfate and KCl (which is thought to open the cell wall matrix) the β-glucanase activity in vivo showed a two-fold increase, so that possibly up to 80% of the enzyme might be cell wall bound. No β-(1→6)-glucanase or α-(1→3)-glucanase activities were detected in any of his preparations. Tests for the latter enzyme were made because S. *pombe* walls contain α-(1→3)-glucan (*26*). Barras also noted the phenomenon of cell wall autohydrolysis. When washed cell wall preparations of S. *pombe* were incubated in the presence of buffer, they underwent hydrolysis resulting in the release of proteins, β-(1→3)-linked oligosaccharides, and high-molecular-weight, soluble glycans. At the same time a "solubilization" of the wall-associated β-(1→3)-glucanase was observed. Recoveries of the enzyme in the solubilized form, however, were low and variable because of instability of the enzyme, and attempts at further purification of this glucanase were unsuccessful.

The association of endo-β-(1→3)-glucanases with yeast cell walls (*S. cerevisiae* and *S. carlsbergensis*) has been mentioned briefly also by Bacon et al. (*43*) and by Matile and co-workers (*44, 45*). However, these workers did not isolate or determine the properties of these enzymes further.

Fleet and Phaff (*46, 47, 48*) also studied the glucanases associated with the fission yeasts. During an initial screening of the four recognized species of the genus *Schizosaccharomyces, Sch. pombe, Sch. versatilis, Sch. octosporus,* and *Sch. malidevorans,* it was found that *Sch. versatilis* contained much higher levels of β-(1→3)- and β-(1→6)-glucanase activities than the other species. No α-(1→3)-glucanase could be demonstrated in any of these four organisms. The stability of the β-glucanases of *Sch. versatilis* was much greater than reported by Barras (*42*) for *Sch. pombe*; this was confirmed by Fleet and Phaff (*47*). A detailed study of the β-glucanases associated with *Sch. versatilis* revealed that the cell-free extract and the culture fluid in which the yeast had been grown contained only exo-β-glucanase activity, which had properties comparable with the enzyme isolated from other yeasts (38). The properties of the intracellular and extracellular enzymes were identical. An electrophoretically homogeneous enzyme preparation showed a molecular weight of 43,000 daltons. The enzymes followed Michaelis-Menten kinetics, and the V_{max} values for soluble laminarin and pustulan hydrolyses were 350 and 52 μmol of glucose released/min/mg of protein, respectively. The corresponding K_m values were 6.25 mg/ml for laminarin and 166 mg/ml for pustulan. The pH optimum for either substrate was 5.0. Although the enzyme showed good activity on soluble laminarin and pustulan, its activity on alkali-insoluble glucan from *Sch. pombe* was rather weak and limited. After about 10% of the total possible hydrolysis had taken place, the reaction came to a halt. Only negligible activity was noted on cell walls or whole cells of *Sch. pombe*. This enzyme therefore does not seem to play a major role in cell wall lysis.

When cell walls of *Sch. versatilis* were stored in buffer at pH 5.0, a slow autohydrolysis took place with release of β-(1→3)-linked oligosaccharides, glucose, and β-glucanase activity (*47*). Under appropriate conditions, up to 80% of the cell-wall-associated β-glucanase could be obtained in soluble form. During the autolytic process the originally thick walls became very thin in appearance under the microscope. The solubilized enzyme was separated into an exo-β-glucanase with properties similar to those of the cytoplasmic enzyme [87% of the total β-glucanase (laminarinase) activity] and an endo-β-(1→3)-glucanase, accounting for 13% of the total activity. The endo-β-(1→3)-glucanase was purified until it was electrophoretically homogeneous. It had a molecular weight of 97,000 daltons and hydrolyzed laminarin in a random manner via various oligosaccharides to a mixture of laminaribiose and glucose. The enzyme had an optimum pH of 5–6, a V_{max} of 5 μmol of glucose equivalents released/min/mg of protein and a K_m value of 0.33 mg of laminarin/ml. This enzyme showed no activity with pustulan. With laminarin it showed sigmoidal kinetics with two cooperative active sites per enzyme molecule. In contrast to the exo-glucanase, the endo-glucanase caused extensive cell wall lysis; complete lysis of *Schizosaccharomyces* cell walls was obtained only after the yeast enzyme was supplemented with a bacterial endo-α-(1→3)-glucanase (*49*).

Evidence for the presence of endo-β-(1→3)-glucanases in budding yeasts also has been forthcoming in recent years. In the bipolarly budding yeasts of the genus *Hanseniaspora* Abd-El-Al and Phaff (*39*) demonstrated in cell extracts and in the culture fluid an endo-β-(1→3)-glucanase. The enzyme was partially purified and was specific for polysaccharides with β-(1→3)-glucosidic bonds. Its lytic activity on cell walls was not tested. Exo-β-glucanases appeared to be absent in these yeasts.

Evidence for the presence of endo-β-(1→3)-glucanase in multilaterally budding yeasts was first supplied in 1970 by Bacon et al. (*43*) who mentioned briefly that such an enzyme was associated with cell walls of baker's yeast. Maddox and Hough (*50*) detected endo-β-(1→3-glucanase activity in chloroform-induced autolyzed brewer's yeast, while

Arnold (*51*) partially purified such an enzyme from cell-free extracts of baker's yeast. Matile et al. (*45*) and Cortat et al. (*44*) demonstrated the existence of glucanase-containing vesicles within the cytoplasm of *Saccharomyces cerevisiae.* These vesicles contained exo- as well as endo-glucanases but the enzymes were not studied in detail. Fleet and Phaff (*47*) obtained qualitative evidence for the occurrence of endo-β-$(1 \rightarrow 3)$-glucanases in the cell walls of *Saccharomyces rosei, Kluyveromyces fragilis, Hansenula anomala, Pichia pastoris,* and *Candida utilis. K. fragilis* and *H. anomala* contained only exo-glucanase in cell extracts (*38*).

Several recent publications indicate that the β-glucanase composition of budding yeasts may be even more complex than originally thought. Farkas et al. (*52*) reinvestigated the β-glucanases of *Saccharomyces cerevisiae.* After establishing that protoplasts of *S. cerevisiae* suspended in a synthetic growth medium excreted β-glucanase into the growth medium (*53*), they fractionated the β-glucanases excreted by protoplasts and intact cells of *S. cerevisiae* by DEAE-cellulose chromatography. The results with both types of cells were similar. The β-glucanase activity could be separated into three activity peaks, two of which partially overlapped. One of the peaks (separate from the other two) represented enzyme which showed activity against laminarin, pustulan, and *p*-nitrophenyl-D-glucoside. Since it also had a low but detectable activity against oxidized laminarin, this enzyme fraction may have represented one form of exo-β-glucanase contaminated with a small amount of endo-β-$(1 \rightarrow 3)$-glucanase. The other two peaks represented a typical endo-β-$(1 \rightarrow 3)$-glucanase (specific for laminarin) and an exo-β-glucanase exhibiting activity for laminarin, pustulan, PNPG, and gentiobiose, typical of other exo-β-glucanases discussed above. The occurrence of two exo-β-glucanase forms in an extract of *Schizosaccharomyces versatilis* was also noted by Fleet and Phaff (*47, 48*), although only one of the forms was studied in detail. Villa et al. (*54*) found three enzymes with laminarinase activity in cell extracts or in the medium in which cells or protoplasts of *Pichia polymorpha* were cultivated. They found separation to be best on columns of DEAE Sephadex A50. In contrast to the findings with *S. cerevisiae* (*52*) the most abundant component (79%) represented a specific exo-β-$(1 \rightarrow 3)$-glucanase without activity on pustulan. This enzyme fraction could be similar in specificity to the exo-β-$(1 \rightarrow 3)$-glucanase from *Candida utilis* (*41*) which was reported not to hydrolyze pustulan. About 20% of the total β-glucanase of *P. polymorpha* was represented by a typical, nonspecific exo-β-glucanase and only 1% by an endo-β-$(1 \rightarrow 3)$-glucanase.

Information about endogenous lytic enzymes affecting the cell walls of basidiomycetous yeasts is less extensive. Villanueva and Gacto (*41*) reported that β-$(1 \rightarrow 3)$-glucanase activity was lacking in cell extracts of

Rhodotorula mucilaginosa, Rh. minuta, Sporobolomyces albidus, and *Sp. salmonicolor,* but present in *Cryptococcus diffluens.* However, Meyer (*49*) clearly demonstrated such activity in the intracellular extract and in the cell walls of *Rhodotorula minuta* var. *texensis, Rh. glutinis, Cryptococcus albidus, Cr. terreus,* and *Cr. laurentii.* The activity in the cell walls of these yeasts was always much higher than that of the intracellular extract. β-($1 \rightarrow 6$)-Glucanase activity of these species was either lacking or extremely low, but chitinase activity was significant in most strains.

Meyer (*49*) also searched for a α-($1 \rightarrow 3$)-glucanase activity in basidiomycetous yeasts and in species of *Schizosaccharomyces* and *Endomyces,* known or suspected to contain α-($1 \rightarrow 3$)-glucan in their walls. He used soluble carboxymethyl-α-($1 \rightarrow 3$)-glucan to increase the sensitivity of the assay. Although he confirmed the absence of α-($1 \rightarrow 3$)-glucanase in species of *Schizosaccharomyces* (*42, 47*), the enzyme was demonstrated in cell walls and, to a limited extent, in the intracellular extract of several *Cryptococcus* and *Rhodotorula* species. Its presence in *Rhodotorula* was surprising since no evidence for the presence of α-($1 \rightarrow 3$)-glucan in this genus could be obtained. *Phaffia rhodozyma,* a species which contains substantial amounts of cell-wall-associated α-($1 \rightarrow 3$)-glucan, lacked α-($1 \rightarrow 3$)-glucanase activity, while *Endomyces tetrasperma* showed only slight activity in its walls. Meyer (*49*) attempted to purify the α-($1 \rightarrow 3$)-glucanase from *Rhodotorula minuta* var. *texensis,* but the intracellular as well as the cell wall-associated enzymes were very unstable to storage at 4°C. Dialysis of the intracellular extract resulted in activation of an inactive fraction of the enzyme, but the activated fraction was also rapidly inactivated. For this reason a study of the properties of this enzyme was not made. Since α-($1 \rightarrow 3$)-glucan is a structural wall component, it is curious that in several of the species containing this polysaccharide no α-($1 \rightarrow 3$)-glucanase activity could be demonstrated in log-phase cells. It is also surprising to find this enzyme in cells not containing this polysaccharide in the cell walls.

Function and Localization of Yeast-associated Glucanases. Abd-El-Al and Phaff (*38, 39*) noted that the β-glucanase levels in *K. fragilis, H. anomala,* and *H. valbyensis* were much higher than the levels of these enzymes in *S. cerevisiae, S. elegans,* and *H. uvarum.* The first three yeasts are examples of species of which the ascus wall lyses rapidly upon maturation of the spores, whereas in the last three species mentioned, the ascus wall remains intact until it is ruptured by a swelling process during germination. Noting this correlation, these workers suggested a role of the yeast glucanases in ascus wall lysis and subsequent spore liberation. Judging by the very limited activity of exo-β-glucanases on isolated cell walls or whole cells of yeast (*46*), it seems most likely that endo-β-(1→3)-glucanases play the major role in such lysis, perhaps supported by exo-β-glucanase action. There is no information regarding the activation forces on the enzymes while the cell wall changes into an ascus wall. One assumption that can be made is that synthetic repair processes of damage caused by lytic enzymes is no longer possible when biosynthesis and activity of repair enzymes is locked away in the ascospores, and the former cell wall, now a nonviable shell for the spores, has become susceptible to breakdown by lytic enzymes.

Johnson (*55*) has proposed that cell wall extension in yeasts takes place by the concerted action of a nucleoside diphosphate glucose glucosyl transferase and a glucanase or glucanases, such that the hydrolytic enzyme(s), presumably of an endo-action pattern, produces breaks in the existing glucan network into which new glucosyl units are incorporated.

In budding yeasts such as S. *cerevisiae*, cell division involves the modification of cell wall properties at the location of new buds. The sudden extrusion of buds seems to depend on the local softening of the existing cell wall (*56*). Cytological studies by Moor with S. *cerevisiae* have shown that the budding process is initiated by a striking localized vesiculation of the endoplasmic reticulum (*57*). The small vesicles produced are discharged into the cell wall at the site of the prospective bud. The secretion of these vesicles continues until growth of the daughter cell is completed. It was postulated that these vesicles may contain cell wall modifying enzymes.

Matile and co-workers (*44, 45*) have demonstrated the association of β-(1→3)-glucanases with vesicular type particles isolated from S. *cerevisiae*. Comparisons were made of the total β-(1→3)-glucanase activity in mechanically prepared crude extracts of S. *cerevisiae* and the intracellular extracts obtained by lysis of prepared protoplasts. The protoplast extracts gave only 20% of the total glucanase found in the mechanically prepared extracts, and part of this activity was associated with vacuoles or sedimentable material. On the basis of these findings, as well as some observations of glucanase activity exhibited by intact cells, Matile and co-workers (*44*) suggested that approximately 85–90% of the β-(1→3)-glucanase associated with whole cells is external to the cytoplasmic membrane and that this represents the soluble enzyme in mechanically prepared cell-free extracts.

Further evidence for the location of cell-associated β-glucanases external to the cytoplasmic membrane has been forthcoming in recent years. Farkas et al. (*52*) confirmed the findings of Cortat et al. (*45*) that about 87% of the total exo-β-(1→3)-glucanase activity of S. *cerevisiae* is released when the cells are converted to protoplasts. They concluded that this enzyme is localized in the periplasmic space and that a part of it is excreted into the growth medium during the budding process. More direct evidence for the location of the β-glucanases in the periplasmic space was offered by Villanueva and Gacto (*41*) with cells of *Candida utilis*. They were able to inactivate selectively the β-(1→3)-glucanase activity of living cells by mild mineral acid treatment without affecting significantly the viability of the cells or the internal β-glucosidase. The same was shown for cells of *Pichia polymorpha* by Villa et al. (*54*). Since during cell breakage the exo-β-glucanase of budding yeasts appears mainly in the soluble extract, the enzyme does not appear to be linked covalently to the cell wall. Mechanical cell disruption must cause extensive damage to the wall-adhering membrane causing release of the enzyme.

Studies with synchronously dividing *S. cerevisiae* cultures have given further support to a functional role of yeast glucanases in cell budding. Cortat et al. (*44*) observed an increase in the glucanase activity at the phase of growth immediately preceding the initiation of budding. The specific activity of glucanase present in exponentially growing cells was nearly twice as high as in stationary phase cells when practically no budding cells were present.

Maddox and Hough (*50*) also noted cyclical changes in the glucanase and mannanase activities of synchronously dividing cultures of *S. carlsbergensis*. The enzyme activities rapidly increased during the phase immediately preceding budding, but they decreased markedly with the onset of budding.

Brock reported (*58*) that conjugation in *H. wingei* required protein synthesis and, noting that β-glucans form the rigid component of yeast cell walls, he postulated that conjugation may involve the induction of cell wall softening enzymes which act on the mating cells. He later showed that during vegetative growth, cells possessed low levels of β-(1→3)-glucanase activity and that this activity increased sharply during conjugation (*37*). A similar sharp increase in β-glucanase activity of *Schizosaccharomyces versatilis* during the early stages of cell agglutination and conjugation was noted by Fleet and Phaff (*46*). It is possible that the mating factors produced by haploid, heterothallic yeasts are directly or indirectly responsible for the induction of glucanases involved in cell wall softening. For example, α-cells of *S. cerevisiae* produce a peptide referred to as α-factor, which causes copulatory outgrowths (so-called "schmoos") of *a-cells.* A similar factor produced by cells of *a* mating type, which affects α-cells, has also been found (*59, 60*).

There is some evidence to suggest that β-glucanases may play a role in the expansion of yeast cells in the absence of cell duplication. Yanagishima (*61*) and Shimoda et al. (*62*) discovered that a plant hormone, auxin (indole-3-acetic acid), induced cell elongation in auxin-responsive mutants of *Saccharomyces ellipsoideus* when added to the culture medium. Since spheroplasts did not respond to this hormone, it was concluded that auxin plays an essential role in the expansion of the cell wall itself. Next, Shimoda, and Yanagishima (*63*) investigated the possibility that enzymic degradation of the glucan portion of the cell wall might be involved in wall loosening and thus cause cell expansion. They used an exo-β-(1 $\rightarrow$ 3)-glucanase of fungal origin and showed that β-(1 $\rightarrow$ 3)-glucanase induced cell expansion more rapidly than auxin, but only in auxin-responsive strains of yeast. In a subsequent paper (*64*), they reported that glucono-δ-lactone, a potent inhibitor of glucanase-type reactions, inhibited auxin-induced cell expansion. With the realization that yeasts possess endogenous β-(1 $\rightarrow$ 3)-glucanases, it was thought that the action of glucono-δ-lactone might be an inhibition of glucanase activity. Crude cell-free extracts of S. *ellipsoideus* exhibited β-(1 $\rightarrow$ 3)-glucanase activity, and it was possible to inhibit this activity by very low concentrations of glucono-δ-lactone. The β-(1 $\rightarrow$ 3)-glucanase activity in extracts of auxin-treated cells was always higher than the activity observed in the extracts of untreated cells. The authors concluded that indole-3-acetic acid induces cell expansion by activation of cell wall-degrading enzymes, including β-(1 $\rightarrow$ 3)-glucanase.

Cell Wall Lysis by Endogenous β-Glucanases. The removal or appreciable weakening of yeast cell walls to facilitate protein recovery from the cells would be most conveniently accomplished by endogenous glucanases. It is now fairly well established that the innermost, alkali-insoluble, micro-fibrillar glucan layer must be hydrolyzed in order to disrupt the integrity of the cell envelope. The critical enzymes needed for this process are endo-β-(1 $\rightarrow$ 3)-glucanases which have been demonstrated in wall preparations from several yeasts, and these enzymes, together with specific or nonspecific exo-β-glucanases, appear to be located in the periplasmic space between the cytoplasmic membrane and the innermost glucan layer. It is also known that the level of these enzymes varies greatly in different species of yeasts and at different stages of the growth cycle. As discussed in the foregoing sections, these enzymes appear to be inactive in resting cells of the stationary growth phase, and show only transient and often localized activity during certain phases of the cell cycle. Little is as yet known about activation and inhibition of these enzymes by cell-associated factors. It would seem desirable to conduct further research with species of yeast with potentially high levels of endo-β-(1 $\rightarrow$ 3)-glucanases and to determine the cell's control mechanisms for their activity. Interference with these control mechanisms by mutation or environmental factors could then lead to β-glucanase activation at a desired point of the growth cycle and, as a consequence, cell wall breakdown and release of protein from the cells.

Yeast Cell Wall Hydrolysis by β-Glucanases from Microbial and Other Sources

Research on the enzymatic digestion of yeast cell walls by exogenous enzymes has been stimulated for three main reasons: (a) the preparation of protoplasts or spheroplasts for physiological studies, (b) studies on the biosynthesis and regeneration of cell walls from protoplasts, and (c) analysis of native and isolated cell walls (*1, 65*). Wall analysis with selected and specific enzymes avoids the often used harsh treatments of cells or cell wall material with strong acids and alkali. Such enzymes can cleave specific linkages or selectively remove certain groups from macromolecules with minimal modification (*65*), but great care must be taken of using only highly purified enzyme preparations to avoid false conclusions.

The methods used to determine cell wall lytic activity of enzyme preparations vary considerably in different laboratories. The following approaches have been used:

1. Since the principal bonds in yeast glucan are β-(1→3) and β-(1→6), the activity of enzyme preparations on the β-(1→3)-glucan laminarin and on the β-(1→6)-glucan pustulan have been determined. However, even though lytic enzymes are characterized by β-(1→3)-glucanase activity, some endo-β-(1→3)-glucanase preparations with laminarinase activity are devoid of lytic activity on cell walls (*66*).

2. The use of purified cell walls as a substrate has provided more reliable results to judge lytic activity. Such lytic activity can be followed either by the cup-plate assay where cell walls are incorporated in agar and clearing is observed visually (*67, 68*), or cell walls are suspended in buffer and the decrease in optical density is followed spectrophotometrically (*69*). Although this procedure demonstrates lytic activity towards isolated cell walls, such preparations may lack activity on living cells (*70*).

3. Determination of changes in optical density of viable cell suspensions has been done both in the presence of osmotic support (*71*) and in its absence (*72*). If an osmotic support is used, dilution of the suspending medium results in rapid lysis of osmotically sensitive cells and a corresponding decrease in optical density. This procedure can also be used in combination with microscopic inspection of the treated cells.

Gastric Juice of the Garden Snail, *Helix pomatia*. Giaja (*73, 74 75*) observed many years ago that the digestive juice of *Helix pomatia*, which is found in a small vesicle of the alimentary canal, possesses the ability to digest yeast cells. His main interest was cell wall removal for cytological studies. Eddy and Williamson (*76, 77*) used gastric snail juice in osmotically stabilized media for the purpose of producing protoplasts.

They found that log-phase cell walls were much more susceptible to hydrolysis than walls of stationary phase cells, while Holter and Ottolenghi (*78*) noted considerable variation in susceptibility to protoplasting between different strains of *Saccharomyces.* The latter observation has been confirmed by others (*79, 80*). In general, sulfhydryl reagents such as cysteine, 2-mercaptoethylamine, or dithiothreitol promote the ease with which cells are converted to protoplasts (*81, 82, 83*). This effect is presumably caused by the fission of disulfide bonds of cell wall proteins (*14*), which thereby opens up the wall structure making the underlying glucan layer more susceptible to the enzyme action. There is also evidence in the literature that there are great differences in susceptibility to protoplasting among various species of yeast (*1*).

Hydrolysis of the wall material during protoplasting of various yeasts by snail enzyme is not uniform over the whole cell surface, indicating differences in thickness, composition, or both. Initial breaks in the wall are usually seen first in the equatorial and the polar regions of the cells; the spheroplasts usually leave the cell through these initially formed ruptures. In *Candida utilis* spheroplasts usually leave the residual hull through an equatorial rupture (*84*); in *Schizosaccharomyces pombe* (*80*) and in *Endomyces magnusii* (*85*) they usually leave near one of the poles of these cylindrical cells; in *Saccharomycodes ludwigii* (a bipolarly budding yeast), at one of the poles; and in *Sporobolomyces,* at the location of ballistospore discharge (*80*).

Darling et al. (*86*), who used a strain of *S. cerevisiae* with elongated cells, observed that the formation of spheroplasts by snail enzymes went through two stages. In the first one the cells lost their thick walls but retained their elongate shape if osmotically protected. Cells of this stage were termed prospheroplasts. The prospheroplasts then assumed a spherical shape during the second stage, presumably shedding a very thin remnant of the alkali-insoluble, microfibrillar glucan layer adjacent to the cytoplasmic membrane. The attack on the wall probably starts with hydrolysis of the intermediate, alkali-soluble glucan layer, resulting in shedding of most of the wall.

It is interesting that when yeast cells transform into asci, the walls become much more susceptible to enzymic digestion. As mentioned in the previous section ascus walls of some species are lysed by endogenous glucanases, liberating the ascospores. Use has been made of the ability of snail enzyme to digest ascus walls in the isolation of single ascospores for genetic studies, from yeasts in which the ascus wall does not lyse spontaneously (*87*). The walls of ascospores are not affected by snail enzyme.

Snail digestive fluid has been reported to contain some 30 enzymes, including glucanases, mannanase, and chitinase, but it appears to be very low in proteolytic activity (*88*) and it lacks α-(1→3)-glucanase entirely (*89*). The enzymes of snail gut juice have been analyzed (*90, 91*). Anderson and Millbank (*91*) fractionated the enzyme complex by gel filtration over Sephadex G-100, and they tested the fractions on log-phase cells of *Saccharomyces carlsbergensis* and on model substrates. Two fractions with β-glucanase activity showed very limited activity on yeast cells or cell walls. These were shown to contain exo-β-(1→3)- and exo-β-(1→6)-glucanase activities. A third fraction caused extensive degradation of walls of the test yeast, and enzymes in this fraction hydrolyzed laminarin and the β-(1→6)-glucan, lutean, with the formation of mono-, di-, and trisaccharides. This provided evidence (a) for the presence of both endo-β-(1→3)- and endo-β-(1→6)-glucanase in snail enzyme, and (b) that these two endo-glucanases are essential factors in yeast cell wall degradation. Lampen and co-workers (*92, 93*) have postulated that in addition to an endo-β-glucanase system, a phosphomannanase is also required for effective protoplast formation. This enzyme was thought to open the outer mannan layer by its partial hydrolysis. However, since highly purified β-glucanase preparations from both yeast (*47*) and bacterial origin (*94, 95*), which lacked mannanase or phosphomannanase activity, were lytic to whole cells, it is questionable that phosphomannanase is an absolutely essential enzyme in the lytic process.

Lysis of Walls by Microbial Enzymes. Salton (*96*) first isolated a number of actinomycetes and myxobacteria which showed lytic activity when streaked on an agar plate containing autoclaved cells of *Candida pulcherrima.* The lytic organisms included *Cytophaga johnsonii, Myxococcus fulvus,* and strains of *Streptomyces.* Later Webley et al. (*70*), with another strain of *Cytophaga johnsonii,* showed that this myxobacterium also hydrolyzes isolated cell walls or the walls of autoclaved intact baker's yeast and that it excretes an active glucanase. However, no activity could be demonstrated on the walls of live yeast cells. This lack of activity was thought to be caused by an outer barrier of the wall that may be destroyed upon autoclaving. In a brief note (*97*) Bacon and co-workers reported attempts at chromatographic fractionation of the β-glucanases produced in the culture fluid by this myxobacterium. Three fractions (A, B, and C) which exhibited laminarinase activity were obtained. Fractions A and B both lysed yeast cell walls, but fraction C, representing most of the laminarinase activity, did not lyse cell walls, either alone or in the presence of an endo-β-(1→6)-glucanase. Fractions A, B, and C were all endo-glucanases, but whereas fraction C produced mainly glucose, laminaribiose, and laminaritriose from laminarin, fractions A and B did not degrade this polysaccharide beyond the pentasaccharide stage. These observations were later expanded in a full-length paper (*43*) which reported that the nonlytic and lytic endo-β-(1→3)-glucanases had been highly purified and their properties studied in more detail. The lytic enzyme (Fractions A and B) was able to hydrolyze effectively only long-chained glucans. These were important observations since they showed that an endo-laminarinase [endo-β-(1→3)-glucanase] does not necessarily act as a lytic enzyme when it is added to a highly insoluble, microfibrillar substrate such as alkali-insoluble yeast glucan. Apparently, besides having endo-β-(1→3)-glucanase activity, a lytic enzyme must have other properties enabling it to hydrolyze this type of glucan.

Aside from the above-mentioned myxobacterium *Cytophaga,* numerous species of streptomycetes, true bacteria as well as higher fungi, have been shown to produce enzymes which are lytic to yeast cell walls. The most extensive and definitive work has been done with *Arthrobacter* sp. (*98, 69, 99, 100, 101, 102*), *Bacillus circulans* (*103, 104, 105, 106, 66, 67, 94, 95*), *Flavobacterium* sp. (*107, 108*), *Micromonospora* sp. (*109*), *Oerskovia* (*71, 110*), *Streptomyces* sp. (*111, 112*), *Alternaria* sp. (*113*), *Basidiomycete* sp. (*114*), *Corticium* sp. (*115*), *Deuteromycete* sp. (*116*), *Gliocladium* sp. (*117*), *Rhizopus* sp. (*118, 119*), and *Trichoderma* sp. (*120*). It is evident from the rather vague descriptions of some of these organisms that their taxonomic designations are not firmly established in all cases. There are considerable differences in the mechanism by which the enzymes from these sepcies hydrolyze cell wall glucan. These differences manifest themselves, at least in part, by the products of the enzymatic hydrolysis. In the following part the enzymes elaborated by the species listed above will be discussed on the basis of the products they produce, which is the direct result of their action patterns. The following five groups will be considered: (a) enzymes which cause random hydrolysis of β-(1→3)-glucans to glucose and laminaribiose; (b) enzymes which produce oligosaccharides of the laminarin series with a degree of polymerization of five or larger. These enzymes may have an exo- or an endo-action pattern or can be debranching enzymes, or a combination of both; (c) yeast glucan debranching enzymes; (d) enzymes which produce glucose and gentiobiose from laminarin or from yeast glucan by an exo-mechanism; (e) enzymes with action patterns which have not been clearly defined.

ENDO-ENZYMES PRODUCING DISACCHARIDES AND GLUCOSE FROM POLYSACCHARIDES. The most thoroughly studied enzymes of this group are produced by strains of *Bacillus circulans.* Horikoshi and Sakaguchi (*121*) isolated a strain of this bacterium from soil, which showed lysis of *Aspergillus oryzae* and *Saccharomyces sake* cell walls. The lytic enzyme complex was induced by growing the bacterium on *Aspergillus* mycelium as sole carbon source. The crude enzyme lowered the turbidity of a heated suspension of *S. sake*, but a suspension of living cells was attacked only poorly. Proteolytic activity was negligible or lacking in the preparation. Later Horikoshi et al. (*103*) separated a β-(1→3)-glucanase from the culture liquid of *B. circulans* grown on *Aspergillus oryzae* mycelium. The enzyme, which hydrolyzed laminarin to glucose and laminaribiose by a random mechanism, also hydrolyzed cell walls of a *Fusarium* species and of *Aspergillus oryzae* (*104*).

Tanaka and Phaff (*67*) isolated another strain from soil, which was designated *B. circulans* WL-12. This strain produced lytic enzymes when grown aerobically in a mineral medium with baker's yeast cell walls as the inducer. They separated the enzyme complex by chromatography over a DEAE-cellulose column into an endo-β-(1→3)-glucanase and an endo-β-(1→6)-glucanase. The former enzyme produced complete clearance by the cup-plate assay in a baker's yeast cell wall-agar plate. The clear zones were of small diameter and had sharp edges. The latter enzyme produced rather large zones of partial clearing with diffuse edges under the same conditions (*67, 68*). The ability of the separate or combined glucanases from *B. circulans* to produce spheroplasts from baker's yeast was extremely limited and was only slightly improved by the addition of sulfhydryl compounds (*68*). However, certain yeast-like fungi (e.g. *Eremothecium ashbyi* and *Ashbya gossypii*) yielded spheroplasts readily with a mixture of the two glucanases. Tanaka et al. (*68*) also studied the lysis of yeast cell walls by the so-called cross-induction technique. *B. circulans* was inoculated on a rectangular agar block containing the walls of a certain species of yeast. Small square blocks of agar containing walls of other species of yeasts were then placed in contact with the rectangular block. The enzymes produced in response to the inducer species diffused into the adjacent blocks and produced clearing of varying degrees and intensities. These experiments showed the great variation in polysaccharide composition of various yeasts as well as the high potential of *B. circulans* to adapt to different inducers. This last conclusion was based on the complete clearing of the blocks containing the inducer cell wall substrate.

The β-glucanase complex of *B. circulans* WL-12, however, has proven to be considerably more complex than originally thought. Fleet and Phaff (*66*) grew *B. circulans* on baker's yeast walls (*46*) and subjected the β-(1→3)-glucanase (tested on laminarin) and β-(1→6)-glucanase (tested on pustulan) to a more thorough purification, involving Sephadex G-100, DEAE-cellulose and carboxymethyl cellulose. The two highly purified β-glucanases caused only a very limited hydrolysis of baker's yeast cell walls, not sufficient to observe any clearance in cell wall-agar plates (cup-plate assay). Since β-(1→3)-bonds are the main linkages in the structural glucan of the yeast cell wall (Figure 1), this finding was unexpected. Since the enzyme had access to the inner as well as the outer wall surface, it seemed unlikely that spatial inaccessibility of the enzyme to the glucan component(s) could be responsible. Considering that the crude culture fluid of *B. circulans* and partially purified β-glucanases from this bacterium (*67, 68*) caused extensive hydrolysis of yeast cell walls, it seemed most likely that some additional enzymes of the crude culture fluid, which were lytic to yeast cell walls, had been removed during the more extensive purification of the two β-glucanases.

To test this possibility Rombouts and Phaff (*94, 95*) followed the purification of the β-glucanase complex of *B. circulans* grown on baker's yeast cell walls by two separate procedures. In addition to the usual tests for β-(1→3)- and β-(1→6)-glucanases by reducing group increase on their respective soluble substrates, laminarin and pustulan, each fraction from the fractionating colums, was also tested for lytic activity on a cell wall suspension. They made two additional improvements in enzyme purification.

First, by using alkali-insoluble glucan rather than cell walls as enzyme inducer, enzyme purification was greatly facilitated since yeast mannan (which is not attacked by *B. circulans*) does not accumulate as a very viscous component in concentrated crude enzyme solutions to be applied to chromatography columns. Yeast glucan is at least as effective as an enzyme inducer as are yeast cell walls.

Secondly, it was found that the lytic activity could be selectively adsorbed (affinity adsorption) from the dialyzed crude culture fluid onto alkali-insoluble glucan. Laminarinase and pustulanase activities remained for the most part unadsorbed in the supernatant liquid. After autodigestion of the centrifuged glucan pellet the concentrated lytic enzyme preparation was subjected to purification by column chromatography (*94, 95*).

In addition to the nonlytic β-(1→3)- and β-(1→6)-glucanase discussed above, the concentrated lytic fraction could be resolved by hydroxyl apatite chromatography into one lytic β-(1→6)-glucanase (*94*) and two lytic β-(1→3)-glucanases (I and II) (*95*). Further purification of the three lytic enzymes was done over columns of DEAE-agarose and carboxymethyl cellulose, and their kinetic properties, action patterns, and molecular weights were determined. The lytic β-(1→6)-glucanase (*94*) with an optimum pH of 6.0 hydrolyzed pustulan via a series of oligosaccharides to a mixture of gentiotriose, gentiobiose, and glucose. The enzyme produced a small amount of gentiobiose from laminarin, which contains a few single glucose residues attached to the β-(1→3)-linked glucan. Since no oligosaccharides of the laminarin series were produced, the enzyme cannot be considered contaminated with a β-(1→3)-glucanase. When applied to a cup in a cell wall-agar plate, the enzyme caused a rather large zone of partial clearing (*see* Ref. *68*). On the basis of the ability of the enzyme to excise gentiobiose from laminarin and its ability to cause partial lysis of yeast cell walls, its lytic action is thought to be caused by a debranching of the alkali-insoluble glucan component. The enzyme differs in essence from the nonlytic β-(1→6)-glucanase of the same organism (*66*) by its positive action on yeast cell walls and yeast glucan, which may be caused in part by a high affinity for these substrates.

The lytic β-(1→3)-glucanase I (*95*) caused complete lysis of cell walls; in cup-plates it caused a small zone of clearing with a sharp border, while the lytic β-(1→3)-glucanase II showed a moderately spreading zone of lysis with a fuzzy outer border. Lysis of the cell walls by the latter enzyme proceeded slowly to completion. Properties of only the first enzyme were studied in detail. It hydrolyzes laminarin randomly to laminaribiose and glucose. Although its action pattern on laminarin is similar to that of the nonlytic β-(1→3)-glucanase (*66*), it differs from the latter in essence by its powerful lytic activity on insoluble yeast glucan which may be related to its high affinity for this substrate. A synergism with lytic β-(1→6)-glucanase was observed during the later stages of yeast glucan hydrolysis. This suggests that the proposed debranching activity of the lytic β-(1→6)-glucanase on insoluble yeast glucan facilitates the activity of the lytic-β-(1→3)-glucanase. It seems likely that the two nonlytic β-glucanases and the lytic β-(1→3)-glucanase II play additional roles in the rapid degradation of yeast cell walls by this bacterium. Action of the three lytic enzymes from *B. circulans* WL-12 on living yeasts has not been studied so far.

Recently Kobayashi et al. (*105*) and Tanaka et al. (*106*) also explored the production of multiple β-glucanases by *B. circulans* WL-12. Their activity profile, as shown by polyacrylamide gel electrophoresis of the enzyme mixture produced on S. *cerevisiae* cells, revealed four different β-glucanases active on soluble laminarin. These enzymes have not as yet been purified and studied further, but most likely they will prove to be identical with the two lytic β-(1→3)-glucanases, the nonlytic β-(1→3)-glucanase, and the lytic β-(1→6)-glucanase which releases gentiobiose from laminarin (*see* above). *B. circulans* was found to produce six laminarinase activities when *Piricularia oryzae* mycelium was used as the inducer substrate (*105*); one of these, which was not produced in significant concentration on *S. cerevisiae* cells, was further purified by electrophoresis, and some of its properties (*72*) resemble to some degree those of the lytic β-(1→3)-glucanase I of Rombouts and Phaff.

Bacillus circulans WL-12 is a versatile organism, since it is also able to produce high levels of endo-α-(1→3)-glucanase when grown on

α-(1→3)-glucan (pseudonigeran) as substrate (*49*). Since this glucan is an important structural component of species of *Cryptococcus,* of the fission yeast *Schizoscharromyces* and of several filamentous fungi (*26*), *B. circulans* is a suitable organism for the production of lytic enzymes able to cause hydrolysis of the walls of such yeasts and fungi. The α-(1→3)-glucanase must be supplemented with β-glucanases to be effective in wall lysis. This is accomplished by growing *B. circulans* on *Schizosaccharomyces* cell walls or a similar substrate containing both β-glucans and α-(1→3)-glucan. The α-(1→3)-glucanase of *B. circulans* WL-12 was purified (*49*) by a combination of affinity adsorption of the enzyme to water-insoluble α-(1→3)-glucan (followed by digestion of the glucan pellet by the adsorbed enzyme), DEAE-agarose chromatography, and polyacrylamide P-150 gel filtration. Its molecular weight was estimated to be 135,000 $\pm$ 10% based on its relative mobility in an SDS-polyacrylamide gel. The optimum pH was approximately 7.7 using soluble carboxymethyl-α-(1→3)-glucan as substrate. Pseudonigeran is hydrolyzed by a random mechanism to glucose and nigerobiose. Several heavy metal ions are inhibitory, and their effect can be reversed by thiol compounds; this suggests involvement of sulfhydryl groups that are essential for activity. The kinetic constants of the enzyme vary significantly with the substrate being tested, i.e., colloidal α-(1→3)-glucan or its soluble carboxymethyl derivative. A second α-(1→3)-glucanase was found in the culture filtrate, accounting for approximately 15% of the total α-(1→3)-glucanase activity. This enzyme was partially purified, but the limited amount of protein available prevented a detailed study of its properties. Since the lytic activity of the main enzyme on *Schizosaccharomyces* walls is rather limited in comparison with the unfractionated culture fluid, the lytic activity may be associated with the minor enzyme component.

ENZYMES PRODUCING OLIGOSACCHARIDES OF D.P. 5 AND LARGER FROM β-(1→3)-LINKED GLUCANS. Doi et al. (*98*) observed two types of β-(1→3)-glucanases during fractionation of the culture fluid of a yellow-colored *Arthrobacter* species when grown in the presence of baker's yeast cells. Yeast glucan was rapidly solubilized by the glucanase type I of their preparations and the reaction was accompanied by the accumulation of an oligosaccharide identified as laminaripentaose. Type II glucanase partially solubilized yeast glucan liberating laminaribiose and glucose. The latter enzyme was compared to the β-(1→3)-glucanase produced by *Bacillus circulans* (*67*), an enzyme later found to contain more than one component (*66, 94*). Although glucanase type I could be resolved into several protein bands by electrophoresis on polyacrylamide gel (*98*), the various components appeared to have enzymatic properties similar to the original glucanase I. In later publications (*69, 122, 123*)

Doi et al. reported further information on the lytic ability of the *Arthrobacter* species. They suggested that the laminaripentaose is not merely a product that accumulates as a *residue* of endo-hydrolysis but that the glucanase I cuts out this oligosaccharide as such from the interior of the long β-(1$\rightarrow$3)-linked portions of insoluble yeast glucan. Further work by disc-gel electrophoresis confirmed that the lytic β-glucanase I component is heterogeneous. The enzyme fractions could be grouped into those which exhibited relatively high lytic activity on yeast cell walls and those which showed much lower activity, but the action pattern of the various fractions was apparently the same. They also reported that preparations from young cultures had higher lytic activity than preparations from older cultures because the latter had a lower proportion of the most active lytic component. A clear interpretation of the properties of the various components is not yet possible.

Another purified enzyme preparation which produces laminaripentaose from insoluble laminarin and from heat-treated pachyman is produced by a strain of *Arthrobacter luteus* (*100, 101, 102*) when grown on yeast cells or β-(1$\rightarrow$3)-glucan. The enzyme, which was named Zymolase (also referred to as Zymolyase) appeared to be homogeneous by electrophoresis in a Tiselius apparatus and by ultracentrifugation. The molecular weight of the enzyme was estimated from ultracentrifugation to be ca. 20,500. The optimum pH for lysis of viable yeast cells was 7.5. The optimum temperature was 35°C. The optimum pH for heat-treated pachyman hydrolysis was 6.5, and the optimum temperature was 45°C. A Lineweaver–Burk plot with heat-treated pachyman yielded a K_m value of 0.04% when the solubilized carbohydrate was assayed by the phenol-sulfuric acid method. Zymolase lost all its activity after incubation at 60°C for 5 min.

Zymolase specifically hydrolyzed linear glucose polymers with β-(1$\rightarrow$3)-linkages, releasing laminaripentaose and leaving some higher oligosaccharides (DP $\geq$ 8) with laminarin as the substrate. Pachyman was resistant to hydrolysis unless it was first heated. The resistance of unheated pachyman appears to be related to its crystallinity. Zymolase hydrolyzed insoluble laminarin from *Laminaria* but not soluble laminarin from *Eisenia*. These data indicate that the enzyme has a specific requirement for long sequences of β-(1$\rightarrow$3)-linked residues. It can cause endo- or exo-hydrolysis (producing laminaripentaose fragments) depending of the chain length or solubility of the substrate.

Kaneko, Kitamura, and Yamamoto (*99*) studied the susceptibilities to lysis by crude enzyme from *Arthrobacter luteus* broth of 26 strains of yeast, representing 18 species and 8 genera, although most species belonged to *Saccharomyces* and *Candida*. Each yeast was tested during logarithmic and stationary phase of growth in different media. The effects of various treatments, such as heating or treatment with 2-mercaptoethanol and sodium dodecyl sulfate, on their susceptibility were also examined. Most strains of *Candida* tested were less susceptible than *Saccharomyces* yeasts, but became as susceptible after treatment with 2-mercaptoethanol. Species of *Rhodotorula* and *Sporobolomyces* did not show any susceptibility, whereas *Schizosaccharomyces pombe* showed low susceptibility to the enzyme. *Candida* yeasts grown on *n*-paraffin did not differ greatly in susceptibility from cells cultured in yeast extract.

Another lytic β-(1→3)-glucanase, which produces as end products from yeast glucan, pachyman, or laminarin mainly oligosaccharides with five or more glucose residues, has been isolated from the culture fluid of *Cytophaga johnsonii* by Bacon and co-workers (*97*). This enzyme acts on only long β-(1→3)-glucan chains, including alkali-extracted cell walls, but it has little effect on living yeast cells. Autoclaved whole cells, however, were susceptible to its action.

Mann et al. (*71*) and more recently Jeffries (*110*) have reported on a soil actinomycete, *Oerskovia xanthineolytica,* which produces a mixture of several β-(1→3)-glucanases capable of lysing various viable yeasts. The enzyme mixture was fractionated and the individual components purified after selective adsorption of the lytic enzymes to alkali-extracted yeast glucan, followed by autodigestion of the glucan (*95*). Biogel P-60 gave three protein peaks with laminarinase activity, only two of which exhibited lytic activity towards viable yeast. Further purification of the combined lytic fractions by chromatography over DEAE Bio-Gel A and on CM Bio-Gel A resulted in two lytic fractions with very low laminarinase activity. Two additional enzymes with laminarinase activity but negligible lytic activity were also purified. Recombination of lytic and nonlytic fractions gave a pronounced synergistic effect on cell lysis. The two lytic enzymes (termed IIIA and IIIB), when allowed to act on laminarin, produced from the beginning of the reaction laminaripentaose as the major product together with a minor product of DP 9. Thus, the two lytic enzymes were thought to act as exo-enzymes, removing five glucose residues at a time from the nonreducing terminals of β-(1→3)-glucan chains. When the residual substrate reached a DP of nine, the enzyme appeared unable to cause further degradation. Enzymes IIIA and IIIB apparently cannot bypass β-(1→6)-branch points in yeast glucan. This is evidenced by the fact that when yeast glucan was used as substrate, approximately 40% of the glucan was converted to laminaripentaose and DP-9 oligosaccharides; 20% was solubilized to products with DP between 15 and 35; and the remainder was of higher molecular weight and partially insoluble.

Yeast Glucan Debranching Enzymes. Rombouts and Phaff (*95*) postulated that the lytic β-(1→6)-glucanase which they purified from the culture fluid of *Bacillus circulans* grown on alkali-insoluble yeast glucan might have a debranching action on such glucan (*see* section on endo-β-glucanases which produce glucose and disaccharides from polymers). Their conclusion was based on the partial clearing of baker's yeast cell walls in the absence of laminarinase activity and on the excision of gentiobiose from laminarin and pachyman. The latter activity was thought to be caused by the ability of the lytic β-(1 → 6)-glucanase to cleave β-(1 → 3)-bonds next to a glucose residue substituted in the 6-position (*65, 124*). The partial hydrolysis of yeast glucan was thought to be caused by the cleavage of one or more β-(1 → 3)-bonds adjacent to the β-(1 → 6)-linkage involved in branching of the alkali-insoluble glucan. The remaining long chains of β-(1 → 3)-linked glucose residues are considered responsible for the residual opacity in cell wall plates treated with this enzyme. A similar type of specificity has been found in the hydrolysis of *Eisenia* laminarin [a glucan with mixed β-(1 → 3)- and β-(1 → 6)-linkages] by a *Gibberella* β-(1 → 6)-glucanase (*125*). In this case evidence was obtained to show that the β-(1 → 6)-glucanase split glycosidic linkages of the 6-substituted glucosyl units in the *Eisenia* laminarin. Hydrolysis of β-(1 → 3)-linkages in this glucan was confirmed by the hydrolysis of 3^2 gentiobiosyl-gentiobiose to gentiobiose by the β-(1 → 6)-glucanase.

Yamamoto et al. (*118*) described the purification and crystallization of a glucanase from the culture fluid of a species of the fungal genus *Rhizopus*. This enzyme did not hydrolyze laminarin, pachyman, or luteose, but it did hydrolyze alkali-insoluble yeast glucan and walls of living yeast with the predominant production of oligosaccharides containing 5–7 glucose residues. Although the nature of the glycosidic bonds hydrolyzed was not determined, these authors suggested that the enzyme may be an endo-β-glucanase of the debranching type, removing side chains of yeast glucan. If the products of this enzyme are indeed laminarin oligosaccharides of DP 5–7, it is difficult to see how this enzyme could act as a debranching enzyme, since the β-(1 → 3)-linked linear portions of the alkali-insoluble glucan are much longer than seven glucose residues (*3*). Later Yamamoto et al. (*126*) identified their fungus as *Rhizopus chinensis* and found that the lytic β-(1 → 6)-glucanase of this mold produced some laminaribiose from yeast glucan. This disaccharide could have originated from the β-(1 → 6)-glucan component (*6*) in which relatively short segments of β-(1 → 6)-linked glucose units are linked to similar units by β-(1 → 3)-bonds. The laminaribiose found as a product points to a difference in substrate specificity of the enzymes from *B. circulans* and *R. chinensis*.

CELL WALL LYTIC ENZYMES WHICH ACT BY THE SEQUENTIAL REMOVAL OF SINGLE GLUCOSE RESIDUES FROM β-(1 → 3)-GLUCANS. Exo-enzymes of this type are well known as endogenous enzymes from yeasts (*38, 48*). However, these enzymes are unable to cause lysis either of cell walls or of alkali-insoluble glucan to a significant extent. On the other hand, an exo-β-(1 $\rightarrow$ 3)-glucanase produced by a basidiomycetous fungus designated as QM 806 has the ability to produce protoplasts from several yeasts (*127*). The purification (*128*), action pattern and specificity (*129, 130*), and the hydrolytic mechanism (*131*) of this enzyme have been thoroughly studied. The exo-enzyme removes single glucose residues sequentially from the nonreducing ends of β-(1 $\rightarrow$ 3)-linked glucan chains. It cannot hydrolyze β-(1 $\rightarrow$ 6)-bonds as do the exo-enzymes from yeast, but, in contrast to the yeast exo-β-glucanases, the QM 806 enzyme can bypass β-(1 $\rightarrow$ 6)-branch points yielding gentiobiose quantitatively from laminarin or from baker's yeast cell walls. The enzyme showed decreasing K_m values and increasing V_{max} values on oligomers with increasing degree of polymerization from two to eight. The extent of binding and the rate of hydrolysis did not appear to reach maximum values even at a chain length of 15–20 glucose residues (*128*). Its ability to bypass β-(1 $\rightarrow$ 6)-linked side chains or branch points probably enhances its lytic potential. The enzyme was shown to hydrolyze cell walls of *Fusarium* and of *Aspergillus oryzae* (*104*) and of *S. cerevisiae* and *Wickerhamia fluorescence* (*127*). The protoplasts were observed to emerge from the walls of the cells and they did not exhibit any residual adhering material when examined by electron microscopy.

Lytic Enzymes with Action Patterns Which Have Not Been Clearly Defined. Yamamoto et al. (*116*) purified and crystallized an enzyme from a Deuteromycete which degrades yeast glucan and log-phase yeast cells. The crystallized enzyme appeared to be homogeneous and was shown to have a molecular weight of ca. 24,500 d. Log-phase cells of *S. cerevisiae* were disrupted by the enzyme preparation, although addition of 2-mercaptoethanol or phosphomannanase enhanced its effectiveness as a lytic agent. Nagasaki et al. (*132*) studied the enzymic and structural properties of the crystallized enzyme. The enzyme was fairly strongly inhibited by laminarin, but not by mannan, phosphomannan, luteose, starch, inulin, pectin, or dextran. Among the sugars and derivatives glucose, sucrose, fructose, mannose, and glucono-δ-lactone inhibited its action completely at appropriate concentrations, while galactose, maltose, melibiose, raffinose, ribose, and xylose had little effect.

The action of the Deuteromycete enzyme on yeast glucan was characterized by a rapid reduction in optical density of a glucan suspension (up to 60%), with a concomitant release of carbohydrate. However, the release of reducing equivalents (as glucose) was slow and very small. The reaction products from laminarin, as determined by Sephadex G-15 chromatography, consisted of a mixture of oligosaccharides ranging from laminaritriose to laminaridecaose. Since there was no evidence that laminarin oligosaccharides with a DP of less than ten could be hydrolyzed further, the smaller oligosaccharides found as products probably arose as remnants of the polymeric β-(1 → 3)-glucan (*116*).

Chesters and Bull studied extensively the enzymatic degradation of laminarin by various fungal species, and reported on the multicomponent nature of these enzymes (*34, 35*). The various types of laminarinases were separated by a combination of chromatographic procedures. One of their fungal species (*Myrothecium verrucaria* IMI25291) produced at least six laminarinase components: (a) an exo-β-(1 → 3)-glucanase producing laminaribiose, 1-0-β-glucosyl-D-mannitol, and gentiobiose from laminarin; (b) an endo-β-(1 → 3)-glucanase producing short-chained oligosaccharides (laminaribiose, -triose, and -tetraose) from laminarin; (c) and (d), imperfectly separated endo-β-(1 → 3)-glucanases producing only long-chained oligosaccharides (DP > 5) from laminarin; (e) and (f), β-glucosidases of low specificity hydrolyzing laminaribiose and gentiobiose. Evidence was obtained for a synergistic effect of these enzyme components, and a working scheme for laminarin hydrolysis was proposed (*133*). The endo-β-(1 → 3)-glucanases (c) and (d) act specifically on laminarin and long-chain fragments and were presumed to be important during the initial stages of hydrolysis. Their action would furnish substrates for the exo-β-(1 → 3)-glucanase (a) and the endo-enzyme (b), which can attack short-chain fragments in addition to the high molecular weight substrates. β-Glucosidase probably acts by hydrolyzing possible limit hydrolysis products such as laminaribiose and/or gentiobiose.

Sugimori et al. (*115*) screened strains of a number of molds (*Aspergillus, Botryotinia, Corticium, Fusarium, Giberella, Mucor, Penicillium, Rhizopus,* and *Sclerotium*) and *Streptomyces* for their abilities to release reducing sugars from heat-treated baker's yeast. Among 26 strains tested, *Corticium centrifugum* IAM 9028 showed the highest activity. The crude enzyme from this organism showed a broad pH activity curve from 2.0 to 7.0 with an optimum near pH 2.5. The heat-treated yeast cells were susceptible to the lytic enzymes over a wide range of temperatures and were readily hydrolyzed at 30°C. In contrast, living yeast cells were hardly attacked at that temperature. Appreciable activity against S. *cerevisiae* stationary-phase cells was observed only when the temperature was raised above 47°C, where it may be assumed that cell death occurred.

They tested 21 species of yeast representing nine genera for their susceptibility to the crude *Corticium* enzyme preparations (*115*). The stationary-phase cells of each yeast were incubated at 30°C for 12 hr with the crude lytic enzyme and were observed for the percent lysis (as calculated by a direct count of intact cells before and after incubation) and release of reducing sugars. Complete or nearly complete lysis was observed for *Candida lipolytica*, *C. guilliermondii*, *Endomyces hordei*, *Endomyces fibuliger*, *Endomycopsis capsularis*, *Schwanniomyces*, and *Torulopsis bacillaris*. Several other species showed intermediate degrees of lysis. Little or no significant lysis was observed with stationary-phase cells for strains of *Candida utilis*, *Rhodotorula rubra*, and *S. cerevisiae*. The individual enzymes constituting the lytic system from *C. centrifugum* were not studied in detail, but β-glucanase and proteinase activities were very high. Mannanase activity was not detected.

General Conclusions on Microbial Enzymes Capable of Causing Yeast Cell Wall Lysis. From the foregoing it is evident that because of variation in cell wall composition and ultrastructural arrangement of cell wall components, yeast species differ greatly in their susceptibility to lysis by a particular enzyme preparation. On the other hand, microorganisms have demonstrated great versatility in producing the appropriate inducible enzymes for cell wall lysis of different yeast species if they are grown on cell walls of a particular yeast species.

With the exception of those yeasts which contain substantial proportions of chitin in their walls (e.g., *Cryptococcus*) the principally β-(1→3)-linked, alkali-insoluble glucans are the structural components of most species of yeasts. In some species, such as *Schizosaccharomyces*, these are supplemented by α-(1→3)-glucans.

Enzymatic studies have shown that a β-(1→3)-glucanase, with demonstrated activity on soluble laminarin as the substrate, is not necessarily endowed with lytic activity on cell walls or on alkali-insoluble β-(1→3)-glucan. Neither is there any correlation between total laminarinase activity of crude preparation and their lytic activity. It is now known that cell lysis is caused by specific β-glucanase components, many of which have very low laminarinase activity. All microorganisms thus far examined appear to form complex mixtures of β-glucanases, of which only some have lytic ability. However, many authors have observed pronounced synergistic activity between lytic and nonlytic components in the process of cell wall digestion.

It is also clear from studies in recent years that there is no universal process by which the alkali-insoluble glucan of *Saccharomyces cerevisiae* or other species can be hydrolyzed enzymatically. Among the lytic β-glucanases the following action patterns have been identified: (a) endo-β-(1→3)-glucanases causing random hydrolysis of β-(1→3)-glucans to laminaribiose and glucose; (b) endo- or exo-β-(1→3)-glucanases which produce as end products laminaripentaose and larger oligosaccharides; (c) yeast glucan debranching enzymes; and (d) exo-β-(1→3)-glucanases which remove single glucose residues in a sequential manner from the nonreducing terminals of β-(1→3)-glucan chains. These glucanases can bypass β-(1→6)-linked branch points.

There are no clear answers to the question as to why some β-glucanases are lytic and others are not. The molecular size of the various enzymes does not appear to play a significant role. An important factor appears to be a high affinity of the lytic enzymes for the insoluble yeast glucan. Most of the K_m values of the β-glucanases have been determined with soluble laminarin as the substrate. Although the lytic enzymes have lower K_m values with this substrate than the nonlytic glucanases, these results do not necessarily apply to insoluble yeast glucan. However, since alkali-insoluble yeast glucan has been used by several investigators to selectively adsorb the lytic activity from crude enzyme preparations, it may be concluded that the nonlytic enzymes have a significantly lower affinity for this substrate. In this connection Clarke and Stone (*134*) suggested that the affinity of a particular glucanase varies with the degree of polymerization of the polysaccharide substrate. Thus the occurrence of many endoglucanases in nature is logical in the sense that where one glucanase (e.g., a lytic one) ceases to have significant affinity for intermediate products, the degradative process is taken over by other glucanases (such as the nonlytic ones) as well as by β-glucosidases.

Use of Beta-Glucanases as an Aid in Protein Recovery from Yeast

The enormous interest in augmenting the world's supply of protein from microbial sources (*135*) has focussed much attention on this source of protein as a nutrient for animals as well as humans. However, there is some evidence that if intact yeast cells are included in a diet, the tough polysaccharide-containing cell wall may constitute at least a partial barrier in the effective utilization of the protein of the cytoplasm. One reason for this is the absence of enzymes in the digestive tract of humans and most warm-blooded animals capable of hydrolyzing the microfibrillar β-glucan component of the cell wall (*135, 136*). Other factors may include the inability of the proteolytic enzymes of the digestive track to make effective contact with the cell's protein.

Although the utility of microorganisms as a source of protein, minerals, and vitamins is restricted by a number of other factors—e.g., a high nucleic acid content (for humans only), lack of texture or a slimy texture, strong color, and an unattractive taste and odor—the purpose of this review is limited to the restriction imposed by the cell wall and the possibilities of its removal or weakening by enzymatic means. The latter approach, namely weakening of the cell wall as a result of partial hydrolysis of the alkali-insoluble glucan, has not been systematically explored. It is based on the idea that protoplast emergence from cells is quite localized rather than caused by complete cell wall digestion in most species that have been investigated (*80, 84, 85*). It may therefore be possible to combine partial cell wall hydrolysis, using a low enzyme concentration, with mechanical rupture of the weakened cells. This approach may greatly reduce the high energy requirements necessary for the mechanical breakage of untreated fresh cells.

Wall Lysis by Endogenous Enzymes. As discussed earlier in this review, studies in recent years have revealed cell-wall-associated lytic enzymes of the endo-β-$(1 \rightarrow 3)$-glucanase type. If the activity of these enzymes could be controlled and stimulated at a particular point during the growth of the yeast (preferably the stationary phase), it would be the most economical and efficient way to rupture or weaken the cell wall (*137*). However, almost nothing is known about the factors which control their activity, either locally or generally, in the living cell. Much research is needed to determine if activating factors are required or removal of inhibitors of preexisting enzymes, or whether de novo synthesis and the factors that control it are necessary. It is known already that the level of such endo-β-$(1 \rightarrow 3)$-glucanases varies greatly from species to species and is also dependent on the particular stage of the growth cycle (*47*).

Wall Lysis by Exogenous Enzymes. Because of the limited availability and high cost of snail digestive fluid, extracellular microbial enzymes appear to offer the best means to weaken or rupture cell walls.

Many workers have observed that log-phase cells are much more susceptible to enzymatic wall lysis than cells of the stationary phase of growth. However for enzyme preparations from *Arthrobacter* or *Oerskovia* the difference in susceptibility does not appear to be great. Since it is more convenient to harvest yeast in the stationary than in the exponential phase of growth, enzymes produced by species of these two genera would appear to have advantages over those from other organisms. Increasing the susceptibility of yeast to lysis may be accomplished by pretreating the harvested yeast with sulfhydryl compounds.

If a particular yeast strain is chosen for the production of single cell protein, it is desirable to grow the chosen microbe on this same yeast to induce lytic enzyme production. The rationale for this is the considerable variation in cell wall composition between yeast species, and the most effective lytic enzyme complex is obtained by using the same strain as enzyme inducer.

Finally it may be mentioned that purification of the crude lytic enzymes produced in a culture broth is not necessary. In view of a marked synergism observed by several investigators between various components of the complex of lytic and nonlytic glucanases produced by a microbe, purification is likely to make an isolated lytic enzyme less effective than it would be in a complex mixture. Most of the cell wall lytic microbes which have been investigated appear to be very low or negligible in proteolytic activity, so that hydrolysis of yeast protein is not a major problem.

For enzyme economy it may be possible to immobilize the lytic enzymes on some type of solid carrier and to pass a yeast slurry over the immobilized enzyme. Macmillan et al. (*138*) experimented with collagen as the enzyme carrier, but since the enzymes in this system were not covalently linked they were slowly leached from the carrier.

Future Research Needs

Although research in the area of enzymatic yeast cell wall degradation has been very active in the last 10–15 years and great progress has been made, many problems remain unresolved and much remains to be done.

An important area where further work is needed deals with the purity of single enzyme components isolated by different investigators and the use of such fractions in studies of cell wall structure. If selective hydrolysis of specific bonds in the complex polysaccharides making up the cell walls of various yeasts is to lead to a better understanding of their structures, it is essential to employ pure enzymes. This criterion has not been adhered to rigorously in some laboratories. All organisms studied thus far produce mixtures of various enzymes involved in wall hydrolysis. The action of individual enzyme components should be determined before an assessment can be made of the synergistic action known to take place with two or more members of an enzyme complex. This approach in turn may lead to a solution of the unanswered question why stationary phase cells are in general more resistant to lysis than are log-phase cells and why some strains or species of yeast are more resistant than others. It is evident that for large scale cell wall lysis stationary phase cells are more convenient and economical to harvest than log-phase cells.

More work is also needed on the subject of lytic properties of β-glucanases. This was emphasized by the discovery that both lytic and nonlytic β-$(1 \rightarrow 3)$- and β-$(1 \rightarrow 6)$-glucanases are produced by microorganisms. Explanations for such differences in enzyme activities and specificities are inadequate at this time.

From the applied point of view where cell wall lytic enzymes are to be used for improvement in protein recovery and protein processing from yeast cells, a number of areas are much in need for further research. One of the problems of using microbial enzymes for this purpose is their inducible nature. Investigations have shown that yeast cell walls are the best inducers, followed by β-$(1 \rightarrow 3)$- and β-$(1 \rightarrow 6)$-glucans, while simple and inexpensive substrates induce little or no enzyme activity. Research is therefore needed to obtain cheaper and better inducers for enzyme production or compounds which enhance enzyme synthesis and release in the culture medium. Such research should be combined with a genetic approach to attempt by mutation to modify the genetic control of enzyme synthesis and allow the organisms to produce constitutive lytic β-glucanases. Efforts should also be directed at establishing conditions for greater stability of lytic β-glucanases, since several of these purified enzymes have been shown to be highly unstable in comparison with some nonlytic β-glucanases. This should include further work on immobilizing lytic β-glucanases on a solid carrier so that yeast slurries can be treated by passing them over such materials. Finally, in view of the difficulty of obtaining complete cell wall hydrolysis, the author feels that partial wall hydrolysis or softening and weakening of the alkali-insoluble glucan layer may facilitate mechanical cell rupture in a yeast slurry for the purpose of extraction and recovery of the cellular protein. Further studies on the control and activation of endogenous yeast endo-β-glucanases for this same purpose, particularly in yeast species where such enzymes are formed in high concentration during part of the cell cycle, are needed and likely to be rewarding.

Literature Cited

1. Phaff, H. J., "Structure and Biosynthesis of the Yeast Cell Envelope," *in* "The Yeasts, Vol. 2, Physiology and Biochemistry of Yeasts," A. H. Rose, J. S. Harrison, Eds., Academic, London, 1971, pp 135–210.
2. Bacon, J. S. D., Davidson, E. D., Jones, D., Taylor, I. F., *Biochem. J.* (1966) **101,** 36C–38C.
3. Manners, D. J., Masson, A. J., Patterson, J. C., *Biochem. J.* (1973) **135,** 19–30.
4. Bacon, J. S. D., Farmer, V. C., *Biochem. J.* (1968) **110,** 34P–35P.
5. Bacon, J. S. D., Farmer, V. C., Jones, D., Taylor, I. F., *Biochem. J.* (1969) **114,** 557-567.
6. Manners, D. J., Masson, A. J., Patterson, J. C., Björndal, H., Lindberg, B., *Biochem. J.* (1973) **135,** 31–36.
7. Misaki, A., Johnson, J., Jr., Kirkwood, S., Scaletti, J. V., Smith, F., *Carbohydr. Res.* (1968) **6,** 150–164.
8. Bell, D. J., Northcote, D. H., *J. Chem. Soc.* (1950) 1944–1947.
9. Manners, D. J., Masson, A. J., Patterson, J. C., *J. Gen. Microbiol.* (1974) **80,** 411–417.
10. Atkins, E. D. T., Parker, K. D., *J. Polym. Sci.,* Part C (1969) **28,** 69–81.
11. Rees, D. A., *MTP Int. Rev. of Sci.: Org. Chem.,* Ser 1, Vol. 7, G. O. Aspinall, Ed., Butterworth, London, 1973, p 251.
12. Kopecka, M., Phaff, H. J., Fleet, G. H., *J. Cell Biol.* (1974) **62,** 66–76.
13. Manners, D. J., Masson, A. J., *FEBS Lett.* (1969) **4,** 122–123.
14. Falcone, G., Nickerson, W. J., *Science* (1956) **124,** 272–273.
15. Eddy, A. A., Woodhead, J. S., *FEBS Lett.* (1968) **1,** 67–68.
16. Kessler, G., Nickerson, W. J., *J. Biol. Chem.* (1959) **234,** 2281–2285.
17. Ballou, C. E., *Adv. Enzymol.* (1974) **40,** 239–270.
18. Ballou, C. E., *Adv. Microb. Physiol.* **14,** A. H. Rose, J. F. Wilkinson, Eds., Academic, New York, 1976.

19. Stewart, T. S., Ballou, C. E., *Biochemistry* (1968) **7**, 1855–1863.
20. Rosenfeld, L., Ballou, C. E., *Biochem. Biophys. Res. Commun.* (1975) **63**, 571–579.
21. Thieme, T. R., Ballou, C. E., *Biochemistry* (1971) **11**, 1115–1119.
22. Nakajima, T., Ballou, C. E., *J. Biol. Chem.* (1974) **249**, 7679–7684.
23. Sentandreu, R., Northcote, D. H., *Carbohydr. Res.* (1969) **10**, 584–585.
24. Nakajima, T., Ballou, C. E., *J. Biol. Chem.* (1974) **249**, 7685–7694.
25. Nakajima, T., Ballou, C. E., *Biochem. Biophys. Res. Commun.* (1975) **66**, 870–879.
26. Bacon, J. S. D., Jones, D., Farmer, V. C., Webley, D. M., *Biochim. Biophys. Acta* (1968) **158**, 313–315.
27. Horisberger, M., Lewis, B. A., Smith, F., *Carbohydr. Res.* (1972) **23**, 183–188.
28. Zonneveld, J. M., *Biochim. Biophys. Acta* (1971) **249**, 506–514.
29. Johnston, I. R., *Biochem. J.* (1965) **96**, 659–664.
30. Kreger, D. R., *Biochim. Biophys. Acta* (1954) **13**, 1–9.
31. McMurrough, I., Rose, A. H., *Biochem. J.* (1967) **105**, 189–203.
32. Mill, P. J., *J. Gen. Microbiol.* (1966) **44**, 329–341.
33. Moore, E. O., *J. Sci. Food Agric.* (1955) **6**, 611–621.
34. Chesters, C. G. C., Bull. A. T., *Biochem. J.* (1963) **86**, 28–31.
35. Chesters, C. G. C., Bull, A. T., *Biochem. J.* (1963) **86**, 31–38.
36. Brock, T. D., *Biochem. Biophys. Res. Commun.* (1965) **19**, 623–628.
37. Brock, T. D., *J. Cell Biol.* (1964) **23**, 15A.
38. Abd-El-Al, A. T., Phaff, H. J., *Biochem. J.* (1968) **109**, 347–360.
39. Abd-El-Al, A. T., Phaff, H. J., *Can. J. Microbiol.* (1969) **15**, 697–701.
40. Tingle, M. A., Halverson, H. O., *Biochim. Biophys. Acta* (1971) **250**, 165–171.
41. Villanueva, J. R., Gacto, M., *Proc. Int. Specialized Symp. Yeasts, 3rd,* Otaniemi/Helsinki (1973) H. Suomalainen, C. Waller, Eds., Part II, 261–283.
42. Barras, D. R., *Antonie van Leeuwenhoek* (1972) **38**, 65–80.
43. Bacon, J. S. D., Gordon, A. H., Jones, D., Taylor, I. F., Webley, D. M., *Biochem. J.* (1970) **120**, 67–78.
44. Cortat, M., Matile, P., Wiemken, A., *Arch. Mikrobiol.* (1972) **82**, 189–205.
45. Matile, P., Cortat, M., Wiemken, A., Frey-Wyssling, A., *Proc. Nat. Acad. Sci.* (1971) **68**, 636–640.
46. Fleet, G. H., Phaff, H. J., *in* "Yeast, Mould and Plant Protoplasts," J. R. Villanueva, I. Garcia-Acha, S. Gascon, F. Uruburu, Eds., Academic, New York, 1973, pp 33–59.
47. Fleet, G. H., Phaff, H. J., *J. Biol. Chem.* (1974) **249**, 1717–1728.
48. Fleet, G. H., Phaff, H. J., *Biochim. Biophys. Acta* (1976) **401**, 318–332.
49. Meyer, M. T., "α-(1 → 3)-Glucanases from *Bacillus circulans* WL-12 and from Certain Yeasts," Ph.D. Dissertation, University of California, Davis, 1975.
50. Maddox, I. S., Hough, J. S., *J. Inst. Brew.* (1971) **77**, 44–47.
51. Arnold, W. N., *J. Biol. Chem.* (1972) **247**, 1161–1169.
52. Farkaš, V., Biely, P., Bauer, Š., *Biochim. Biophys. Acta* (1973) **321**, 246–255.
53. Biely, P., Farkaš, V., Bauer, Š., *FEBS Lett.* (1972) **23**, 153–156.
54. Villa, T. G., Notario, V., Villanueva, J. R., *Arch. Microbiol.* (1975) **104**, 201–206.
55. Johnson, B. F., *J. Bacteriol.* (1968) **95**, 1169–1172.
56. Beran, K., *Adv. Microbiol. Physiol.* (1968) **2**, 143–171.
57. Moor, H., *Arch. Mikrobiol.* (1967) **57**, 135–146.
58. Brock, T. D., *J. Gen. Microbiol.* (1961) **26**, 487–497.
59. MacKay, V., Manney, T. R., *Genetics* (1974) **76**, 255–271.
60. Hartwell, L. H., *Bacteriol. Rev.* (1974) **38**, 164–198.
61. Yanagishima, N., *Plant Cell Physiol.* (1963) **4**, 257–264.

62. Shimoda, C., Massada, Y., Yanagishima, N., *Physiol. Plant.* (1967) **20,** 299–305.
63. Shimoda, C., Yanagishima, N., *Physiol. Plant.* (1968) **21,** 1163–1169.
64. Shimoda, C., Yanagishima, N., *Physiol. Plant.* (1971) **24,** 46–50.
65. Marshall, J. J., Adv. in Carbohyd. Chem. and Biochem., R. S. Tipson, D. Horton, Eds. **30,** Academic, New York, 1974, pp 257–370.
66. Fleet, G. H., Phaff, H. J., *J. Bacteriol.* (1974) **119,** 207–219.
67. Tanaka, H., Phaff, H. J., *J. Bacteriol.* (1965) **89,** 1570–1580.
68. Tanaka, H., Phaff, H. J., Higgins, L. W., "Symposium on Yeast Protoplasts," *Abh. Dtsch. Akad. Wiss. Berlin, Kl. Med.* (1966) **6,** 113–129, 353–357.
69. Doi, K., Doi, A., Fukui, T., *Agric. Biol. Chem.* (1973) **37,** 1619–1627.
70. Webley, D. M., Follett, E. A. C., Taylor, I. F., *Antonie van Leeuwenhoek* (1967) **33,** 159–165.
71. Mann, J. W., Heinz, C. E., Macmillan, J. D., *J. Bacteriol.* (1972) **111,** 821–824.
72. Kobayashi, Y., Tanaka, H., Ogasawara, N., *Agric. Biol. Chem.* (1974) **38,** 973–978.
73. Giaja, J., *C. R. Soc. Biol.* (1914) **77,** 2–4.
74. Giaja, J., *C. R. Soc. Biol.* (1919) **82,** 719–720.
75. Giaja, J., *C. R. Soc. Biol.* (1922) **86,** 708–709.
76. Eddy, A. A., Williamson, D. H., *Nature* (London) (1957) **179,** 1252–1253.
77. Eddy, A. A., Williamson, D. H., *Nature* (London) (1959) **183,** 1101–1104.
78. Holter, H., Ottolenghi, P., *C. R. Trav. Lab. Carlsberg* (1960) **31,** 409–422.
79. Millbank, J. W., Macrae, R. M., *Nature* (London) (1964) **201,** 1347.
80. Rost, K., Venner, H., *Arch. Mikrobiol.* (1965) **51,** 122–129.
81. Burger, M., Bacon, E. E., Bacon, J. S. D., *Biochem. J.* (1961) **78,** 504–511.
82. Duell, E. A., Inoue, S., Utter, M. F., *J. Bacteriol.* (1964) **88,** 1762–1773.
83. Sommer, A., Lewis, M. J., *J. Gen. Microbiol.* (1971) **68,** 327–335.
84. Svihla, G., Schlenk, F., Dainko, J. L., *J. Bacteriol.* (1961) **82,** 808–814.
85. Zvjagilskaja, R. A., Afanasjewa, T. P., "Symposium on Yeast Protoplasts," *Abh. Dtsch. Akad. Wiss. Berlin, Kl. Med.* (1966) **6,** 27–29.
86. Darling, S., Theilade, J., Birch-Andersen, A., *J. Bacteriol.* (1969) **98,** 797–810.
87. Johnston, J. R., Mortimer, R. K., *J. Bacteriol.* (1959) **78,** 292.
88. Holden, M., Tracey, M. V., *Biochem. J.* (1950) **47,** 407–414.
89. de Vries, O. M. H., Wessels, J. G. H., *J. Gen. Micirobol.* (1972) **73,** 13–22.
90. Myers, F. L., Northcote, D. H., *J. Exp. Biol.* (1958) **35,** 639–648.
91. Anderson, F. B., Millbank, J. W., *Biochem. J.* (1966) **99,** 682–687.
92. McLellan, W. L., Jr., Lampen, J. O., *J. Bacteriol.* (1968) **95,** 967–974.
93. McLellan, W. L., Jr., McDaniel, L. E., Lampen, J. O., *J. Bacteriol* (1970) **102,** 261–270.
94. Rombouts, F., Phaff, H. J., *Eur. J. Biochem.* (1976) **63,** 109–120.
95. Rombouts, F., Phaff, H. J., *Eur. J. Biochem.* (1976) **63,** 121–130.
96. Salton, M. R. J., *J. Gen. Microbiol.* (1955) **12,** 25–30.
97. Bacon, J. S. D., Gordon, A. H., Webley, D. M., *Biochem. J.* (1970) **117,** 42.
98. Doi, K., Doi, A., Fukui, T., *J. Biochem.* (1971) **70,** 711–714.
99. Keneko, T., Kitamura, K., Yamamoto, Y., *Rep. Res. Lab. Kirin Brew. Co.* (1973) **16,** 77–85 (*see also Agric. Biol. Chem.* (1973) **37,** 2295–2302).
100. Kitamura, K., Yamamoto, Y., *Arch. Biochem. Biophys.* (1972) **153,** 403–406.
101. Kitamura, K., Kaneko, T., Yamamoto, Y., *Arch. Biochem. Biophys.* (1971) **145,** 402–404.

102. Kitamura, K., Kaneko, T., Yamamoto, Y., *J. Gen. Appl. Microbiol.* (1972) **18,** 57–71.
103. Horikoshi, K., Koffler, H., Arima, K., *Biochim. Biophys. Acta* (1962) **73,** 268–275.
104. Horikoshi, K., *in* "Yeast, Mold and Plant Protoplasts," J. R. Villanueva, I. Garcia-Acha, S. Gascon, F. Uruburu, Eds., Academic, London, 1973, pp 25–32.
105. Kobayashi, Y., Tanaka, H., Ogasawara, N., *Agric. Biol. Chem.* (1974) **38,** 959–965.
106. Tanaka, H., Kobayashi, Y., Ogasawara, N., *Agric. Biol. Chem.* (1974) **38,** 967–972.
107. Harada, T., Moori, K., Amemura, A., *Agric. Biol. Chem.* (1972) **36,** 2611–2613.
108. Yamamoto, S., Nagasaki, S., *J. Ferment. Technol.* (1972) **50,** 117–135.
109. Monreal, J., Uruburu, F. D., Villanueva, J. R., *J. Bacteriol.* (1967) **94,** 241–244.
110. Jefiries, T. W., "Purification and Characterization of the Yeast Lytic Enzyme System from *Oerskovia Xanthineolytica,*" Ph.D. Dissertation, Rutgers University, New Brunswick, N. J., 1975.
111. Mendoza, C. G., Villanueva, J. R., *Nature* (London) (1962) **195,** 1326–1327.
112. Elorza, M. V., Ruiz, E. M., Villanueva, J. R., *Nature* (London) (1966) **210,** 442–443.
113. Yamoto, S., Yadomae, T., Miyazaki, T., *J. Ferment. Technol.* (1974) **52,** 706–712.
114. Bauer, H., Bush, D. A., Horisberger, M., *Experientia* (1972) **28,** 11–13.
115. Sugimori, T., Uchida, Y., Tsukada, Y., *Agric. Biol. Chem.* (1972) **36,** 669–675.
116. Yamamoto, S., Fukuyama, J., Nagasaki, S., *Agric. Biol. Chem.* (1974) **38,** 329–337.
117. Yokotsuka, K., Goto, S., Yokotsuka, I., Kushida, T., *J. Ferment. Technol.* (1974) **52,** 701–705.
118. Yamamoto, S., Shiraishi, T., Nagasaki, S., *Biochem. Biophys. Res. Commun.* (1972) **46,** 1802–1809.
119. Horitsu, H., Satake, T., Tomoyeda, M., *Agric. Biol. Chem.* (1973) **37,** 1007–1012.
120. Toyama, N., Ogawa, K., *J. Ferment. Technol.* (1968) **46,** 626–633.
121. Horikoshi, K., Sakaguchi, K., *J. Gen. Appl. Microbiol.* (1958) **1,** 1–11.
122. Doi, K., Doi, A., Ozaki, T., Fukui, T., *J. Biochem.* (1973) **73,** 667–670.
123. Doi, K., Doi, A., Ozaki, T., Fukui, T., *Agric. Biol. Chem.* (1973) **37,** 1629–1633.
124. Parrish, F. W., Perlin, A. S., Reese, E. T., *Can. J. Chem.* (1960) **38,** 2094–2104.
125. Shibata, Y., *J. Biochem.* (1974) **75,** 85–92.
126. Yamamoto, S., Kobayashi, R., Nagasaki, S., *Agric. Biol. Chem.* (1974) **38,** 1493–1500.
127. Bauer, H., Bush, D. A., Horisberger, M., *Experientia* (1972) **28,** 11–13.
128. Huotari, F. I., Nelson, T. E., Smith, F., Kirkwood, S., *J. Biol. Chem.* (1968) **243,** 952–956.
129. Nelson, T. E., Johnson, J., Jantzen, E., Kirkwood, S., *J. Biol. Chem.* (1969) **244,** 5972–5980.
130. Eveleigh, D. E., *Bacteriol. Proc.* (1969) p 130.
131. Nelson, T. E., *J. Biol. Chem.* (1970) **245,** 869–872.
132. Nagasaki, S., Fukuyama, J., Yamamoto, S., Kobayashi, R., *Agric. Biol. Chem.* (1974) **38,** 349–357.
133. Bull, A. T., Chesters, C. G. C., *Adv. Enzymol.* (1966) **28,** 325–364.
134. Clarke, A. E., Stone, B. A., *Biochem. J.* (1965) **96,** 793–801.
135. Kihlberg, R., *Annu. Rev. Microbiol.* (1972) **26,** 427–466.

136. Labuza, T. P., "Cell Collection: Recovery and Drying for SCP Manufacture," *in* "Single Cell Protein II," S. R. Tannenbaum and D. I. C. Wang, Eds., MIT Press, Cambridge, Mass., 1975, pp 69–104.
137. Kröning, A., Egel, R., *Arch. Microbiol.* (1974) **99**, 241–249.
138. Macmillan, J. D., Cuffari, G. L., Jeffries, T. W., Wilber-Murphy, J., *Proc. Int. Symp. Yeasts, 4th, Vienna, Austria,* (1974) Part I, pp 25–26.
139. Fleet, G. H., Manners, D. J., in preparation.

RECEIVED January 26, 1976.

9

Removal of Naturally Occurring Toxicants through Enzymatic Processing

IRVIN E. LIENER

Department of Biochemistry, College of Biological Sciences,
University of Minnesota, St. Paul, Minn. 55108

Many plants contain glycosides which, upon hydrolysis by endogenous enzymes, release substances which may be toxic. Examples of such glycosides are the goitrogens of rapeseed and the cyanogens of cassava, lima beans, and fruit kernels. Suitable processing techniques or traditional methods of preparation often take advantage of the enzymatic release of such toxic constituents in order to detoxify such foods. The destruction of a number of other toxic constituents may be facilitated by enzymes of microbial origin as exemplified by: the removal of flatulence-producing factors in beans, the hydrolysis of the phytate of soybeans and wheat, and the destruction of free gossypol in cottonseed. The lactose of milk can be reduced effectively by enzymatic hydrolysis to provide an excellent source of protein to lactose-intolerant individuals.

Although proteins of plant origin offer considerable promise for alleviating the shortage of food protein now facing many segments of the world's population—a situation which will certainly become more acute as the expansion of people continues unabated—many plants are known to contain substances which have an adverse effect on the nutritional properties of the protein (*1, 2*). In many cases such substances may remain relatively innocuous as long as they are not hydrolyzed by enzymes which frequently accompany these substances in the plant tissues. Paradoxically, advantage can sometimes be taken of the action of those endogenous enzymes to effect the removal of these toxicants by appropriate methods of food preparation or processing techniques. In the discussion that follows, examples will be cited whereby enzymatic detoxification has been applied in a practical fashion to the production of sources of protein that can be used with relative impunity in the diet of man or animals.

Goiterogenic Glycosides

Many cruciferous plants such as the rapeseed (*Brassica napus*), mustard seed (*Brassica juncea*), and Abyssian kale (*Crambe abyssinica*) provide a potentially valuable source of protein, particularly for feeding animals. The protein content of the meal remaining after extraction of the oil is relatively high (30–50%), and the amino acid composition compares favorably with that of soybeans (*3, 4*). The use of such oilseeds, however, is limited by the fact that they contain thioglycosides, also referred to as glucosinolates, which upon hydrolysis yield products that, unlike the parent glycoside, are goiterogenic and act as growth depressants. These goiterogenic products, primarily isothiocyanates and oxazolidine-thiones, are liberated from the parent glycosides by enzymes (β-thioglycosidases) normally present in the plant tissue, and these act on their substrates when the seed is crushed and moistened. Although a wide variety of thioglycosides are present in cruciferous seeds (*5*), the principal one present in rapeseed and crambe is progoitrin which, when acted upon by myrosinase, yields an isothiocyanate that spontaneously cyclizes to a vinyloxazolidine-thione (Figure 1).

The toxicity of rapeseed can be reduced in one or a combination of several different ways: (a) destruction of the enzyme by moist heat treatemnt (*6*); (b) removal of the thioglycosides by extraction with hot water (*7, 8*), dilute alkali (*9*), or acetone (*10*), or by decomposition with iron salts (*11*) or soda ash (*12*); or (c) removal of goiterogenic end products.

Although procedure (a) prevents the further breakdown of the thioglycosides in the meal, the possibility remains that the glycosides which are allowed to remain in the meal could be subsequently hydrolyzed to toxic end products after ingestion by enzymes produced in the intestinal tract (*13, 14*). In one instance at least, enzyme-inactivated mustard seed cake was found to be toxic to cattle because kohlrabi, another cruciferous plant which contains myrosinase, was fed at the same time to the animals (*15*). There is some indication, however, that in ruminants the goiterogenic oxazolidine-thiones which may be released by microbial activity in the rumen may be isomerized by the microflora to a nontoxic derivative characterized as thiazolidines (*16*) as shown in Figure 2. Although procedure (b), especially in combination with procedure (a), effects a marked reduction in toxicity, it does not eliminate any goiterogenic substances which may have been produced prior to processing.

In procedure (c), all of the thioglycosides are deliberately permitted to undergo conversion to gointerogenic end products which are

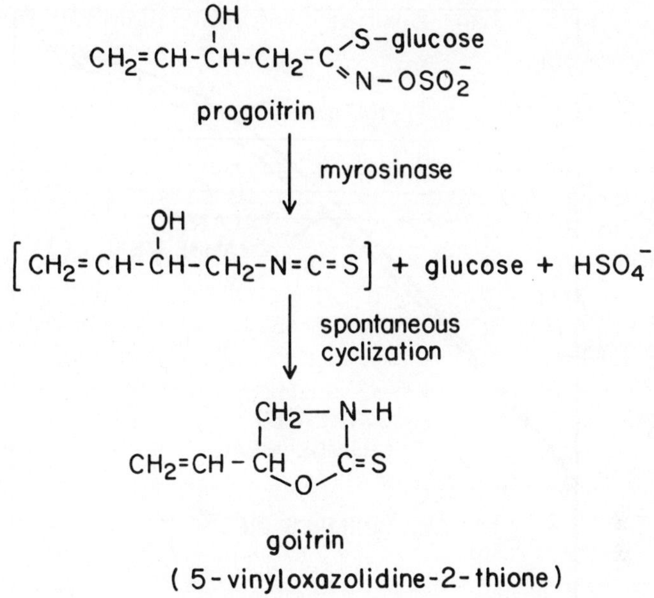

OH
$CH_2{=}CH{-}CH{-}CH_2{-}C$ S−glucose / N−OSO_2^-
progoitrin
myrosinase
OH
[$CH_2{=}CH{-}CH{-}CH_2{-}N{=}C{=}S$] + glucose + HSO_4^-
spontaneous
cyclization
CH_2— N-H
$CH_2{=}CH-CH$ C=S
O
goitrin
(5-vinyloxazolidine-2-thione)

Figure 1. Structure of the thioglycoside progoitrin present in rapeseed and the products produced by the action of the enzyme myrosinase

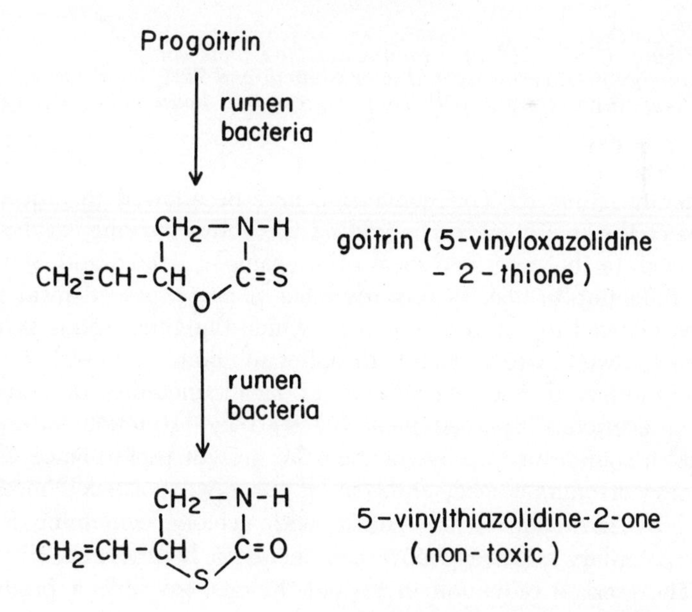

Journal of the American Oil Chemists' Society

Figure 2. Conversion of progoitrin to a nontoxic derivative in the rumen of sheep (16)

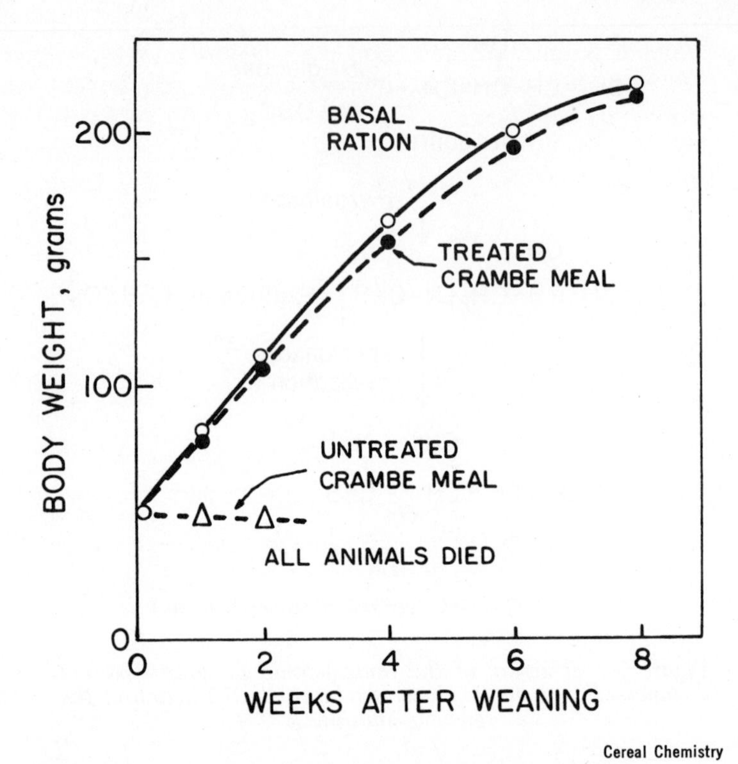

Figure 3. Growth of weanling rats fed diets containing 28% crambe meal, either treated or untreated, and 72% basal ration. Comparison is made with control group fed basal ration alone (19).

subsequently removed. This conversion may be allowed to proceed by autolysis of the raw meal, or by adding the active enzyme, in the form of raw meal, to the processed meal. For example, Belzile and Bell (*17*) noted a reduction in toxicity towards mice of raw rapeseed meal which had been allowed to autolyze and from which the goiterogenic products had been removed by extraction with buffer solutions (pH 6–9). Goering et al. (*18*) observed that the addition of a small amount of raw rapeseed to solvent-extracted rapeseed meal followed by extraction with water produced a significant improvement in the growth performance of rats. In the case of crambe seed, autolysis of the raw, moistened meal was followed by extraction of the goitrin with acetone containing 2–12% water or a ternary mixture of 53% acetone, 44% hexane, and 3% water (*19*). The marked reduction in toxicity in rats fed such a product is shown in Figure 3. It is evident from these results that enzymatic hydrolysis followed by extraction with aqueous acetone provides a very effective means of detoxifying crambe seed meal.

In the case of mustard seed meal the principal hydrolytic product of the thioglycoside, sinigrin, is allylisothiocyanate (Figure 4) which, unless removed from the meal, imparts an extremely pungent flavor and thus limits the amount that can be used for feeding purposes. In a process developed by Mustakas and co-workers (*20, 21*), advantage is

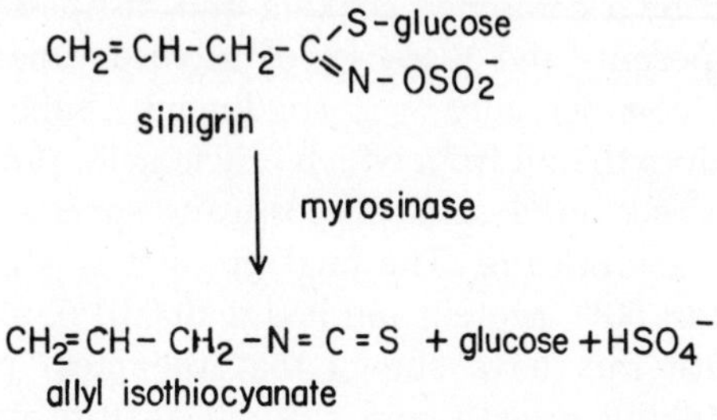

$$CH_2{=}CH{-}CH_2{-}C\begin{smallmatrix}\nearrow S\text{-glucose} \\ \searrow N\text{-}OSO_2^-\end{smallmatrix}$$

sinigrin

$$\downarrow \text{ myrosinase}$$

$$CH_2{=}CH{-}CH_2{-}N{=}C{=}S + \text{glucose} + HSO_4^-$$

allyl isothiocyanate

Figure 4. Enzymatic hydrolysis of sinigrin, the principal thioglycoside of mustard seed

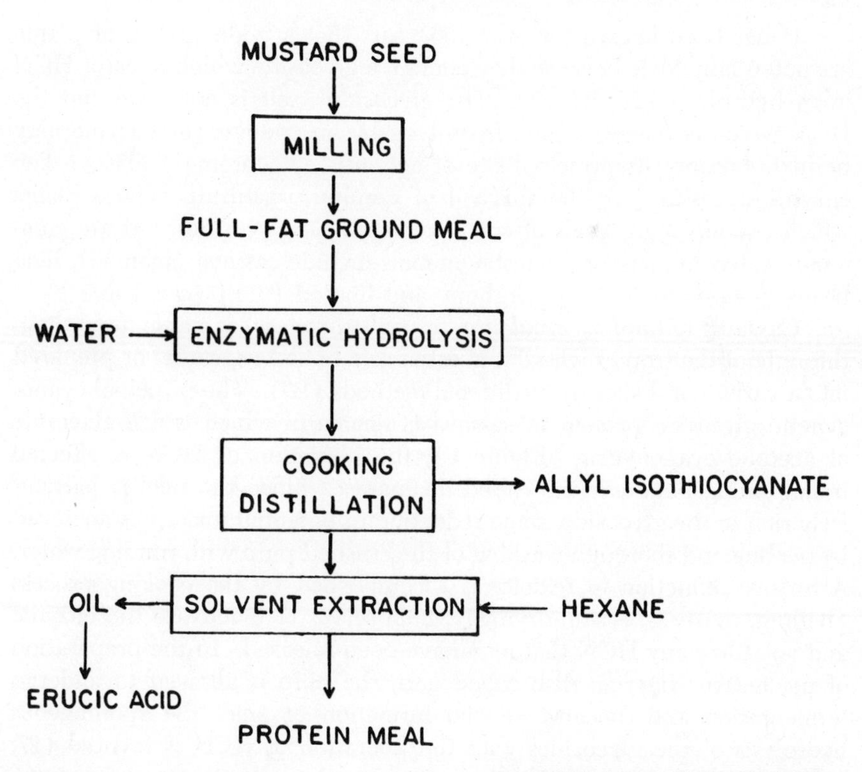

Journal of the American Oil Chemists' Society

Figure 5. Flow sheet for the enzymatic detoxification of mustard seed (20)

taken of the fact that allylisocyanate is not only water-soluble but volatile as well, so it can be removed by distillation after enzymatic hydrolysis of the glycoside. A flow sheet of the complete process which incorporates this feature is shown in Figure 5. In this process the raw ground meal is moistened to 30% and the conversion of the thioglycoside to allylisothiocyanate by endogenous enzymes is allowed to proceed for 15–45 min at 55°C, followed by a combined cooking and distillation step at 100°C. The latter step permits the recovery of allylisocyanate (mustard oil) which is utilized commercially as a condiment. Subsequent extraction with hexane removes the oil from which erucic acid, present to the extent of 25%, may be recovered since it possesses special properties which make it useful as a lubricant. The final product is a bland meal which now contains almost 50% protein and less than 0.01% allylisothiocyanate. Feeding tests with rats have shown that the meal produced by this method did not inhibit growth, and a 50:50 mixture with soybean meal gave as good growth as soybean meal alone (*22*).

Cyanogenetic Glycosides

It has been known for over 150 years that a wide variety of plants are potentially toxic because they contain a glycoside which releases HCN upon hydrolysis (*23, 24, 25*). The glycoside itself is not toxic, but the HCN which is released upon hydrolysis by an endogenous enzyme may be toxic because its principal site of action is cytochrome oxidase, a key enzyme necessary for the survival of aerobic organisms. Those plants which contain high levels of such cyanogenetic glycosides and are commonly eaten by man or domestic animals include cassava (manioc), lima beans (*Phaseolus luatus*), sorghum, and linseed (flax) (*see* Table I).

Cassava (*Manihot esculenta* or *utilissima*) is a staple food item throughout the tropics where it is eaten as a boiled vegetable or prepared into a variety of dishes by traditional methods (*26*). The principal cyanogenetic glycoside present in cassava is linamarin which is a β-glucoside of acetone cyanohydrin (Figure 6); the liberation of HCN is effected by an endogenous enzyme called linamarase. Since the peel is particularly rich in the glycoside, some reduction in potential toxicity is achieved by peeling and thorough washing of the crushed pulp with running water. A further reduction in toxicity is accomplished by the cooking process (boiling, roasting, or sun-drying) which serves to inactivate the enzyme and volatilize any HCN that may have been released. In the preparation of the native Nigerian dish called gari, the pulp is allowed to undergo fermentation and, because of the formation of acid, the spontaneous hydrolysis of the glycosides with the liberation of HCN is favored (*27, 28*). Subsequent frying effects almost complete elimination of any remaining HCN.

Table I. Cyanide Content of Plants Commonly Eaten by Man or Domestic Animals (*23, 24*)

Plant	*HCN Yield, (mg/100 g)*
Lima bean (*Phaseolus lunatus*)	
Samples incriminated in fatal human poisoning	210–312
Normal levels	14.4–16.7
Sorghum	250
Cassava	113
Linseed meal	53
Black-eyed pea (*Vigna sinensis*)	2.1
Garden pea (*Pisum sativum*)	2.3
Kidney bean (*Phaseolus vulgaris*)	2.0
Bengal gram (*Cicer arietinum*)	0.8
Red gram (*Cajanus cajans*)	0.5

West Indian Medical Journal

Tapioca is the product obtained from cassava after thorough washing. Because of its high starch content (about 65%) and its high digestibility, tapioca has been recommended for human diets and for animal nutrition. Tapioca, however, may still contain sufficiently high levels of cyanide to preclude its use at levels higher than 10% for poultry feeding (*29*). Because tapioca has not been subjected to any heat treatment, the hydrolytic enzymes still remain. It has been claimed, however, that these enzymes catalyze the condensation of HCN with aldehydic compounds to form cyanohydrins, so that the addition of glucose to unprocessed cassava causes the disappearance of HCN (*30*). It is therefore suggested that in the preparation of foods from unprocessed cassava liberal amounts of glucose should be added.

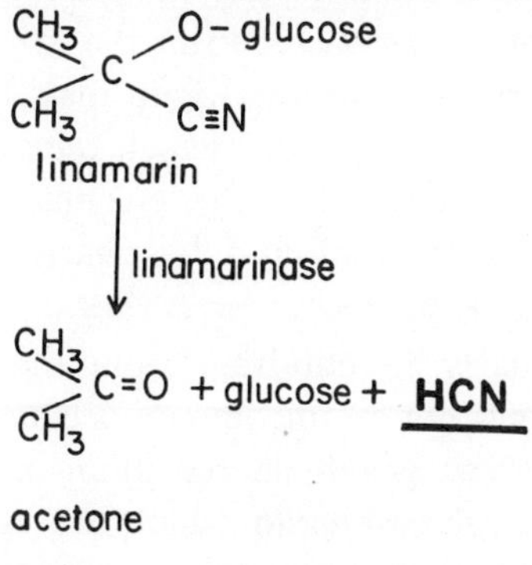

Figure 6. Enzymatic hydrolysis of linamarin, the cyanogenetic glycoside of cassava and lima beans

Despite the apparent effectiveness of traditional methods of preparing cassava dishes in reducing the cyanide content, there is evidence to indicate that a tropical disease known as ataxic neuropathy may be associated with the consumption of cassava (*31, 32, 33*). One of the metabolic routes whereby cyanide is detoxified involves the formation of thiocyanate as shown in Figure 7. Elevated levels of thiocyanate in blood plasma have been observed in individuals who suffered from ataxia neuropathy and who had a history of high consumption of cassava. Rats fed boiled or fermented cassava also showed high levels of thiocyanate in plasma and developed clinical signs of ataxia neuropathy. The mechanism whereby cyanide or its metabolic product, thiocyanate, are involved in the pathogenesis of this disease remains to be elucidated.

Serious outbreaks of poisoning in man and animals have frequently been attributed to the consumption of certain varieties of lima beans (*23, 24*). The particular glycoside involved is identical to the one present in cassava (Figure 6) but is sometimes referred to as phaseolunatin. Rahman et al. (*34*) have described a simple procedure for reducing the cyanide content of lima beans to a negligibly low, nontoxic level. Water is added to the ground bean flour to a level of 25–50% which favors the autolytic release of cyanide. Subsequent drying and cooking of the flour with water for 20 min reduces the cyanide contents to levels where it is physically impossible to eat enough to reach an estimated toxic dose of 50 mg of HCN for man.

It is by no means clear why human intoxication has been occasionally reported in cases where the lima beans had been cooked and the hydrolytic enzymes presumably destroyed (*35, 36*). Significant amounts of intact glycoside might still be present, for example, if conditions were such that very little autolysis had been allowed to occur prior to cooking. Since cyanide was observed in the urine (*36*), it is possible that enzymes secreted by the intestinal flora may be responsible for the release of cyanide from any glycosides which remain after cooking (*37*). Auld (*38*), however, contends that the enzymic hydrolysis of cyanogenetic glycosides is greatly inhibited by nearly all of the conditions prevailing in the digestive tract.

Poisoning, particularly in children, from the consumption of the kernels of various fruits is not an uncommon occurrence (*39*). The seeds of such fruits as the apricot, peach, cherry, plum, and almonds are known to contain cyanogenetic glycosides in which the aglycone is D-mandelonitrile (*40*) as exemplified by amygdalin whose structure is shown in Figure 8. Amygdalin itself is harmless as such, but when the seed is crushed and the pulp moistened, enzymatic hydrolysis rapidly ensues with the concomitant release of HCN.

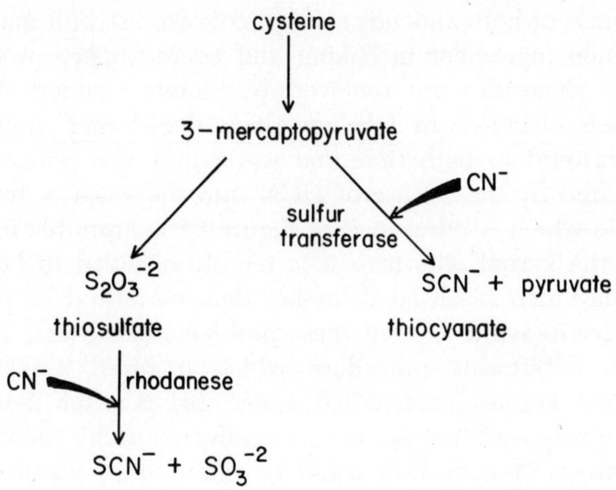

British Medical Journal

Figure 7. Enzymatic detoxification of cyanide as it occurs in vivo (31)

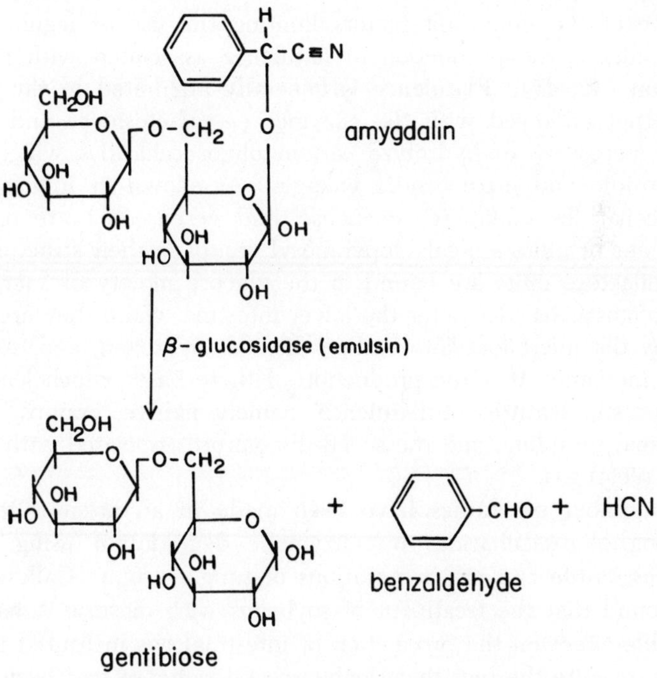

Figure 8. Structure of amygdalin, the cyanogenetic glucoside in the kernel of many species of fruits

The kernels of both almonds and apricots are used for making marzipan, a common ingredient in baking and confectionery products. The cyanogenetic glycosides are removed by simply soaking the coarsely ground kernels for 24–48 hr (the so-called "debittering" process). This process is wasteful in both time and water and also poses a pollution problem created by the release of HCN into the wash water. Also the benzaldehyde which is released (*see* Figure 8) is insoluble in water and dissolves in the kernel oil where it is readily oxidized to benzoic acid. The latter may then reach levels higher than permitted by food regulations. In order to avoid some of these problems Schab and Yannai (*41*) developed a debittering procedure which involved soaking coarsely ground apricot kernels in acidified water, pH 6.5, for 2 hr at 55°C. Complete hydrolysis of the glycoside was effected under these conditions, and the products of hydrolysis could be removed by steam distillation. Marzipan made from apricot kernels which had been processed in this manner was found to be as organoleptically acceptable as the product made from almonds prepared in the conventional fashion.

Flatulence-producing Factors

One of the important factors limiting the use of legumes in the human diet is the production of flatulence associated with their consumption (*42, 43*). Flatulence is generally attributed to the fact that man is not endowed with the enzymes (α-galactosidase and β-fructosidase) necessary to hydrolyze certain oligosaccharides which contain α-galacosidic and β-fructosidic linkages. As shown in Figure 9, these oligosaccharides (raffinose, stachyose, and verbascose) are related by having one or more α-D-galactopyranosyl groups in their structure, where the α-galactose units are bound to the glucose moiety of sucrose. The intact oligosaccharides enter the lower intestine where they are metabolized by the microflora into carbon dioxide, hydrogen, and, to a lesser extent, methane. It is the production of these bases which leads to the characteristic features of flatulence, namely nausea, cramps, diarrhea, abdominal rumbling, and the social discomfort associated with the ejection of rectal gas.

A number of studies have been made in an attempt to remove offending oligosaccharides by enzymatic degradation using, in most instances, crude enzyme preparations of fungal origin. Calloway et al. (*42*) found that the treatment of soybeans with diastase A had only a negligible effect on the production of intestinal gas measured in human subjects despite the fact that stachyose and raffinose had been virtually eliminated. Although several studies have shown that the oligosaccharide content of soybeans can be reduced to negligible levels by treatment

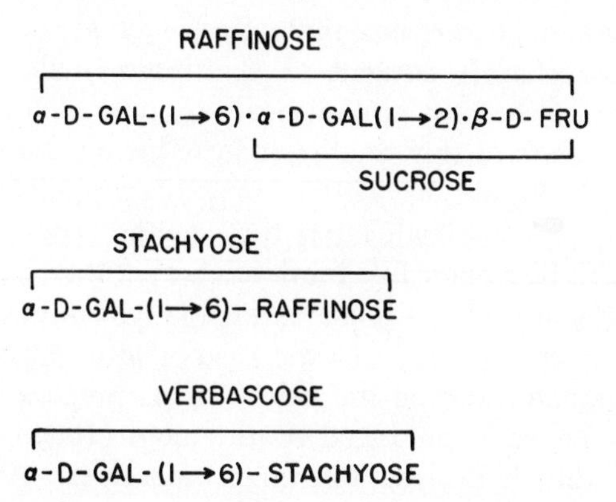

Figure 9. Structural relationship between the oligosaccharides believed to be involved in the production of flatulence

with mold enzymes (*42, 44*), the flatulence-producing properties of the bean as measured in human subjects were not significantly altered by such treatment (*42*). The treatment of soymilk by an enzyme purified from *Aspergillus saitoi* also resulted in the complete hydrolysis of its constituent oligosaccharides (*45*), but the effectiveness of this treatment in reducing flatulence was not evaluated. Significant reduction in the oligosaccharide content of the soybean can also be effected by a combination of soaking, germination, and resoaking (*46*), although in this case it is difficult to assess how much of this reduction is caused by leaching or

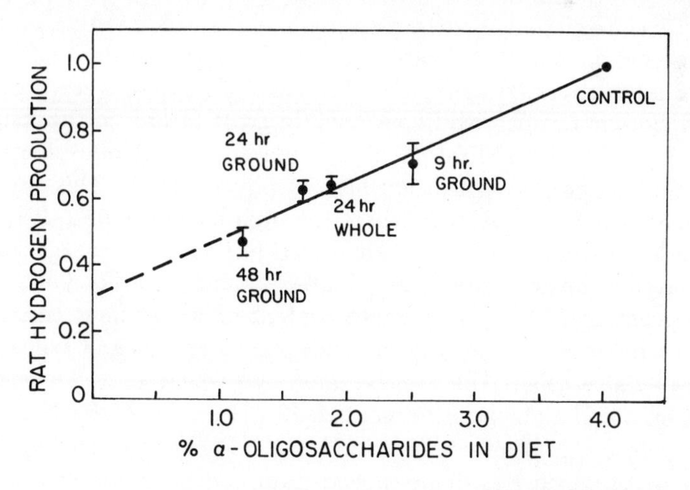

Journal of Food Science

Figure 10. Flatulence in rats produced by California small white beans which had been allowed to autolyze for different times at pH 5.2 and 55°C (47)

by autolysis by endogenous enzymes. Becker et al. (*47*) were able to lower the oligosaccharide content of California small white beans by allowing a slurry of the raw, ground bean to undergo autolysis at pH 5.2 and 55°C. The effect of this treatment in reducing the flatulence effect in rats is shown in Figure 10. If this curve is extrapolated to 0% oligosaccharide, there is some indication that another component (or components) may also be responsible for flatulence, at least in the case of rats.

Such traditional soybean foods as tofu (soybean curd) and tempeh have little flatus activity (*42*). In the case of tofu, oligosaccharides are presumably eliminated during the course of its preparation, and, in the case of tempeh, enzymes produced by the mold (*Rhizopus*) during fermentation probably hydrolyze the oligosaccharides. As might be expected because of their low carbohydrate content ($< 1\%$), soy protein isolates are devoid of flatus activity (*43*). It follows that textured meat analogs made from isolated soy protein are most likely free of flatus activity.

Phytic Acid

Phytate, a cyclic compound (inositol) containing six phosphate radicals (*see* Figure 11), readily chelates such di- and trivalent metal ions as calcium, magnesium, zinc, and iron to form poorly soluble compounds that are not readily absorbed from the intestines (*48*). The ability of phytic acid to bind metal ions is lost when the phosphate groups are hydrolyzed through the action of the enzyme phytase. Although phytase activity has been shown to be present in the small intestine of various experimental animals (*49*), its presence in the intestines of humans remains a controversial issue (*50, 51*). Even if phytase were present, its activity does not appear to be sufficiently great to prevent phytate from interfering with the utilization of calcium, zinc, and phosphorus from cereal-containing diets which are high in phytate (*51*). This was found to be particularly true in Iran where the high incidence of a disturbance in mineral metabolism has been attributed to their life-long consumption of unleavened bread made from a high-extraction rate flour with a high phytate content (*52*). The phytate content of wheat flour can be considerably reduced by allowing the endogenous enzymes of wheat to act on the phytates (*53*). This can be done by making an aqueous suspension of bran and allowing it to incubate at pH 4.5, 50°C, for 6 hr. Using human subjects, McCance and Widdowson (*53*) found that bread made from flour to which this dephytinized bran had been added provided a much more readily available source of calcium, magnesium, and phosphorus than untreated controls. Reinhold (*54*) found that the fermentation of whole wheat breads with baker's yeast caused a significant

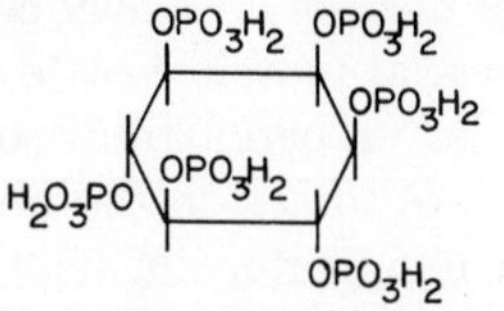

Figure 11. Structure of phytic acid (myo-*inositol 1,2,3,4,5,6-hexadihydrogen phosphate)*

reduction in phytate content in three different whole wheat flours, representing different extraction rates, available in Iran (*see* Figure 12). Yeast fermentation also caused a marked increase in the solubility of zinc which was accompanied by an increase in the uptake of this metal by isolated intestinal strips (*55*). These effects may be attributed in large measure to the action of yeast phytase resulting in the hydrolysis of phytic acid.

It has been estimated that two thirds of the phosphorus of soybean oil meal is bound as phytate and, as such, is unavailable to the chick and furthermore interferes with the availability of calcium (*56*). The treatment of soybean meal with the culture filtrate of *Aspergillus ficcum,* however, caused a 90% release of the phytate-bound phosphorus and reduced the chick's requirement for calcium by at least one third (*57*).

Journal of the American Dietetic Association

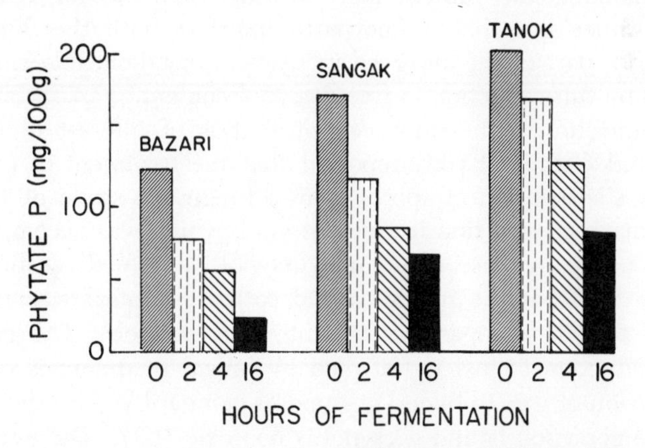

Figure 12. Destruction of phytate in whole wheat meals by action of yeast. Bazari: 75–85% extraction rate (ER); Sangok: 85–90% ER; Tanok: 95–100% ER (54).

In soybean concentrates and isolates much of the phytate remains associated with the protein; in fact, phytate may constitute as much as 2–3% of the weight of a commercial protein isolate (*57*). A low-phytate soybean protein isolate can be prepared from soybean flour, however, by allowing endogenous phytase to act on the phytate in a 6% suspension of the flour at pH 5 at a temperature of 65°C (*58*). Hydrolysis of the phytate facilitates its separation from the bulk of the soybean protein which is then concentrated by ultrafiltration using a membrane which is permeable to phytate and its hydrolysis products but impermeable to protein. The product obtained by this method contains over 90% protein and only about 0.3% phosphorus.

The phytate content of California small white beans (*Phaseolus vulgaris*) can be greatly reduced by allowing a suspension of raw beans to autolyze (pH 5.2, 35°–55°C, 20–48 hr) or by adding a preparation of wheat germ phytase under the same conditions (*47, 59*).

Rojas and Scott (*60*) used a culture filtrate of *Aspergillus ficcum* as a source of phytase in an attempt to reduce the phytate content of solvent-extracted cottonseed meal. As might be expected, phytase treatment of cottonseed meal served to increase the availability of phosphorus and zinc for the growing chick. This treatment also resulted in a substantial increase in the metabolizable energy of the meal for chicks, an effect which was attributed to the hydrolysis of phytate and its release from the protein, thereby making the latter more readily digestible. In addition, phytase treatment effected a reduction in the level of free gossypol. This effect may also be due to the breakdown of the phytate-protein complex, making the protein more readily available for binding the gossypol. Since gossypol is known to interfere with the digestion of protein (*61*), part of the observed increase in metabolizable energy may be caused by this reduction in free gossypol content.

In connection with cottonseed meal it is of interest to note that Baugher and Campbell (*62*) reported that the treatment of cottonseed meal with a suspension of spores from a fungus (a strain of *Diplodia*) caused a marked reduction in free gossypol, which was accompanied by a corresponding increase in bound gossypol. Rat feeding studies confirmed the fact that the fungal-treated cottonseed meal was essentially devoid of toxicity compared with untreated controls. The conclusion that the reduction in toxicity effected by fungal treatment is caused by the conversion of free to bound gossypol is compatible with the fact that, unlike free gossypol, bound gossypol is nontoxic (*63*). The exact nature of the enzymatic process responsible for this conversion is not known, but the ability of a phytase to release protein from a phytate–protein complex thus permitting the protein to bind gossypol, as postulated by Rojas and Scott (*60*), could explain the fungal detoxification of gossypol.

Lactose Hydrolysis

It is estimated that 70–80% of non-white adults and about 20–30% of white adults have some degree of lactose intolerance (*64*). The latter term is used to describe individuals who, by virtue of having a genetically acquired deficiency of the lactose-hydrolyzing enzyme (lactose or β-D-galactosidase) in their intestines, are unable to split lactose into glucose and galactose. This then leads to an influx of fluid into the intestinal lumen, the consequences of which are abdominal distention, cramps, and an accumulation of gas in the lower intestines. Lactose intolerance is most frequently found among Negroes of North America and Africa, South American Indians, Asians, and other population groups of low economic standards.

Since lactose intolerance is most often associated with those segments of the world's population which need the nutrients provided by milk, considerable interest has been manifested in the possible development of techniques which would permit the selective removal of lactose from milk. One such approach is to subject milk to a prehydrolysis treatment whereby lactose could be converted into absorbable monosaccharides prior to general distribution. This has been accomplished by treating fresh raw whole milk with an immobilized form of lactase derived from *Saccharomyces lactis* (*65*). A preparation of this enzyme from *Saccharomyces fragilis* has also been used for this purpose without immobilization (*66*). As much as 90% of the lactose of milk can be hydrolyzed in this fashion.

Paige et al. (*65*) conducted an experiment with 32 black children in Baltimore, ranging in age from 13 to 19 years. Of these, 22 were judged to be lactose-intolerant as measured by the lack of a rise in blood glucose following the ingestion of a test load of 8 oz of untreated milk (equivalent to 12 g lactose). These children were then administered untreated milk as well as milk in which 50% and 90% of the lactose had been enzymatically hydrolyzed. The ability to tolerate these various milk products was evaluated by measuring the rise in blood sugar after a given period of time following ingestion. The results of this experiment are shown in Table II. The lactose-intolerant children showed a progressive increase in blood sugar, that is, greater tolerance towards lactose, with increasing degrees of lactose hydrolysis. The hydrolysis of lactose made little difference as far as the lactose-tolerant children were concerned. Although the lactose-intolerant children had lower blood sugar on raw milk or milk in which the lactose had been hydrolyzed 50%, these differences were eliminated when the lactose-intolerant children were given milk in which 90% of the lactose had been hydrolyzed. From these results it would appear that the enzymatic removal of lactose from milk offers considerable promise for providing much-needed nutrients to large

Table II. Effect of Lactose-hydrolyzed Milk on Rise in Blood Sugar in Lactose-tolerant and -intolerant Children (*66*)[a]

	Milk Tested		
	Untreated	*50% Hydrolysis*	*90% Hydrolysis*
Lactose-intolerant ($n = 22$)	4.4[†]	8.8[†]	14.5[‡]
Lactose-tolerant ($n = 10$)	12.2[‡]	13.3[‡]	13.7[‡]

[a] Expressed as mg/100 ml blood. The difference between groups having different superscripts (†, ‡) is significant at $P < 0.05$.

Zeszytz Naukowe Akademii Rolniczo-Technecznej w Olsztynie, Technologia Zywnosci

segments of population, both domestic and abroad, who are unable to tolerate the lactose which milk normally contains.

Conclusions

In this paper an attempt has been made to cite examples of how toxic substances which are natural constituents of many plant tissues may be removed or at least reduced through the action of specific enzymes. These enzymes may be present endogenously to release the toxic substancs from an innocuous precursor; processing techniques or traditional modes of food preparation are then employed to eliminate the toxicants so formed. Such is the case of goitrogenic glycosides of rapeseed and the cyanogenic glycosides of cassava, lima beans, and fruit kernels. A most promising approach to the problem of detoxification, and one which will probably be used more commonly in the future, is the use of microbial enzymes which selectively attack natural toxicants. This method has been employed successfully to reduce the flatulence-producing factors in beans, the phytate content of soybeans and wheat, and the gossypol in cottonseed. The use of immobilized enzymes represents an area of technological advancement that will no doubt find wider application in the future for the enzymatic detoxification of foods. This technique has already proved to be a practical means whereby the lactose content of milk can be reduced dramatically by enzymatic hydrolysis, thus providing a valuable source of protein to a large segment of the population which is otherwise incapable of tolerating milk in the diet.

Literature Cited

1. Liener, I. E., Ed., "Toxic Constituents of Plant Foodstuffs," Academic Press, New York, 1969.
2. "Toxicants Occurring Naturally in Foods," Nat. Acad. Sci., Wash., D. C., 1973.
3. Bell, J. M., *Can. J. Agric. Sci.* (1955) **35,** 242–251.

4. Miller, R. W., Van Etten, C. H., McCrew, C., Wolff, I. A., Jones, Q. J., 1962. "Amino Acid Composition of Seed Meals from Forty-one Species of Cruciferae," **10**, 426–430.
5. Van Etten, C. H., Wolff, I. A., "Natural Sulfur Compounds," *in* "Toxicants Occurring Naturally in Foods," Nat. Acad. Sci., Wash., D. C., 1973, pp 210–234.
6. Eapen, K. E., Tape, N. W., Sims, R. P. A., *J. Am. Oil Chem. Soc.* (1968) **45**, 194–196.
7. Eapen, K. E., Tape, N. W., Sims, R. P. A., *J. Am. Oil Chem. Soc.* (1969) **46**, 52–55.
8. Agren, G., Eklund, A., *J. Sci. Food Agric.* (1972) **23**, 1451–1462.
9. Kozlowska, H., Sosulski, F. W., Youngs, C. G., *Can. Inst. Food Sci. Technol. J.* (1972) **5**, 149–154.
10. Van Etten, C. H., Daxenbichler, M. E., Peters, J. E., Wolff, I. A., Booth, A. N., *J. Agric. Food Chem.* (1965) **13**, 24–27.
11. Kirk, L. D., Mustakas, G. C., Griffin, E. J., Jr., Booth, A. N., *J. Am. Oil Chem. Soc.* (1971) **48**, 845–850.
12. Mustakas, G. C., Kirk, L. D., Griffin, E. L., Jr., Clanton, D. C., *J. Am. Oil Chem. Soc.* (1968) **45**, 53–57.
13. Oginsky, E. L., Stein, A. E., Greer, M. A., *Proc. Soc. Exp. Biol. Med.* (1965) **119**, 360–364.
14. Marangos, A., Hill, R., *Proc. Nutr. Soc.* (1974) **33**, 90A.
15. Poulsen, E., *Nord. Veterinaermed.* (1958) **10**, 487–489.
16. Lanzani, A., Piana, G., Piva, G., Cardillo, M., Rastelli, A., Jacini, G., *J. Am. Oil Chem. Soc.* (1974) **51**, 517–518.
17. Belzile, R. J., Bell, J. M., *Can. J. Anim. Sci.* (1966) **46**, 165–169.
18. Goering, K. J., Thomas, O. O., Beardsley, D. R., Curran, W. A., Jr., *J. Nutr.* (1960) **72**, 210–216.
19. Tookey, A. L., Van Etten, C. H., Peters, J. E., Wolff, I. A., *Cereal Chem.* (1965) **42**, 507–514.
20. Mustakas, G. C., Kirk, L. D., Griffin, E. L., Jr., *J. Am. Oil Chem. Soc.* (1962) **39**, 372–377.
21. Mustakas, G. C., Griffin, E. L., Jr., Gastrock, E. A., D'Aquina, E. L., Keating, E. J., Patton, E. L., *Biotechnol. Bioeng.* (1963) **5**, 27–39.
22. Mustakas, G. C., *Chemurgic Digest* (1964) Jan.-Feb., pp 9–11.
23. Montgomery, R. D., *West Indian Med. J.* (1964) **13**, 1–11.
24. Montgomery, R. D., *Am. J. Clin. Nutr.* (1965) **17**, 103–113.
25. Montgomery, R. D., "Cyanogens," *in* "Toxic Constituents of Plant Foodstuffs," Liener, I. E., Ed., Academic, New York, 1969, pp 143–157.
26. Favier, J.-C., Chevassas-Agnes, S., Gallon, G., *Ann. Nutr. Aliment.* (1971) **25**, 1–59.
27. Collard, P., Levi, S., *Nature* (1959) **183**, 620–621.
28. Akinrele, I. A., *J. Sci. Food Agric.* (1964) **15**, 589–594.
29. Vogt, H., *World's Poult. Sci. J.* (1966) **22**, 113–125.
30. Oke, O. L., *Nature* (1966) **212**, 1055–1056.
31. Osuntokun, B. O., Durowoju, J. E., McFarlane, H., Wilson, J., *Brit. Med. J.* (1968) **3**, 647–649.
32. Osuntokun, B. O., *Brit. J. Nutr.* (1970) **24**, 797–800.
33. Osuntokun, B. O., Aladetoyinbo, A., Adauja, A. D. G., *Lancet* (1970) Aug. 15, 372–373.
34. Rahman, S. A., De, S. S., Subrahmanyan, V., *Curr. Sci.* (1947) **11**, 351–352.
35. Davidson, A., Stevenson, T., *Practitioner* (1884) **32**, 435–439.
36. Gabel, W., Kruger, F., *Muench. Med. Wochenschr.* (1920) **67**, 214–215.
37. Winkler, W. O., *J. Assoc. Off. Agric. Chem.* (1958) **34**, 541–547.
38. Auld, S. J. M., *J. Agric. Sci.* (1912–1913) **5**, 409–417.
39. Sayre, J. W., Kaymaksalan, S., *N. Engl. J. Med.* (1964) **270**, 113–114.
40. Conn, E. E., *J. Agric. Food Chem.* (1969) **17**, 519–526.

41. Schab, R., Yannai, S., *J. Food Sci. Technol.* (1973) **10,** 51–59.
42. Calloway, D. H., Hickey, C. A., Murphy, E. L., *J. Food Sci.* (1971) **36,** 251–255.
43. Rackis, J. J., *Adv. Chem. Ser.* (1975) **15,** 207–222.
44. Sherba, S. E., *Chem. Abstr.* (1970) **74,** 2858n.
45. Sugimoto, H., Van Buren, J. P., *J. Food Sci.* (1970) **35,** 655–660.
46. Kim, W. J., Smit, C. J. B., Nakayama, J. O. M., *Lebensm.-Wiss. + Technol.* (1973) **6,** 201–204.
47. Becker, R., Olsen, A. C., Frederick, D. P., Kon, S., Gumbmann, M. R., Wagner, J. R., *J. Food Sci.* (1974) **39,** 766–769.
48. Oberleas, D., "Phytates," *in* "Toxicants Occurring Naturally in Foods," Nat. Acad. Sci., Wash., D. C., 1973, pp 363–371.
49. Spitzer, R. R., Phillips, P. H., *J. Nutr.* (1945) **30,** 183–192.
50. McCance, R. A., Widdowson, E. M., *Biochem. J.* (1935) **29B,** 2694–2700.
51. Bitar, K., Reinhold, J. G., *Biochim. Biophys. Acta* (1972) **268,** 442–447.
52. Reinhold, J. G., Nedayati, H., Lahimgarzadeh, A., Nasr, K., *Ecol. Food Nutr.* (1973) **2,** 157–162.
53. McCance, R. A., Widdowson, E. M., *J. Physiol. (London)* (1942) **101,** 304–313.
54. Reinhold, J. G., *J. Am. Diet. Assoc.* (1975) **66,** 38–41.
55. Reinhold, J. G., Parsa, A., Kariman, N., Hammick, J. W., Ismail-Beigi, F., *J. Nutr.* (1974) **104,** 976–982.
56. Nelson, T. S., McGillivray, J. J., Shieh, T. R., Wodzinski, R. J., Ware, J. H., *Poult. Sci.* (1968) **47,** 1985–1989.
57. Nelson, T. S., Shieh, J. N., Wodzinski, R. J., Ware, J. H., *Poult. Sci.* (1968) **47,** 1842–1848.
58. Okubo, K., Waldrop, A. B., Iacobucci, G. A., Myers, D. V., *Cereal Chem.* (1975) **52,** 263–271.
59. Kon, S., Olson, A. C., Frederick, D. P., Eggling, S. B., Wagner, J. R., *J. Food Sci.* (1973) **38,** 215–220.
60. Rojas, S. W., Scott, M. L., *Poult. Sci.* (1969) **48,** 819–835.
61. Hill, F. W., Totsuka, K., *Poult. Sci.* (1964) **43,** 362–370.
62. Baugher, W. L., Campbell, T. C., *Science* (1969) **164,** 1526–1527.
63. Berardi, L. C., Goldblatt, L. A., "Gossypol," *in* "Toxic Constituents of Plant Foodstuffs," Liener, I. E., Ed., Academic, New York, 1969, pp 212–266.
64. Bayliss, T. M., Paige, D. M., Ferry, G. D., *Gastroenterology* (1971) **60,** 605–608.
65. Paige, D. M., Bayless, D. M., Huang, S.-S., Wexler, R., *Am. Chem. Soc. Symp. Ser.* (1975) **15,** 191–206.
66. Surazynski, A., Poznanski, S., Chojnowski, W., Mrozek, Z., Rogala, L., *Zesz. Nauk. Akad. Roln.-Tech. Olsztynie, Technol. Zywn.* (1975) **4,** 77–86; *Chem. Abstr.* (1975) **83,** 95178p.

RECEIVED January 26, 1976.

INDEX

B

C

D

E

N

O

P

T

U

V

W

X

The text of this book is set in 10 point Caledonia with two points of leading. The chapter numerals are set in 30 point Garamond; the chapter titles are set in 18 point Garamond Bold.

The book is printed offset on Text White Opaque, 50-pound. The cover is Joanna Book Binding blue linen.

Jacket design by Norman Favin.
Editing and production by Joan Comstock.

The book was composed by the Service Composition Co., Baltimore, Md., printed and bound by The Maple Press Co., York, Pa.